"Dr Slyman has drawn on her empirical research to explore a common issue that affects many children and students. A 'blockage' when it comes to mathematics and statistics is addressed through socio-cognitive, socio-cultural and psychological factors. The issue demands our attention and educators, parents, researchers, as well as undergraduate and postgraduate students will benefit from this thought provoking and inspirational text."

Dr. Richard Race, *Teesside University, UK*

"In an era where traditional teaching methods often fall short, Unlocking the Power of Simulation Games offers a bold new vision for Mathematics and Statistics education. The author's pioneering approach to Acumenous Game-Based Learning is both innovative and practical, demonstrating how games can be used to foster a deeper understanding and appreciation of complex subjects. This book is a treasure trove of insights for educators, researchers, and game designers who are passionate about reshaping the future of education. It provides the tools and inspiration needed to harness the power of games as a force for positive change. A transformative read that will leave a lasting impact on educators and students alike."

Dr. Julio Garay, *Bronx Community College, New York City, USA*

"This book provides an innovative interdisciplinary exploration of what applying game-based learning to mathematics and statistics in HE might look like. Bridging the fields of mathematics and statistics education with approaches to learning as social practice, it uniquely foregrounds students' learning experiences, opportunities and challenges. It introduces the acumenous game-based learning framework in simulation games and aptly illustrates how it can enhance student agency, ownership and autonomy and increase confidence and motivation. Written in accessible language, it opens new avenues to reconceptualise 21st century HE, proposing exciting implications for policy, practice and employability."

Dr. Vally Lytra, *Goldsmiths, University of London, UK*

"With this insightful book, the author invites us to rethink our approach to teaching Mathematics and Statistics. Unlocking the Power of Simulation Games provides a fresh perspective on how the integration of game-based learning can dramatically improve student outcomes. By focusing on the novel Acumenous Game-Based Learning framework, the author offers educators practical strategies for creating more engaging and effective learning experiences. This book is an essential guide for anyone committed to enhancing student engagement and achievement in mathematical education. Prepare to be inspired by the transformative potential of game-based learning!"

Prof. Sylvester Arnab, *Coventry University, UK*

Acumenous Game-Based Learning in Simulation Games

Drawing on a comprehensive theoretical foundation encompassing socio-constructivism, social learning, and motivation, this book offers a new theoretical contribution and sheds light on the transformative power of acumenous game-based learning to reengage learners with Mathematics and Statistics Education.

Foregrounding six years of research into simulation games, and exploring inquisitional and acumenous theory, the book ultimately explores the power of science and game-based technologies, applying theoretical knowledge to practice in order to reframe negative connotations associated with mathematical and statistical learning. The chapters explore maths anxiety and resilience, as well as understanding dyscalculia and its ramifications, introducing a novel theoretical framework that will redefine traditional teaching paradigms. Ultimately exploring the powerful potential of simulation games as motivational and pedagogical tools to redress the complex societal problem of negative emotions toward Mathematics and Statistics Education, readers will gain invaluable insights into unlocking new learning opportunities, paving the way for a future where mathematics and statistics become accessible, engaging, and empowering for all.

This book is of benefit to postgraduate students, scholars, and academics in the fields of educational technology, mathematics education, and teaching and learning in higher education more broadly.

Souad Slyman is a senior lecturer who served as interim Deputy Head of MOSI (Marketing, Operations, Supply Chain, and Innovation) cluster group, University of Roehampton, UK, as well as a highly experienced business computing educator and researcher.

Routledge Research in Digital Education and Educational Technology

This series provides a forum for established and emerging scholars to discuss the latest debates, research, and practice in the field of digital education and educational technology globally, including distance and open learning.

Overcoming Challenges in Online Learning
Perspectives from Asia and Africa
Edited by Areej ElSayary and Abdulrasheed Olowoselu

The New Digital Education Policy Landscape
From Education Systems to Platforms
Edited by Cristóbal Cobo and Axel Rivas

Digitalization and Digital Competence in Educational Contexts
A Nordic Perspective from Policy to Practice
Edited by Sara Willermark, Anders D. Olofsson and J. Ola Lindberg

Learning, Teaching, and Social Media
A Generational Approach
Andrew McWhirter

Designing Learning with Digital Technologies
Perspectives from Multimodality in Education
Edited by Fei Victor Lim and Mercedes Querol-Julián

Charting the AI Transition in Education and Business Environments
Navigating the Generative Inflection Point for Industry 4.0 Success
James Hutson and TJ Rains

Acumenous Game-Based Learning in Simulation Games
Transforming Fears of Mathematics and Statistics Education
Souad Slyman

For more information about this series please visit: https://www.routledge.com/Routledge-Research-in-Education/book-series/RRDEET

Acumenous Game-Based Learning in Simulation Games

Transforming Fears of Mathematics and Statistics Education

Souad Slyman

Routledge
Taylor & Francis Group
LONDON AND NEW YORK

First published 2025
by Routledge
4 Park Square, Milton Park, Abingdon, Oxon OX14 4RN

and by Routledge
605 Third Avenue, New York, NY 10158

Routledge is an imprint of the Taylor & Francis Group, an informa business

British Library Cataloguing-in-Publication Data
A catalogue record for this book is available from the British Library

ISBN: 978-1-032-71726-5 (hbk)
ISBN: 978-1-032-72341-9 (pbk)
ISBN: 978-1-032-72342-6 (ebk)

DOI: 10.4324/9781032723426

Typeset in Times New Roman
by SPi Technologies India Pvt Ltd (Straive)

Contents

Illustrations

Figures

Tables

Abbreviations

2D GBL	2Dimensional Game-Based Learning.
3D GBL	3Dimensional Game-Based Learning.
AC_3	Acumen, Creativity, Critical Thinking, and Collaboration.
ACME	Advisory Committee on Mathematics Education.
AGBL	Acumenous Game-Based Learning.
AI	Artificial Intelligence.
BERA	British Education Research Association.
CGT	Constructivist Grounded Theory.
DEGs	Digital Educational Games.
DfE	Department of Education.
E-Learning	Electronic Learning or learning through the internet online.
ESGs	Educational Simulation Games.
FE	Further Education College.
FGIs	Focus-Group Interviews.
ESGAE	Educational Simulation Games Analysis and Evaluation Model.
GAP	Game Playability Approachability.
GBL	Game-Based Learning.
GEAM	Game Elements Attributes Model.
HCI	Human Computer Interaction Design.
HE	Higher Education.
HEA	Higher Education Academy.
HEP	Heuristics Evaluation for Playability.
HRF	Hopelab Research Foundation.
IGBL	Inquisitional Game-Based Learning.
ISO	International Standards Organizations.
IT	Information Technology.
Logo	A programming language (see Glossary).
MDA	Mechanics, Dynamics, and Aesthetics.
MEGs	Mathematics Educational Games.
MMORPGs	Massive Multiple Online Roleplay Games.
NACCCE	National Advisory Committee on Creative and Cultural Education UK.

NGH	Networked Game Heuristics.
PC	Personal Computer.
PLAY	Playability Heuristics for Mobile Games.
SPSS	Statistical Package for the Social Sciences.
SGs	Simulation Games.
SGs-BL	Simulation Games-Based Learning.
STEM	Science, Technology, Engineering, and Mathematics.
T&L	Teaching and Learning.
Vs	Versus.
UI	User Interface.
UK	United Kingdom.
ZDP	Zone of Proximal Development.

Key Terms

Acumen means the ability to understand, reason, and decide.

Acumenify from acumen.

Acumenous Game-Based Learning (AGBL) is a new theoretical framework conceived from Chapter 7 (and was further developed in Chapter 9). It is the ability to understand, reason, and decide.

AC_3 is a proposed teaching and learning framework which stands for acumen, creativity, critical thinking, and collaboration (Final Conclusions).

Applied Statistics can mean applying statistics in different disciplines/situations, such as improving quality, production, and making better decisions, and therefore it becomes applicable (Chapter 3).

Attitude is how individuals think or approach everyday tasks. It can be associated with good or bad thinking depending on the individual's level of thinking (Slyman, 2014), and it is related to feelings and emotions (such as anger and happiness). Positive attitudes are linked to good thinking, and negative attitudes are associated with bad thinking (De Bono, 1992). Throughout this book, attitudes are going to be referred to as "negative emotions" or "preconceptions" (see preconceptions and feelings; Chapter 3).

Constructivist Grounded Theory (CGT) is an inquiry method adopted to explore social processes, manage, and construct original meaning and theories from the data (Charmaz, 2006; Chapter 6).

Cyberdam is a 2D simulation game developed by Mayer et al. (2013) (Chapter 3).

Data Analysis is a process of analyzing the collected data, and it can be qualitative or quantitative data analysis (Chapter 6).

Debriefing is a brief explanation about the gameplay session and what it involves (Chapter 3).

Epistemology is the process of knowing/knowledge (Chapter 2).

Experience is an event or an observation occurring and having effects on a someone's life. In this book, "experience" is defined in terms of confidence, feelings, views, behavior, perceived usefulness, and fun, and how entertaining, un/interesting, easy, difficult, or dis/engaging an event is occurring and having effects on someone.

Feelings (or emotions) are how we feel and respond to a situation, a thing, or a person, and it is a synonym to emotions. Throughout this book, feelings and emotions are used interchangeably to mean the same concept.

Focus-Group Interview (FGI) is a way of collecting qualitative data through asking participants a set of questions as in a small/large group.

ESGAE (Educational Simulation Games Analysis and Evaluation) is a proposed model for analyzing and evaluating educational simulation games. This was the outcome of Chapter 6 guided by Chapter 4.

Game is regarded as a fun playful dynamic learning activity with a fixed set of rules and limited time, and can simulate real-world problems. It has specific rules, usually ends in either win or lose, and involves manipulation, curiosity, secrecy, and mystery (Chapter 2).

Game-Based Learning (GBL) is a type of game where one can learn and play. It can have learning outcomes where players may investigate real-life examples in a risk-free environment (Chapters 2 and 3).

GameMaker is a free commercial game designing program to construct own games, using drag-and-drop and semantic (interpretation of coding) coding (Habgood, 2007; Chapter 3).

Gamestar Mechanic Game is a platform for creating, learning, and designing games (Gee et al., 2008; Chapter 3).

Gen "Alpha" is a generation that comes after generation Z (Mccrindle, 2018; Chapter 3).

Generation "Z" is the post-millennium students who use the internet and social media with flair (Dede, 2005).

Human Computer Interaction Design (HCI) is the study of human and computers interaction with each other (Chapters 3, 4, and 7).

Inquisitional Game-Based Learning (IGBL) or "quiz like" is a quiz type of learning that can occur in conversations or discourses such as investigation in court/police station: Question-answer format type of learning, and it is the opposite of Acumenous Game-Based Learning (see Chapters 7, 9, and 10).

Interpretivist/Constructivist Paradigm is associated with qualitative research methods and embraced to see and understand the world.

JavaScript is a programming language for creating and developing webpages (Chapter 3).

Kahoot is a method of assessment where teachers/students designed their own questions in "kahoot" to test their knowledge; teachers can also create their own assessment (Chapter 2).

Logo is a programming language where students were given an opportunity to design a mathematical game of their own choice (Chapter 3).

Madison 2200 is a simulation game in which urban planners redesign a local downtown pedestrian mall (Chapter 3).

MDA (Mechanics, Dynamics, and Aesthetics) is a design methodology for guiding the creative process of thinking. It is for building prototypes and architecture software (Chapter 4).

MinecraftEdu is a type of Minecraft game with an educational focus targeted for schools (Chapter 3).

Numbas is a Mathematics shooting game where players are required to select right answers using a tab of numbers as in calculators (Chapter 5).

Pairwise T-Test is a statistical procedure known as dependent sample t-test where a variable is measured twice. The test is used to find out whether the mean is different or the same in two related groups.

Positivist Paradigm relates to quantitative research methods, which are associated with scientific measurements and numbers.

Playtesting is a term used in gaming. It means to test the games while playing them (i.e., testing the quality and design of the games prior to final release, etc.; Chapters 4 and 6).

Pre- and Post-Test is a self-efficacy questionnaire that was employed before and after the Statistics lecture gameplay scenario with Excel to test if there are any differences in participants' self-efficacy prior and post (Chapter 9).

Preconception means an opinion, an attitude, or a concept that someone has prior to learning or experience. It is a social construct linked to emotions and include how students are socially constructed in Mathematics and/or Statistics classrooms (as "good/bad" mathematicians). Throughout this book, preconception is used to mean attitudes, that is, negative preconceptions. To avoid repetition, I refer to "negative preconceptions" as "preconceptions" (see also Chapters 7, 9, and 10).

Questionnaires are a tool/set of questions designed with the purpose of finding out more information about a specific relevant topic and it is part of quantitative measures (Dwyer and Slyman, 2016).

Re-Mission is a shooter fantasy game where players control a superhero robot named "Roxxi," who fights and shoots bad evils, in this case cancer cells, using antibiotic weapons (Granic et al., 2014; Chapter 3).

Self-efficacy is a psychological process that lies within one's personality or self-belief. It is the belief in one's own ability to succeed in or accomplish something specific, such as producing a Mathematics and Statistics outcome (see Bandura, 1986, 1997; Zimmerman, 2000; Chapter 2). Throughout this book, self-efficacy and confidence are used interchangeably to mean the same concept (self-efficacy).

SimCity is a 3D SG of an entertainment type that was developed by Will Wright (Chapter 5).

Simoleons is a term used in the "SimCity" game for extra cash (Chapter 7).

Simulation Games (SGs) are an experiential learning method that can simulate an environment or an aspect of it. They can consist of choices, fantasy, rules/goals, challenge, mystery, reward and control, and immediate feedback. For example, "SimCity" or "Minecraft" games (Chapter 2).

Statistics is a field of Mathematics that is associated with data analysis (Dwyer and Slyman, 2016; Chapter 3).

Sudoku is a puzzle game where the player inputs numbers in a grid from one to nine (Chapter 2).

XML is an extensible Markup language or a set of rules for encoding documents aimed at defining the syntax of other languages (Chapter 3).

Preface

"Unlocking the Power of Simulation Games: Transforming Negative Emotions of Mathematics and Statistics Education." Discover an exploration into the realm of game-based learning, aimed at revolutionizing the way we approach Mathematics and Statistics Education. In this book, the author delves into a new fascinating world of "Acumenous" and "Inquisitional" game-based learning, introducing a novel theoretical framework that redefines traditional teaching paradigms.

Within these pages, the reader will embark on a journey through the transformative potential of Acumenous Game-Based Learning (AGBL) and witness the profound impact it can have on students' understanding, reasoning, decision-making, and overall attitudes toward Mathematics and Statistics. Drawing upon research findings, the author sheds light on the captivating aspects of game design, exploring the ways in which usability heuristics and principles of Human-Computer Interaction (HCI) can be employed to enhance players' experiences.

Challenging the commonly held belief that Mathematics Educational Games (MEGs) are inherently disengaging, the author unveils a transformative approach to overcoming preconceptions and negative emotions surrounding Mathematical learning. Through empirical evidence, the reader will witness the significant impact of AGBL in nurturing students' understanding of Mathematics and Statistics, all while bolstering their self-efficacy and confidence in these subjects.

This book serves as an inspiring call to action, urging educators, researchers, and game enthusiasts alike to embrace the power of games as a catalyst for positive change in the realm of Mathematics and Statistics education. By embracing the principles and insights shared within these pages, we can reshape the landscape of learning and empower students to unlock their full potential in the fascinating world of numbers and data.

All case studies/interviewees and associated data within this book have been anonymized in accordance with the University of London's ethical protocols. To ensure the privacy and confidentiality of participants, all identifiable information was anonymized. This approach not only protects the identity of those involved but also adheres to ethical guidelines that promote respect and integrity in research. By implementing these measures, we strive to maintain the trust of our participants and uphold the ethical standards set forth by the academic community.

In addition, informed consent was obtained from all participants prior to their involvement in this research. Participants were thoroughly briefed on the purpose of the study, the use of their data, and their rights to withdraw at any time without consequence. This process was designed to empower participants, allowing them to make informed decisions about their participation. Our commitment to ethical research practices ensures that the data presented in this book are both responsibly sourced and respectfully handled, reflecting the highest standards of academic integrity.

About the Author

Dr. Souad Slyman, is a Principal/Senior Lecturer who has served as an Interim Deputy head of the MOSI (Marketing, Operations, Supply Chain and Innovations) cluster group. She is a distinguished business computing educator, researcher, and business co-owner. Her journey is marked by a rich multidisciplinary background that encompasses academia, business, and extensive experience in the clothing and food retail industry.

Dr. Slyman's academic journey in teaching and learning started 20 years ago working at various colleges and universities, such as Goldsmiths, University of London, Birkbeck, University of London, and Glion HE institution. She has a first class Honors degree in Mathematics and Computing (BSc), followed by Master's of Art in Education (MA), showcasing her commitment to both technical and educational domains. Her academic pinnacle was achieved with a PhD in Computer Science from Goldsmiths, University of London, where her research focused on game studies, design, and development and how we can use games as a motivational pedagogical tool in teaching Mathematics and Business Statistics at both undergraduate and postgraduate levels.

Before her tenure at Roehampton, Dr. Slyman played diverse roles, ranging from Business Manager and Visiting Lecturer to Mathematics & Computing Teacher and Community Interpreter. This rich tapestry of experiences, including co-owning businesses in clothing and food retail, has endowed her with a practical understanding of industry dynamics, complementing her academic pursuits. Dr. Slyman is a Program Leader for three master programs including MSc in Digital Marketing and MSc in Global Marketing. She also plays a pivotal and strategic role at the university level where she serves as a student retention champion.

Dr. Slyman's research is a testament to her dedication to improving learning outcomes and skills through innovative approaches. Her expertise spans quantitative, qualitative, and mixed research methods. Notable contributions include her research on business marketing simulation games, game-based learning in HE, consumer behavior/experience, business leadership, management, and social change. Her current research is on AI in education and how we can harness AI skills to improve teaching, learning, and student experience in HE. Dr. Slyman is also a TED speaker, an author, and active researcher who has written many research papers, presenting them at different venues all over the world.

In her recent projects, Dr. Slyman collaborated on "Entrepreneurs creating entrepreneurs," a study on retailers supporting the formation of micro-businesses, presented at the 18th European Conference on Innovation and Entrepreneurship. Another noteworthy contribution is the development of an Evaluation Framework for Educational Simulation Roleplay Games, presented at the 16th European Conference on Games Based Learning.

Dr. Slyman is also an advocate for the effective use of technology in education, evident in her research on game-based learning and the application of digital technology to enhance graduate employability skills. Her engagement extends to professional affiliations, including being a Member of the Centre for Sustainability and Research Management and a Senior Fellow of Higher Education Academy (SFHEA).

Dr. Slyman's commitment to education is further reflected in her extensive teaching portfolio. She has delivered courses on MBA Applied Research and Consultancy project, MSc in Data and Analytics for Marketers, MSc in Digital Marketing Practice, MSc in Social Media and Influences, BSc in Business Data Analysis, AI in education and more, across undergraduate and postgraduate levels.

Beyond academia, Dr. Slyman actively engages in consultancy work, working with partners all the over the world from Europe, the USA to Africa. She also supervises MSc students and final year undergraduates. Her consultancy projects include working as an External Examiner for different HE institutions like Kingston, Sheffield Hallam, and Canterbury Christ Church Universities. Other projects include examining the impact of gender inequality on industries (e.g., Ryanair), assessing the effects of video games as an educational learning tool, AI and emerging technologies, and analyzing the impact of social media on business growth.

In summary, Dr. Souad Slyman's remarkable journey, characterized by academic excellence, diverse industry experiences, and impactful research, underscores her commitment to advancing education, empowering students, and contributing valuable insights to the intersection of academia and industry.

Acknowledgments

This voyage has been a tremendous and stimulating journey. I would like to take this opportunity to express my gratefulness to those who have accompanied me during this journey.

This book would not have been possible without the advice, guidance, and patience of my colleagues, their extensive knowledge, and experience in this multidisciplinary topic of research have facilitated the birth and flourishing of this book.

I am also truly indebted and thankful to Dr. Marco Gillies, Dr. John Jessel, and Dr. Vally Lytra at Goldsmiths, University of London, Prof. Norbert Pachler, from UCL, and Prof. Sylvester Arnab, from Coventry University, and to my colleagues Dr. Richard Race and Dr. John Dwyer from the University of Roehampton, for their advice, motivation, and support at all times.

I would also like to thank my beloved parents for their wisdom and encouragement, and to my sons (Yassin, Ayoub, and Youness): The driving force behind this unique book for their motivation and support throughout this intellectual journey.

Part I

Math Dread, Self-Efficacy Empowerment and Negative Emotions through Game-Based Learning

This book delves into the realm of game-based learning and its potential as a tool for enhancing the teaching and learning of Statistics topics typically covered in the undergraduate Higher Education (HE) Psychology, Education or Business/ Computing syllabus.

In this initial section, Part I, I will commence by providing an overview of the empirical research (Chapter 1) in this book and the rationale behind its execution. Subsequently, I will delve into an exploration of prior research related to learning, self-efficacy, and games (Chapter 2), followed by an examination of existing educational literature regarding Game-Based Learning (GBL) and the utilization of Simulation Games (SGs) in the context of Mathematics and Statistics (Chapter 3).

Mathematics, Statistics, and Computing are vast and different disciplines in their own rights. The nature of Mathematics, for instance, is that it is a combination of 'abstract' concepts, patterns, symbols, and structure. Statistics is a field of Mathematics that is associated with data analysis. I initiated this book project looking at Mathematics, and then I narrowed it down to Statistics (my field of interest as a Business Computing Senior lecturer). Hence the word 'Mathematics and Statistics.' It means Mathematics in general and Statistics in particular. To avoid repetition, I refer to this as 'Mathematics and Statistics.' Applied Statistics is a term I will be referring to throughout this book and it can mean applying statistics in different disciplines/ situations such as improving quality, production and making better decisions, as in business, education or computing and therefore it becomes applicable, i.e., applied. This book focuses specifically on how we can improve students' self-efficacy in Mathematics and Statistics using SGs (Simulation Games) (Chapter 1). Self-efficacy is the belief in one's own ability to succeed in or accomplish something specific, such as producing a Mathematics and Statistics outcome (Chapter 2). Throughout this book, I use the terms 'self-efficacy' and 'confidence' interchangeably to mean the same concept (self-efficacy), because self-efficacy is also known as confidence and most people would use the word 'confidence' rather than

DOI: 10.4324/9781032723426-1

'self-efficacy.' There are many varieties of Personal Computer (PC) SGs available, including both entertainment and educational games (Chapter 2). I am focusing -explicitly- on PC SGs like 'SimCity,' a widely used entertainment game. I am interested in this type of SG because it has many activities one can construct/manage within a simulation; for instance, as a mayor/ leader, one can plan, build, and manage a city.

1 Introduction

The Mathematics Dilemma: Overcoming Negative Emotions through Game-Based Learning

The primary objective of this book is to unpack these negative emotions, meticulously examining their origins and implications. By harnessing the power of game-based learning, we aim to transform these emotions, fostering a positive learning environment that enhances productivity, decision-making skills, and overall life outcomes. Through a comprehensive exploration of both theoretical frameworks and practical strategies, we bridge the gap between research and application, offering tangible solutions for overcoming the Mathematics dilemma.

Before concluding this chapter, we provide an overview of the subsequent ten chapters, offering a glimpse into the breadth and depth of the insights to come. Furthermore, we explore the theoretical perspectives that underpin our work, grounding our discussions in robust scholarly foundations.

Prepare to embark on a transformative journey, where negative emotions/preconceptions are challenged, and the potential for meaningful learning is unleashed. This book not only illuminates the path toward Mathematics and Statistics proficiency but also equips educators, parents, and learners with the tools to create a positive and empowering educational experience. Join us as we unravel the complexities, unlock the potential, and embrace the possibilities of game-based learning in the realm of Mathematics and Statistics.

In this section, I will commence by providing an overview of the research at hand, delving into my personal research background, and elucidating the motivation behind the creation of this book, encompassing the genesis of my journey in this endeavor. Subsequently, I will expound upon the research investigations I have conducted, offer insights into the research objectives, and furnish an exposition on the organizational structure of this book.

Grounded in my extensive experience as a senior educator and researcher, which has seen me collaborate with numerous esteemed UK universities and engage in both national and international projects, as well as present my research findings on the global stage at various conferences, this book serves as a natural extension of the empirical research that commenced during my pursuit of a Master of Arts in Education in 2013–2014 and a PhD in Computer Science in 2015.

DOI: 10.4324/9781032723426-2

My overarching perspective on education perceives it as a pivotal catalyst for personal, intellectual, and social development, echoing the assertion by Jarvis (1987) that epistemology, the process of acquiring knowledge, is inherently socially constructed and can flourish under the right environmental conditions, nourishment, and contextual milieu.

My affinity for the fields of Mathematics and computing, both as a learner and senior lecturer throughout my academic journey, instilled in me a profound appreciation for the significance of these subjects when explored at the university level because "adult[s] learn what they want to learn and what is meaningful for them to learn" (Illeris, 2007:208–210). This journey was marked by an evolving comprehension, embracing the notion of deep understanding as posited by Gardner (1993) and Boaler (2009), accompanied by a remarkable ascent in achievement and personal satisfaction.

It was during this voyage that I began to ponder the underlying motives fueling negative attitudes toward learning, particularly the adverse sentiments often associated with Mathematics (Slyman, 2014), as I assumed the role of a senior lecturer, ushering in new programs and engaging with undergraduates and postgraduates. The prospect of mitigating these unfavorable emotions, especially concerning Mathematics—a subject of paramount importance to our society and economy, due to its fundamental nature—seemed elusive.

The nature of Mathematics, for instance, is that it is a combination of "abstract concepts," "rules," "terms," and "language," patterns, symbols, and structure (De Bono, 1992). It is a creative and highly interconnected discipline, which has become fundamentally crucial to everyday life and employment, as well as being a key to success in the modern business world as its significance continues to grow.

The Mathematics Dilemma: Overcoming Negative Emotions through Game-Based Learning

Mathematical and Statistical learning have become increasingly fundamental in science and technology today (Smith Report, 2004, 2017). Many learners perceive Mathematics (in particular, Statistics) as a complex subject and express negative emotions toward it (Slyman, 2014).

Research indicates that there are many socio-cognitive, socio-cultural, and psychological factors behind "blockage" about Mathematics and Statistics (Skemp, 1986:15; Illeris, 2007; Boaler, 2009; Lewis, 2013) exacerbating this "social problem," a real problem in our society, that operates as a "stimulus" and leads me to wonder about issues surrounding learners that hinder learning. However, these issues (negative emotions) are still present and have led many researchers such as Dweck (2006), Illeris (2007), and Lewis (2013) to examine why for many undergraduates Mathematics and Statistics bring fear and anxiety or even pain, "neural pain" (Lyons and Beilock, 2012:1).

Research suggests how "fear" and "anxiety" as emotional factors are contributors to disaffect Mathematics and Statistics (Early, 1992). This is known as Maths anxiety or phobia (Maloney et al., 2013). Nevertheless, Jameson and Fusco (2014:309) demonstrate that the reasons behind Maths anxiety are still unclear and could be related to "self-efficacy," that is, low confidence. Thus, this influences learners' experiences with this numerical subject. But how can we make Mathematics and Statistics stimulating and experiment with it in a playful creative way? Are there any strategies? There is an increasing necessity to discern the effects of game-based learning (GBL).

GBL has gained much interest in many educational settings, as a building tool for increasing self-efficacy and motivation, and higher education is no difference. GBL is widely integrated in Statistics, computing, and business education as an effective pedagogical tool especially in business management, strategy, and international business (Faria and Wellington, 2004; Vos, 2014). Research studies show that GBL has many cognitive and social benefits from improving students' self-efficacy to increasing motivation and engagement (Faria, 2001; Dondlinger, 2007; Moreno-Ger et al., 2008; Li et al., 2013; Malaka, 2014; McGonigal, 2015; Vlachopoulos and Makri, 2017; see Chapter 3 for further details).

Advocates of GBL argue that educational games, in particular simulation games (a subset genre of GBL), can transform the way students learn, motivate, and engage generations "Z" or "Alpha" to learn a great deal, and it can also increase their self-efficacy (e.g., Prensky, 2002; Gee, 2003a, 2003b; Dede, 2005; Mayer et al., 2013; Vos, 2014). For example, Mayo (2009) suggests that educational games can be beneficial to students over classroom teaching because games give learners control, immediate feedback, and support. However, others would argue that if entertainment fails in educational games, the learning experience will suffer (Kirriemuir and McFarlane, 2004; Moreno-Ger et al., 2008). In contrast, some researchers regard GBL as an effective tool for building self-efficacy if it is linked to learning objectives. For instance, McGonigal (2015) argues that games can have many emotional and social benefits, such as increasing self-efficacy, yet when it comes to the length of time individuals should spend playing games, she advises players to have a "balance" (i.e., a time balance) and a goal in mind of why they are playing the game (McGonigal, 2015).

The use of simulation games (SGs) in teaching in HE is not a new concept. For example, some HE institutions in the United States of America have utilized SGs for over 34 years (Wellington and Faria, 1995, cited in Tao et al., 2012). In the USA, for instance, about 64.1 percent of 1085 university academics were utilizing marketing SGs (Faria and Wellington, 2004; Jagger et al., 2016). Examples of SGs include flight and military simulators (Chapter 2). SGs are important because they allow students to engage in a real-life experience and use decision-making skills to resolve real-world problems in a safe environment (Vos, 2014; Vlachopoulos and Makri, 2017). A number of research studies suggest that SGs could be a facilitating tool for assisting learning and building self-efficacy (e.g., Faria and Wellington, 2004; Tao et al., 2012; Li et al., 2013; Vos, 2014; McGonigal, 2015; Jagger et al., 2016). Although SGs have been used for teaching complex ideas and solving problems (see Cordova and Lepper, 1996; Dondlinger, 2007; Cardinot and Fairfield, 2019), there is no research to my knowledge that addresses how we can use SGs to address self-efficacy in Mathematics and Statistics due to negative emotions toward this subject. Can SGs facilitate the teaching and learning of Mathematics and Statistics in HE, given that many students express negative emotions (i.e., pre-conceptions) toward it? The aim of this book therefore is to investigate SGs, in a real-life experience, and explore if students' self-efficacy can be addressed (see Chapters 7, 9, and 10 for more details).

This book looks at GBL, and its use, in learning and evaluates the potential of SG effects on self-efficacy, looking specifically at its use with Education, Psychology, Business, or Business Computing undergraduates in HE.

In order to gain in-depth insights into this transformative journey, and if games, with its game characteristics, game design and development, are of importance, and could be used as a tool for building self-efficacy, there were four research objectives that led to four empirical investigations, as follows:

- **Investigate game players' experiences in SGs** in order to understand their **perceptions of games, gameplay, game design, and development**. This led to the development and application of the ESGAE (Educational Simulation Games Analysis and Evaluation) Model, where I critically analyze and evaluate two games: "Numbas" (educational) and "SimCity" (entertainment) (see Chapter 6), Game Design and Development.
- **Explore connections between Mathematics and GBL with the intention of understanding participants' experiences with Mathematics and GBL**. This led to the development of the theoretical framework: Acumenous and Inquisitional GBL (Chapter 7), Mathematics, and GBL.
- **Explore similarities and differences between Quiz and Simulation GBL** (Chapter 8) to gain further insights into participants' experiences in using the two GBL: Quiz and Simulation GBL.
- **Investigate participants' experiences in learning Mathematics and Statistics using SGs** (Chapter 9): Statistics and GBL, leading to the development of the theoretical framework: Acumenous and Inquisitional Theory (Chapter 10).

The next Chapter, **Chapter 2**, will discuss previous research on learning, self-efficacy, and games, SGs and GBL, as well as looking at the theoretical perspectives that underpin and inform the work discussed in this book.

Bibliography

Boaler, J. (2009) *The Elephant in the Classroom*. London: Souvenir Press.

Cardinot, A. and Fairfield, J.A. (2019) Game-Based Learning to Engage Students with Physics and Astronomy Using a Board Game. *International Journal of Game-Based Learning (IJGBL)*, 9(1), pp. 42–57.

Cordova, D.I. and Lepper, M.R. (1996) Intrinsic Motivation and the Process of Learning: Beneficial Effects of Contextualization, Personalization, and Choice. *Journal of Educational Psychology*, 88, pp. 715–730.

De Bono, E. (1992) *Teach Your Child How to Think*. England: McQuaig Group Inc.
Dede, C. (2005) Planning for Neomillennial Learning Styles. *Educause Quarterly*, 28(1), pp. 7–12.
Dondlinger, J.M. (2007) Educational Video Game Design: A Review of the Literature. *Journal of Applied Educational Technology*, 4(1), Spring/Summer, pp. 21–31.
Dweck, C.S. (2006) *Mindset: The New Psychology of Success*. United States: Ballantine Books Trade Paperback Edition.
Early, R.E. (1992) The Alchemy of Mathematical Experience: A Psychoanalysis of Student Writings. *For the Learning of Statistics*, 12(1), pp. 15–20.
Faria, A.J. (2001) The Changing Nature of Business Simulation/Gaming Research. *Simulation and Gaming*, 32(1), pp. 97–110. DOI: 10.1177/104687810103200108
Faria, A.J. and Wellington, W.J. (2004) A Survey of Simulation Game Users, Former-Users, and Never-Users. *Simulation and Gaming*, 35(2), pp. 178–207. DOI: 10.1177/1046878104263543
Gardner, H. 1993 *Multiple Intelligences: The Theory in Practice*. New York: Basic Books/Hachette Book Group.
Gee, J.P. (2003a) *What Video Games Have to Teach Us About Learning and Literacy*. New York: Palgrave Macmillan.
Gee, J.P. (2003b) Good Video Games and Good Learning. Available at: https://academiccolab.org/resources/documents/Good_Learning.pdf (Accessed 03/01/2021).
Illeris, K. (2007) *How We Learn: Learning and Non-Learning in School and Beyond*. London: Routledge.
Jagger, S., Siala, H. and Sloan, D. (2016) It's All in The Game: A 3D Learning Model for Business Ethics. *Journal of Business Ethics*, 137(2), pp. 383–403.
Jameson, M.M. and Fusco, B.R. (2014) Math Anxiety, Math Self-Concept, and Math Self-Efficacy in Adult Learners Compared to Traditional Undergraduate Students. *Adult Education Quarterly*, 64(4), pp. 306–322.
Jarvis, P., (1987) Meaningful and Meaningless Experience: Towards an Analysis of Learning from Life. *Adult Education Quarterly*, 37(3), pp. 164–172.
Kirriemuir, J. and Mcfarlane, A. (2004) Literature Review in Games and Learning. Available at: https://telearn.archives-ouvertes.fr/hal-00190453/document (Accessed 12/04/2020).
Lewis, G. (2013) Emotion and Disaffection with School Mathematics. *Research in Mathematics Education*, 15(1), pp. 70–86.
Li, Z.Z., Cheng, Y.B. and Liu, C.C. (2013) A Constructionism Framework for Designing Game-Like Learning Systems: Its Effect on Different Learners. *British Journal of Educational Technology*, 44(2), pp. 208–224.
Lyons, I.M. and Beilock, S.L. (2012) When Math Hurts: Math Anxiety Predicts Pain Network Activation in Anticipation of Doing Math. *PLoS One*, 7(10), p. e48076.
Malaka, R. (2014) How Computer Games Can Improve Your Health and Fitness. Gobel, S., and Wiemeyer, J., (eds.) *Games for Training Education, Health and Sports, 4th Conference on Serious Games, GameDays 2014*, Darmstadt, Germany, April 1–5, 2014, Springer.
Maloney, E.A., Schaeffer, M.W. and Beilock, S.L. (2013) Mathematics Anxiety and Stereotype Threat: Shared Mechanisms, Negative Consequences and Promising Interventions. *Research in Mathematics Education*, 15(2), pp. 115–128.
Mayer, I., Warmelink, H. and Bekebrede, G. (2013) Learning in a Game-Based Virtual Environment: A Comparative Evaluation in Higher Education. *European Journal of Engineering Education*, 38(1), pp. 85–106.
Mayo, M. (2009) Video Games: A Route to Large-Scale STEM Education? *Science*, 323, pp. 79–82.
McGonigal, J. (2015) *Superbetter: A Revolutionary Approach to Getting Happier Stronger and Happier, Braver and More Resilient*. New York: Penguin Press.

Moreno-Ger, P., Burgos, D. Martínez-Ortiz, I. Sierra, J.L. and Fernández-Manjón, B. (2008) Educational Game Design for Online Education. *Computers in Human Behavior*, 24(6), pp. 2530–2540. DOI: 10.1016/j.chb.2008.03.012

Prensky, M. (2002) The Motivation of Gameplay: The Real Twenty-First Century Learning Revolution. *On the Horizon*, 10(1), pp. 5–11.

Skemp, R.R. (1986) *The Psychology of Learning Mathematics Great*. Britain: Penguin Books Ltd.

Slyman, S. (2014) I Hate Maths! An Investigation into Students' Attitudes Towards Mathematics. MA Thesis, University of Roehampton, London, United Kingdom.

Smith Report. (2004) Available at: http://www.mathsinquiry.org.uk/report/maths InquiryFinalReport.pdf (Accessed 20/12/2014).

Smith Report. (2017) Available online at: https://www.gov.uk/government/publications/smith-review-of-post-16-maths-report-and-government-response (Accessed 20/12/2018).

Tao, Y.H., Cheng, C.J. and Sun, S.Y. (2012) Alignment of Teacher and Students Perceptions on the Continued Use of Business Simulation Games. *Educational Technology and Society*, 15(3), pp. 177–189.

Vlachopoulos, D. and Makri, A. (2017) The Effect of Games and Simulations on Higher Education: A Systematic Literature Review. *International Journal of Educational Technology in Higher Education*, 14(1), pp. 22. DOI: 10.1186/s41239-017-0062-1

Vos, L. (2014) Marketing Simulation Games: A Review of Issues in Teaching and Learning. *The Marketing Review*, 14(1), p. 6796. DOI: 10.1362/146934714X1394890 9473220

2 Understanding the Power of Theory

Learning, Self-Efficacy, and Games in Education

In Chapter 1, an overview of this research was presented focusing on research background and objectives, and why I set out on this book journey. I articulated that this empirical research is grounded in my experience as a business computing senior lecturer and researcher at a South London University; it is also a continuation to the empirical research that began (my Master of Arts in Education) in 2013–2014. I explained that as a university senior lecturer, I convene many programs and modules like Mathematics and Statistics at UG and PG levels. However, I could not envision how to mitigate what appeared to be negative emotions and attitudes that appeared to be exhibited by undergraduates toward Mathematics (Slyman, 2014). My empirical research in 2014 indicates that there are many factors behind such attitudes (see Chapter 3), but identifying ways to change them was beyond the scope of that investigation. As I explained in Chapter 1, this question led me to embark on this book journey looking for new ways (i.e., GBL) to address Mathematics and Statistics. I then moved on to research investigations, contributions to knowledge, and I concluded the chapter explaining how this book will be structured.

In the subsequent sections, we shift our focus toward self-efficacy theory and its profound implications for learning and motivation. By exploring the extensive literature on self-efficacy and its relationship to games, SGs, and GBL, we illuminate the transformative potential of these learning tools in building learners' confidence and belief in their own abilities.

This chapter also serves as a critical juncture for establishing a common understanding of the terminology employed throughout the book. We carefully define what constitutes a game within the context of our discussions, ensuring clarity and coherence. Furthermore, we showcase compelling evidence that highlights the efficacy of games, SGs, and GBL in fostering self-efficacy and propelling transformative learning experiences.

Prepare to immerse yourself in a theoretical landscape where innovative pedagogies merge with game-based strategies. This chapter sets the stage for the transformative journey that lies ahead, equipping educators, researchers,

DOI: 10.4324/9781032723426-3

and learners with a deep understanding of the theoretical underpinnings that inform subsequent explorations. By leveraging theory, we unlock the true potential of games as powerful tools for learning, self-efficacy, and motivation in the educational realm.

This chapter, Chapter 2, presents previous research on learning, self-efficacy, and games, which underpins and informs the work discussed in this book. This review is broken down into three sections: First, there is a discussion of learning theories that are relevant, highlighting socio-constructivist perspectives and its links to Gee's work in particular. Gee's work will be discussed in detail in Gee's Principles Section. Next, this chapter discusses self-efficacy theory in relation to learning and motivation; finally, there is a discussion of key literature on games, different types/genres of games, GBL and SGs, followed by an explanation of how they are employed as a learning tool.

Previous Research on Learning, Self-Efficacy, and Games

This section introduces and discusses theories of learning that account for the use of GBL in HE and have influenced the research in this book. I will start by discussing socio-constructivist perspectives on learning and teaching, moving on to James Paul Gee's (2003a, 2003b, 2004, 2005a, 2005b, 2007, 2011) 13 principles specifically, since these concepts are of particular relevance to this research because of the way in which these principles are reflected in specific types of games (like SGs see Simulation Games Section).

Socio-Constructivist Perspectives

In this book, I take a broad socio-constructivist perspective to learning. Socio-constructivist perspectives as a learning theory emphasize the notion of social learning where students interact, collaborate, and share ideas in social activities as in group work or under the guidance of a capable adult. This enables students to share knowledge, develop problem-solving abilities, creativity and critical thinking skills, and appreciate a range of different perspectives. Vygotsky's theory (1980) is about language, thought, and their mediation by society. Vygotsky suggests that the process of knowing is affected by other people and is mediated by community and culture.

Vygotsky stresses the social origins of language and thinking unlike other previous thinkers such as Piaget. Vygotsky was leading the way in synthesizing the mechanisms by which culture becomes an integral part of each individual's nature. As humans, Vygotsky (1980) emphasizes that we actively change and adapt according to the various contexts of culture we are exposed to and live in. We have the ability to "externalize" and share our understanding with the same group we belong to. He argues that social interaction is significant to cognitive development. Vygotsky talks about learning occurring in two planes, first the inter-social and then the intra-social. First the child learns from interaction with others and then internalizes this learning. Since, when children are faced with challenging tasks, their emotional use of language increases as well as their efforts to achieve a more intelligent solution. Vygotsky (1980) claims that speech develops from society to the individual (i.e., the way we think (thought) is influenced by environment). He argues that individuals develop learning through culture (beliefs, religions, or social forms of an existing group) and language (mother tongues) (Vygotsky, 1980; Kim, 2001; Fosnot and Perry, 2005; Draper, 2013). In Vygotsky's views, social interaction (environment), communication (language and other modes), and culture are significant processes/key factors to cognitive development, and these can influence individual cognitive development.

Other researchers in the socio-constructivist tradition indicate that learning is a process of constructing meaning, and people themselves make sense of their experience based on their personal interpretations of the world (e.g., Palincsar, 1998). Learning is viewed as a social iterative process where the learner constructs new concepts based on their current understanding, selecting information, making hypotheses and decisions (see Bruner, 1973; Applefield et al., 2000). Learners build and transform knowledge based on their experienced environment through culture and collaborative social interaction. In this vein, this transformation of knowledge, and the way understanding and confidence are gained through previous experience and social interaction, is of significance to this research. This is because through this social iterative process, people can learn and grow. Through playing games (GBL), I propose learners' cognitive and social development can be better stimulated.

Social constructivists regard learning as an active social process where learners discover ideas, concepts, and facts as a group, and hence they promote the guesswork and intuitive thinking in learners (Brown et al., 1989). This is very much linked to Gee's (2003a, 2003b, 2004, 2005a, 2005b, 2007, 2011) work, a socio-constructivist academic and scholar, who uses games in learning (GBL). This book is building on Gee's research on games and learning focusing on the 13 learning principles, as I will explain next.

James Paul Gee's Principles

One of the prominent socio-constructivist thinkers and educators who writes copiously about games and learning, and whose work is influenced by Vygotsky, is Gee (2003a, 2003b, 2004, 2005a, 2005b, 2007, 2011). Gee's work is situated more broadly in an expanded view of new literacies where a language is viewed as one communication system (others include multimodal literacies, visual literacies). Gee's work is situated within a literacy/language as socio-cultural practice perspective. Literacies are not only cognitive outcomes but are connected to social practices. For instance, when learners participate in an activity as a group (from social cultural practice/perspective), they learn literacy of that activity be it Mathematics, problem-solving, or English, etc.

Gee examines what he terms good video games (that meet the 13 principles, discussed next) as a pedagogical tool for learning as well as a critique of traditional schooling. In his view, what children learn at schools is how to pass tests but not how to use what they learn and apply it in a real life. Such an educational system is boring and setting people to failure. Gee suggests that for learning to be meaningful, it needs to be experienced through an activity, an image/symbol, or an action. He explains that some students cannot use or apply the knowledge they develop to pass tests in other contexts. It necessitates to be experienced/played through games, that is, game-based learning. Gee's rationale is that game-based learning, as a system, can be an empowering tool that we could bring to the classroom. A limitation in his work is that he does not discuss bad games but instead he focuses on good games from educational perspectives, as I will explain later.

GBL have many "principles," design features and skills that could enrich teaching and learning. Gee (2003a, 2003b) initially proposes 36 learning principles that make good games entertaining/engaging and later on condense these principles to 16 (Gee, 2004:15–23) as follows:

- **Identity**: Learners take a new identity and make commitment to the new character they set out to play/take.
- **Interaction** is how the player interacts with the game and make decisions at each step of the game.
- **Production** where players produce and co-design at each level and their decisions influence the game and the outcomes.
- **Risk-taking** where players are not afraid of failing but are happy to take risks, make progress, and gain feedback.
- **Customization** where player chooses the attributes and characteristics of their role-playing character. They try new styles of learning and solve different problems.
- **Agent**, a sense of control and ownership.
- **Well-ordered problems** that contain levels where players set up to solve problems in a free space. Level of the complexity increases as the player goes up in levels, they try new problems and become creative to solve problems and then set out to solve new ones.
- **Challenge and consolidation or "cycle of expertise"** where several challenging problems are set, and the player has to solve this and practice until they develop a mastery. Then the game introduces new problems, and the player has to use what they learn to solve new challenges.
- **Just "in time" and on "demand" principle** where information and knowledge is introduced step by step and as required.
- **Situated meanings or meaning as action**, where the player associates words with images, symbols, actions, and dialogues, that is, situated meaning. This helps the learner to understand the meaning of words better in different contexts.
- **Pleasantly frustrating**, challenging with a good doable problem-solving activity where players feel a sense of accomplishments and obtain a solution within a regime of competence-leading to flow since good games create flow (where players forget about time and immerse themselves in gameplay).

- **System thinking**, model-based reasoning in complex system or deep understanding where players think about systems like designers and producers about different variables, and how their decisions and actions might impact the outcomes.
- **Explore, think laterally, rethink goals**, exploring and thinking creatively and revisiting goals.
- **Smart tools and distributed knowledge** where the virtual character lend the skills and knowledge to the player.
- **Cross-functional team**, or identity, the commitment to play the roleplay.
- **Performance before competence principle**, the ability to practice the skills before competence.

Gee (2005a, 2005b:6–14, 2007, 2008, 2011) later on shortens these principles to 13 and groups them under three categories, "empowering learners," "problem-based learning," and "deep understanding." **"Empowering learners"** includes:

- **"Agent/co-design"** where the player takes the role of an agent designing part of the story/game, producing, making decisions, and co-designing at each level and affecting the outcome. Their decisions influence the game and the outcomes too.
- **"Customization"** is how the game invites different players to interact with the game, customizing their experience, giving them opportunity to try out new styles and different strategies, and therefore allowing them to become different types of learners.
- **"Identity"** where players know why they are playing, when, and how. There are clear goals/skills and different styles of doing things.
- **"Manipulation"** allows players to manipulate the game character where they feel empowerment and embodiment within that space, as their mind and body engage with the game. Players feel that they are part of that space and empowered.

While **"problem-based"** learning houses:

- **"Sequenced"**/well-ordered problems that contain levels where players set up to solve problems in a free space. Level of the complexity increases as the player goes up in levels, they try out new challenges in sequences and become creative problem-solvers.
- **"Pleasantly frustrating,"** challenging with a good doable problem-solving activity where players feel a sense of accomplishments and obtain a solution within a regime of competence leading to flow since good games create flow (where players forget about time and immerse themselves in gameplay).
- **"Cycle of expertise"** or challenge and consolidation where a set of challenging problems are set, and the player has to solve this and practice until they develop a mastery. Then the game introduces new problems, and the player has to use what they learn to solve new challenges.
- **Just "in time" and on "demand" principle** where information and knowledge is introduced step by step and as required.

- **"Fish tank"** principle breaking down complexities and systems, exploring a small version of a complex system with the aim of understanding it before moving on to explore larger complex systems.
- **"Sandbox"** principle or risk-free environment where players are not afraid of failing but are happy to take risks, explore, make progress, and gain feedbacks.
- **"Skills under strategies"** mastering the basic skills under a strategy before moving on to harder tasks, knowing reasons why such learning is crucial, and having a goal in mind.

"Understanding" is the third category within Gee's 13 principles and has:

- **"System thinking,"** deep understanding, or model-based reasoning in complex system where players think deeply about different variables and their relationships. They think about systems like designers and producers. They know how their decisions and actions might impact the outcomes.
- **"Situated meanings"** or meaning as action where players associate words with images, symbols, actions, virtual/simulated experiences and dialogues, that is, situated meaning. This helps the learner to understand the meaning of words better and applied it later on in different contexts.

Having highlighted Gee's (2003a, 2003b, 2004, 2005a, 2005b, 2007, 2008, 2011) principles, from learning perspectives, which come under three categories: "empowering" learners, "problem-based" learning, and "deep understanding," I will now focus on Gee's relevant principles to my research in this book, namely empowering learners (agent/co-design and identity), problem-based learning (sequenced/well-ordered problems) and deep understanding (see Chapter 10). I will apply these principles to my data analysis in Chapter 10 (Outline of the Main Findings and My Theoretical Contribution, to avoid repetition).

These concepts of Gee's games learning principles are of relevance to this book because of the way in which these principles are reflected in specific types of games. For instance, games have mental challenges, they are interactive and stimulating. They can create authentic contexts for practicing skills that can be transferred to real life, and they can present a "hub" and context for cognitive social learning, which is fundamental to the social constructivist perspective. Games have rules/goals, feedback, levels, and many cognitive, emotional, and social benefits. In gaming, users go through many cognitive processes while interacting with the game. For example, through gameplay (which involves cognitive skills such as planning, making decisions, and choices), the players think, play, and engage with the game, as an interactive system. This combination of cognitive learning and interaction with the game acts as a driving force in behavior/action (i.e., engagement), where the game player obtains feedback as a winning or losing state (see further details in the Simulation Games Section). This subsequently may have effects on learners' self-efficacy. Self-efficacy is discussed next.

Self-Efficacy Theory

As discussed earlier (Chapter 1), this research focuses on how we could enhance students' self-efficacy in Mathematics and Statistics using SGs (a subgenre of GBL). Research demonstrates that "blockage" about Mathematics and Statistics could be related to self-efficacy (Jameson and Fusco, 2014). In this book, I define self-efficacy and distinguish it from other related terms in the literature. I describe its role in terms of cognition, and how this can affect engagement, behavior, and therefore the learning outcomes.

In this book, I follow Bandura's definition of self-efficacy, I take self-efficacy as a psychological process that lies within one's personality or self-belief. This ability is within one's own self. On this basis, self-efficacy is the capability and ability to believe in one's own performance to execute Mathematics and Statistics tasks and produce an outcome without fear. It can be linked to Gee's (2003a, 2003b, 2004, 2005a, 2005b, 2007, 2011) 13 principles, in particular "agent/codesign," as I will explain later (Chapter 10).

In (1977), Bandura defines self-efficacy in terms of a cause effect relationship, that is, human motivation in relation to an outcome/consequence and an individual's abilities. This suggests that self-efficacy is associated with students' motivation, and it is linked to a previous experience and individuals' perceived abilities. Bandura (1997:25) defines self-efficacy as personal beliefs in one's capabilities and abilities to complete tasks and he assesses it through reporting its "level, generality, and strength" across activities. For example, the "level" of self-efficacy depends on how difficult the Mathematics and Statistics task is, whereas "generality" refers to "transferability of self-efficacy beliefs across activities" (Zimmerman, 2000:83) (i.e., self-belief to perform Statistics tasks across Mathematics activities from specific to general). "Strength" is related to how confidently a person can perform Mathematics and Statistics tasks. These characteristics of self-efficacy are usually measured using questionnaires with specific tasks, various levels of difficulty and confidence rating. Self-efficacy also depends on performance "context" or where learning is taking place (a library/lecture room, for instance) and a mastery of performance (Bandura, 1997). In other words, previous experience with success and whether a learner has achieved previously a success in that task or ability to complete it successfully. On the other hand, previous experience with failure can have negative impacts on learners' ability to complete tasks (Bandura, 1997). Self-efficacy refers to future performance and it is assessed before activities (Bandura, 1997). In the late 1970s, several researchers start to measure self-beliefs in each specific task. This was self-efficacy that Bandura measures. Bandura (1986) identifies it as the "self-efficacy" theory (Jarvis, 2005). Self-efficacy has "emerged as a highly effective predictor of students' motivation and learning" (Zimmerman, 2000:82).

Research shows that self-efficacy correlates highly with success in college, coursework, and perseverance (e.g., Bandura, 1977). Evidence from the literature shows that "self-efficacious" students are more determined, work harder, and put in more effort than others—with low self-efficacy. They are also less likely to have adverse emotional reactions (e.g., negative emotions toward learning) and are more likely to choose hard complex tasks rather than easy tasks. However, this can be only applied to high self-efficacious learners but not to learners with low self-efficacy (see Bandura, 1997). In other studies, Zimmerman (2000:86) suggests that academic motivation is a "choice of activities, level of effort, persistence and emotional reactions." According to Bandura and Schunk (1981), high self-efficacy is correlated with choice of activities. In a meta-analytic review of 70 studies on the relationship between self-efficacy and motivation, Multon et al. (1991) find that self-efficacy has positive significant effects on students' motivation and learning experiences. This indicates that self-efficacy plays a key role in motivation and learning (see Bandura, 1997). According to Bandura (1997) students' self-efficacy to manage learning activities (e.g., Mathematics and Statistics) can influence them emotionally by decreasing their negative emotions. This, thus (I propose), indicates that self-efficacy can have impacts on students' emotions and motivation; where self-efficacy is high, it can lead students to learn more, and where self-efficacy is low, it can lead to negative emotions. There is an increasing necessity to discern the effects of game-based learning (GBL) (Mayer et al., 2013).

From the body of research and theory, the following implications are provided for the current project:

- Educators should pay attention to self-efficacy, since it correlates highly with success at university, coursework, and perseverance. In other words, previous positive experience and whether a learner has achieved previously success in that task or ability to complete it successfully.
- High self-efficacious students are more determined, more resilient when faced with obstacles, work harder and put in more effort than others—with low self-efficacy. They also choose challenging tasks rather than easy ones. They manage learning activities (e.g., Mathematics and Statistics) which can influence them emotionally by decreasing their preconceptions. This can have impacts on students' emotions and motivation; where their self-efficacy is high, it can lead students to learn more, and where self-efficacy is low, it can lead to negative emotions (i.e., preconceptions toward learning).
- The changing profile of today's learners and the use of technology as a method of communication and networking (as well as the great flexibility and wide-availability of internet-based learning) has given generation "Z" great perception (or positive views) in terms of ease of use, fun, and usefulness, that is there is a shift in learning styles (Jagger et al., 2016).

Accordingly, there is a necessity to find new methods to engage generations "Z" (Dede, 2005; Jagger et al., 2016) and "Alpha" (Mccrindle, 2018). One possibility is by employing GBL (as a learning tool), because today's learners are adopting and adapting games as a way of learning. They also prefer stimulating teaching methods (Anderson and Lawton, 2009, cited in Jagger et al., 2016), that is, "stimulated learning" (in other words, engaging learning). Therefore, there is an urgent necessity to adapt and adjust our teaching methods and styles as educators/game researchers in order to facilitate and cater for these learners' needs. Perhaps an interesting problem to explore is how we can employ SGs creatively in HE (as I will discuss later on in the Simulation GBL Approach to Teaching Section).

Overall, the learning theory examined in this section—socio-constructivism has a pedagogic stance; that is empowering learners to develop a better understanding of their learning and build their confidence. The next section discusses the characteristics of games, considers how they can be defined in relation to GBL, and examines theories relating to simulation games and how they can be used as a learning tool.

Games

Having examined a socio-constructivist perspective and the potential benefit of GBL, this section discusses a body of literature review on games in the context of game-based learning. The nature of games and gaming and a working definition is suggested. In addition, the characteristics of games, different game types, and motivations for individuals to play games, both for entertainment and learning, are also considered.

Games can be complex, and researchers present different definitions. Throughout this book, I follow Schell's (2015, 2020) definition of games, I take games as a fun playful dynamic learning activity with a fixed set of rules, and limited time, and can simulate real-world problems. It has specific rules, usually ends in either win or lose, and involves manipulation, curiosity, secrecy, and mystery. For instance, Koster (2005:148) describes games as "well-crafted…art," while Garris et al. (2002) and Kafai (2006) perceive games as a complex model that can solve real-world problems and increase confidence. According to Whitton (2011), games can be described as some visual digital system that accepts input from players and uses this input according to a set of programming rules. Schell (2015, 2020), in contrast, defines a game as a fun playful activity with surprises. Play is also defined in various ways.

On the one hand, play is a "free activity standing quite consciously outside life," not serious but systematic with a set of rules and specific "boundaries of time and space according to fixed rules and in an orderly manner" (Huizinga, 1938, cited in Salen and Zimmerman, 2011:4). In contrast, Koster (2005) considers play as a life learning skill and Resnick (2017:128) associates play with creativity and argues that creativity comes from "experimenting," risk-taking, and "testing boundaries." Pelletier (2009) argues that play contributes to cognitive, moral, and social development. Play is about fun and relaxation or a person desires and feelings. However, different disciplines define play differently. According to Pelletier (2009), in education, play is seen as progress where individuals develop some sort of skills and progress leads to growth.

It is worth noting that a "play" and "game" have different meanings. Games are about making choices and solving puzzles. They consist of specific rules and may have only two outcomes: Either win or lose. This is a luring motivator for game players, because game players tend to wish to avoid losing and so they perform their best. Play, in contrast, is a "manipulation that indulges curiosity" (Schell, 2015:40, 2020), "secrecy," and "mystery." In other words, it is the "make-believe by nature" approach to play, a source of happiness and fun activity within a limited time and a fixed set of rules (Salen and Zimmerman, 2011:5). As a game researcher, this playful and fun activity can lead to enjoyment and pleasure. This enjoyment/pleasure, according to Lazzaro (2004), can occur through brain releasing "endorphins" into the human system naturally (aesthetic appreciation) or via chemical manipulations (taking drugs) (McGonigal, 2015; Schell, 2015, 2020). Drugs in gaming "terminology," according to Koster (2005), are referred to as the "learning" process that goes on inside the brain, the stimulated pleasure one obtains from learning. Lazzaro (2004) describes it as "hard fun" (challenging). Csikszentmihalyi (1997) calls it a "flow" (complete immersion). Fun is the act of mastering a problem mentally. It is the "feedback the brain gives us when we are absorbing patterns for learning purposes" (Koster, 2005:94–97). It is about practicing, learning, and gaining more confidence.

Game Types

Game types are many and diverse. These include simulations, educational, edutainment and first-shooter games, adventure, Statistics, memory, engineering, and strategy games. In terms of modality, game types include paper-based games, for instance noughts and crosses game on papers, digital-based (online or offline) games on PCs, tablets, videos, and mobile phones. They can be played by a single player (oneself) or multiple players with various genres from action, simulations, quiz, adventure, and sports to puzzle games (Gee, 2003a, 2003b; Lazzaro, 2004; McGonigal, 2015; Schell, 2015, 2020). In the context of GBL, games such as simulations, educational or serious games, and edutainment are often included in discussions.

Game-Based Learning (GBL) is a type of learning that can occur when playing games. GBL uses digital technology to deliver and address teaching, learning, and assessment. They can have learning outcomes where players can maintain the enjoyment of play while learning is absorbed (Vos, 2014; Skilbeck, 2017). An area often associated with GBL is simulations (see Chapter 3 for further discussion).

Simulation games are a type of GBL that can simulate an environment or an aspect of it. It can be used as a method of teaching and learning. Simulation is an interactive environment where students can apply previous knowledge and practical skills to solve complex problems (Vos, 2014). The idea is that they investigate tangible scenarios just as in real life with the aim of building on their self-efficacy, employability, and leadership skills (Vlachopoulos and Makri, 2017). One example is "SimCity," which is an entertainment simulation game. "SimCity" will be discussed in detail in Chapter 5, analyzed in Chapter 6 as part of this empirical work, and discussed more in Simulation Games Section, as it is the focus of this book.

Educational games are a type of GBL that can have specific rules and educational (pedagogic) learning outcomes. Educational games focus on education rather than entertainment or fun, and they are intended for other purposes than entertainment (Zyda, 2005). One example is "Numbas," which is an educational game aimed at teaching Mathematics. "Numbas" will be discussed in detail in Chapter 5 and analyzed in Chapter 6, as part of this empirical work.

Edutainment is another type of GBL that integrates education and entertainment at the same time as its main focus besides gamification. Gamification is the use of points or rewards for instance in a non-game context with the purpose of turning iterative activities into motivating tasks (Isaacs, 2015, cited in Crockett, 2018). There is a substantial literature for both edutainment and gamification (Jagger et al., 2016). Both perspectives are beyond the scope of this book, due to a word limitation.

There are many characteristics, elements, features, or activities, associated with games in the context of GBL, which makes games intriguing and captivating. Different researchers present different characteristics, and sometimes they emphasize one element over the other. For instance, Malone (1980) suggests that game characteristics are: Challenge, fantasy, and curiosity, while de Felix and Johnston (1993 cited in Garris et al., 2002) propose that the essential characteristics of games are: Dynamic visuals, interaction, rules, and a goal. Garris et al. (2002), on the other hand, suggest that game characteristics are fantasy, rules/ goals, sensory stimuli, challenge, mystery, and control. Dondlinger (2007) adds to this gameplay, rewards, learning objectives, and storytelling. Kolb and Kolb (2008) conclude that games characteristics comprise interesting choices, fantasy, rules/goals, challenge, mystery, reward and control, immediate and useful feedback. As explained above, there are many game characteristics, for the purpose of understanding, analyzing, and evaluating games (which will be revisited in Chapters 4 and 6). I conclude that game characteristics are, according to Garris et al. (2002), Prensky (2002), and Schell (2015, 2020):

- Game learning objectives.
- Game storylines.
- Game mechanics (the functionalities of the game, the core mechanics and how the game behaves).
- Rules (the rules of the game include players' action, abilities, and constraints).
- Goals and objectives (what motivates players to achieve/gain from the game).
- Game motives: Challenge/competition (this is how the game is played; it could be against oneself as in "Sudoku" or against another player as in "8 Ball Pool"
- Game risks.
- Game design.
- Game interactions (players' interaction with the game and other players).

Based on the game characteristics (above) and to further increase our understanding of how different games function, a review of many games conducted including a sample of 16 existing games (different representative) is shown in Table 2.1.

Having reviewed many and different representative samples of computer games (see an example, Table 2.1) from Sudoku, 8 Ball Pall, to Numbas, I will now focus on simulation vs. quiz games (i.e., SimCity vs. Numbas) in order to gain further insights into the potential value of GBL (in Simulation Games Section; see further details in Chapter 3). I will return to discuss SimCity and Numbas games in detail in Chapters 5 and 6, since these are game design and development chapters.

As noted above, this book focuses on how we can improve self-efficacy in Mathematics and Statistics using SGs. The next (and final) section of this chapter will discuss simulation games, define them, and consider how they can be employed as a learning tool in HE.

Simulation Games

Simulation is a method of teaching/learning. It is an interactive environment where students can apply previous knowledge and practical skills to solve complex problems. Simulation games are an experiential learning method. They have game features and consist of choices, fantasy, rules/goals, challenge, mystery, reward, control, and immediate feedback (Caillois, 1961; Malone, 1980; Jones, 1998; Garris et al., 2002; Gee, 2003a, 2003b; Prensky, 2006). Simulation games are widely used in many fields, such as healthcare, marketing and higher education, and they are significant because students can investigate real-life scenarios, in a safe environment and without involving any high costs (e.g., Faria and Wellington, 2004; Vos, 2014). They are aimed at building students' employability or leadership skills (Vlachopoulos and Makri, 2017) and therefore their transferable skills.

According to Vos (2014), there are three types of simulation games: Introductory, strategic, and specialist games. Introductory games are those aimed at introducing, for instance, principles of business marketing. Strategic games are used for "those in higher level of marketing and business courses" and specialist games focus on specific aspects of "marketing decision-making" (e.g., online marketing). Simulation strategic games, according to Vos, have operational and strategic decisions: Involve decision-making, finance, production, human resources, competitors, performance indicators, and numerical data to interpret and measure decisions against (Vos, 2014:68–69).

Examples of simulation games include flight and military simulators. Flight simulators are types of educational/serious games that teach novice pilots essential flying skills without running the risk of endangering themselves or others, whereas in the military, these game simulators teach the army numerous strategic military techniques from ground combat, first aid to handling weapons. For example, "League of Legends" is a multiplayer strategic game

Table 2.1 A review of 16 different types of games played by the author used to gain insights into the potential value of game-based learning and what skills it can, or could, improve (Slyman, 2022:30).

Examples of different types of games	*Types/genres*	*Game status*	*Game characteristics and its potential GBL value*	
			What it can improve	*What it could improve*
Angry Birds	Comic.	Online/Offline/ Mobile/PC.	Mood, competition, self-efficacy skills, and happiness.	Game design; interactivity; and creativity.
Stack	Mathematics, construction and building.	Online/Offline/PC/ Mobile /Video.	Mood and happiness, memory, concentration, self-efficacy, and strategic skills.	Interactivity, game design, and interaction; game mechanics.
Sudoku	Mathematics; puzzle.	Online/Offline/PC/ Mobile /Video.	Memory, mind, self-efficacy, and Mathematical skills.	Numbers; game design and interaction; game mechanics.
Star Craft	Science fiction; strategy.	Online/Offline/PC/ Mobile /Video.	Problem-solving; increase spatial attention, self-efficacy and concentration skills, determination, and speed.	Time-limit; creativity; storylines; rules.
FIFA 14	Sport game; football;strategy.	Online/Offline/PC/ Mobile /Video.	Socializing, self-efficacy and strategy skills.	Time-limit; game risks; game design.
Call of Duty	Adventure; multi-genreshooting.	Online/OfflinePC/ Video.	Visual attention and decision-making; curiosity; increase spatial attention, self-efficacy and concentration skills, determination, and speed.	Storylines (reduce violence).
8 Ball Pool	Sports; strategy.	Online/Offline/PC/ Mobile /Video.	Visual attention; socializing; competition, self-efficacy, and decision-making.	Storylines; surprise; time-limit.
Chess	Cardboard; strategy.	Online/Offline/PC/ Mobile /Video.	Social interaction, play and socializing; self-efficacy anddecision-making.	Time-limit; game interaction; game design and mechanics.
Minecraft	Adventure.	Online/Offline/PC/ Mobile /Video.	Increase spatial attention, concentration and self-efficacy skills, determination, and speed.	Time-limit; game interaction; game design and mechanics.

(*Continued*)

Table 2.1 (Continued)

Examples of different types of games	*Types/genres*	*Game status*	*Game characteristics and its potential GBL value*	
			What it can improve	*What it could improve*
Super Mario	Arcade; adventure.	Online/Offline/PC/ Mobile /Video.	Mood, happiness; self-efficacy, curiosity, and speed.	Time-limit; game interaction; game design; storylines; surprises.
Bubble	Pattern; shooting.	Online/Offline/PC/ Mobile /Video.	Logic, happiness, visual attention, decision-making and self-efficacy; competition; interactive; reaction time.	Time-limit; Game riskssocial connection to other online sites; customer feedback support; human computer interaction.
Candy Crush	Pattern; puzzle.	Online/Offline/PC/ Mobile /Video.	Visual attention, self-efficacy, and decision-making.	Time-limit; game interaction; game design.
Kahoot	Quiz.	Online/PC/Mobile.	Visual attention; decision-making; problem-solving and self-efficacy; competition/ team competing against each other.	Game interaction; game design and mechanics; surviving skills, mission; features; social interaction; rewards or feedback.
Quizlet	Puzzle; quiz.	Online/PC/Mobile.	Visual attention, self-efficacy, and decision-making; answering questions; multiple choice questions; revising key terms/ curriculum-based activities; competition.	Game interaction; game design and mechanics; creative and critical thinking skills, mission; features; social interaction.
SimCity BuildIt	Simulation; roleplay and strategy.	Online/Offline PC/ Video/ Mobile.	Decision-making, self-efficacy, analytical strategic and creative thinking skills.	Feedback; human computer interaction; online pay system; critical thinking skills.
Numbas	Quiz.	Online/Offline PC/ Video/Mobile.	Mathematical, analytical, self-efficacy and thinking skills.	Feedback; human computer interaction; online pay system. Game interaction; game design and mechanics.

where players play against other team in battles and win some league points. Other examples include "Minecraft," "SimCity," car racing games, fantasy, and sports, for instance FIFA and adventure games. Adventure games consist of storylines and characters that interact with the system to solve puzzles and quizzes. Simulation games can involve "slow pace, reflection, study of the environment and problem-solving" (Moreno-Ger et al., 2008:2535).

Simulation Games as a Learning Tool

Simulation games could be a facilitating tool for assisting learning (Faria and Wellington, 2004; Tao et al., 2012; Li et al., 2013; Jagger et al., 2016; Hamari et al., 2016). There are several reasons behind integrating games as a learning tool, since it allows students to engage in a real-life experience and use decision-making skills to resolve real-world problems in a safe environment (Vos and Brennan, 2010; Vos, 2014). In return, students can develop good time management skills, and the ability to interpret numerical and financial data, thus obtaining a better understanding of how marketing performance is measured. In a study conducted by Li et al. (2013) where they construct a railway simulation model to teach first-year Computer Science undergraduates basic programming and object-oriented skills, the researchers find that simulation games have a positive influence on students' perceptions, self-efficacy, and motivation skills. According to Amory et al. (1999 cited in Dondlinger, 2007), simulation games, adventure, and strategic games are ranked the most popular. This is because players enjoy using their problem-solving and decision-making skills in these simulations.

Research studies suggest that simulation games can bring pedagogical benefits (Kolb and Kolb, 2008). This is because simulation games allow students to investigate real-life complex scenarios suited to work-life situations of intense global-market competition and rapidly evolving technology where students formulate crucial decisions but are allowed to make mistakes and learn from their experience without involving any financial costs or damage to their organization (Vos, 2014). In other words, students gain transferable skills. For instance, games like "Minecraft" (a digital game) are aimed at teaching players how to widen their strategic and creative skills using Mathematical concepts such as cubes, perimeter, and probability. Research has shown that SGs can encourage students to complete a great deal of Mathematical problems due to its interactive simulated features (Dondlinger, 2007; Lee et al., 2012). SGs can play a beneficial role in teaching and learning because they can strategically enable students to see the impacts of their decision-making straightaway, in a friendly environment (Vos, 2014). Some of the learning objectives of the SGs are to teach students to learn to think strategically, invest, and make decisions in iterative cyclic processes, allowing them to refine their decisions over time, thus showing real-life impacts on companies' performances. For example, games such as the "Age of Computers" or the "Leaping Frogs," with its problem/solution scenarios encourage students' problem-solving and decision-making strategies to develop (Sindre et al., 2009; Li et al., 2013), whereas SGs such as the popular SG "SimCity" require players to plan, build, expand, and manage a city within the computer simulation. Players use their analytical and thinking skills to strategically decide each time how to approach tasks and build a city, considering short- and long-term effects on the city. Consequently, I am specifically interested in these types of SGs such as "SimCity" (see Chapter 5 for further discussion of this game) and how I can embed Mathematics and Statistics with technology as a design tool for building engaging GBL experiences. GBL, its impacts, and the potential value of its use are discussed next (Chapter 3).

In all, this chapter has presented previous research on learning, self-efficacy, and games, games types and game characteristics, GBL and SGs, looking at a number of learning theories. It then discussed self-efficacy theory and its implications for this book project. This chapter has considered the theoretical potential for computer games to be an effective tool for teaching and learning.

The next chapter, **Chapter 3**, will explore the potential of game-based learning: Theoretical foundations and case studies, highlighting the significance of GBL on users' experiences from psychological and social lens.

Bibliography

Adams, A., Lunt, P. and Cairns, P. (2008) A Qualitative Approach to HCI Research. In: Cairns, P. and Cox, A. (ed.). *Research Methods for Human-Computer Interaction.* Cambridge, UK: Cambridge University Press, pp. 138–157. Available at: http://www.cambridge.org/catalogue/catalogue.asp?isbn=9780521870122andss=toc (Accessed 20/11/2016).

ADDIE Model (2016) Available at: http://www.learning-theories.com/addie-model.html (Accessed 20/04/2016).

Anderson, S. (2020) Quantitative Analysis. Available at: https://www.investopedia.com/terms/q/quantitativeanalysis.asp (Accessed 20/02/2021).

Applefield, J.M., Huber, R. and Moallem, M. (2000) Constructivism in Theory and Practice: Toward a Better Understanding. *The High School Journal*, 84(2), pp. 35–53.

Arthur, J., Waring, M. Coe, R. and Hedges, L.V. (2012) *Research Methods and Methodologies in Education.* California: Sage Publications Inc.

Bandura, A. (1977) Self-efficacy: Toward a Unifying Theory of Behavioral Change. *Psychological Review*, 2(84), pp. 191–215. Available at: https://edu.ational-innovation.sydney.edu.au/news/pdfs/Bandura%201977.pdf

Bandura, A. (1986) *Social Foundations of Thought and Action: A Social Cognitive Theory*. Englewood Cliffs, NJ: Prentice-Hall.

Bandura, A. (1997) *Self-efficacy: The Exercise of Control.* New York: W.H. Freeman and Company.

Bandura, A. and Schunk, D. (1981) Cultivating Competence, Self-Efficacy, and Intrinsic Interest Through Proximal Self-Motivation. *Journal of Personality and Social Psychology*, 41(3), pp. 586–598.

Barbosa, T., Lopes, S.P. Leao. C. Soares, F. and Carvalho, V. (2019) Serious Game for Teaching Statistics in Higher Education: Storyboard Design. In: *7th EAI International Conference, Braga, Portugal, on 'Interactivity, Game Creation, Design, Learning, and Innovation'* (ed.) Springer, pp. 169–175. DOI: 10.1007/978-3-030-06134-0_18

Bartholomew, H., Darragh, L., Ell, F. and Saunders, J. (2011) "I am Natural, and I Do It for a Love!": Exploring Students' Accounts of Studying Mathematics. *International Journal of Mathematical Education in Science and Technology*, 42(7), pp. 915–924.

Bassey, M. (1999) *Case Study Research in Educational Settings*. Buckingham: Open University Press.

Bellotti, F., Kapralos, B. Lee, K. Moreno-Ger, P. and Berta, R. (2013) Assessment in and of Serious Games: An overview. *Advances in Human-Computer Interaction*, 2013, p. 1.

Belluigi, D.Z. (2009) Exploring the Discourses Around 'Creativity' and 'Critical Thinking' in a South African Creative Arts Curriculum. *Studies in Higher Education*, 34(6), pp. 699–717.

Bennedesen, J. and Caspersen, M.E. (2004) Programming in Context: A Model-First Approach to CS1. In *Proceedings of the 35th SIGCSE Technical Symposium on the Computer Science Education*, pp. 477–481. ACM Press.

Bevan, N. (2001) International Standards for HCl and Usability. *International Journal of Human Computer Studies*, 55(4), pp. 533–552.

Boaler, J. (2009) *The Elephant in the Classroom*. London: Souvenir Press.

Brennan, R. and Vos, L. (2010) Marketing Simulation Games: Student and Lecturer Perspectives. *Marketing Intelligence and Planning*.

Brown, J.S., Collins, A. and Duguid, P. (1989) Situated Cognition and the Culture of Learning. *Educational Researcher*, 18(1), 32–42.

Bruner, J. (1973) *Going Beyond the Information Given*. New York: Norton.

Bryant, A. (2017) *Grounded Theory and Grounded Theorizing: Pragmatism in Research Practice*. New York, NY: Oxford University Press.

Bryman, A. (2008) *Social Research Methods*, 3rd Edition. Oxford: Oxford University Press (first published: 2001).

Bryman, A. (2012) *Social Research Methods*. 4th Edition. United States: Oxford University Press.

Bryman, A. (2016) *Social Research Methods*. 5th Edition. United States: Oxford University Press.

Burton, R.R., Brown, J.S. and Fischer, G. (1984) Skiing as a Model of Instruction. In B. Rogoff and J. Lave (ed.). *Everyday Cognition: Its Development in Social Context*, pp. 139–150.

Caillois, R. (1961) *Man, Play, and Games*. New York: Free Press.

Cardinot, A. and Fairfield, J.A. (2019) Game-Based Learning to Engage Students with Physics and Astronomy Using a Board Game. *International Journal of Game-Based Learning (IJGBL)*, 9(1), pp. 42–57.

Charmaz, K. (2006) *Constructing Grounded Theory*, 1st Edition. London: Sage Publications Ltd.

Charmaz, K. (2014) *Constructing Grounded Theory*, 2nd Edition. London: Sage Publications Ltd.

Chen, M.P., Wong, Y.T. and Wang, L.C. (2014) Effects of Type of Exploratory Strategy and Prior Knowledge on Middle School Students' Learning of Chemical Formulas from a 3D Roleplaying Game. *Educational Technology Research and Development*, 62, pp. 163–185. DOI: 10.1007/s11423-013-9324-3

Clark, D.B., Tanner-Smith, E.E. and Killingsworth, S.S. (2015) Digital Games, Design, and Learning a Systematic Review and Meta-Analysis. *Review of Educational Research*, 86(1), pp. 79–122.

Claxton, G. (2002) *Building Learning Power*, 1st Edition. Bristol: TLO Limited.

Cohen, L., Manion, L. and Morrison, K. (2011) *Research Methods in Education*. London: Routledge.

Cohen, L., Manion, L. and Morrison, K. (2018) *Research Methods in Education*. London: Routledge.

Conlon, C., Carney, G., Timonen, V. and Scharf, T. (2015) Emergent Reconstruction in Grounded Theory: Learning from Team-Based Interview Research. *Qualitative Research*, 15, pp. 39–56. DOI: 10.1177/1468794113495038

Cook, J. (2005) Negotiating the Curriculum: Programming for Learning. In *Negotiating the Curriculum*, pp. 15–31. Routledge.

Cooper, P.A. (1993) Paradigm Shifts in Designed Instruction: From Behaviorism to Cognitivism to Constructivism. *Educational Technology*, 33(5), pp. 12–19.

Corbin, J. and Strauss, A. (2015) *Basics of Qualitative Research: Techniques and Procedures for Developing Grounded Theory*, 4th Edition. Thousand Oaks, CA: Sage.

Cordova, D.I. and Lepper, M.R. (1996) Intrinsic Motivation and the Process of Learning: Beneficial Effects of Contextualization, Personalization, and Choice. *Journal of Educational Psychology*, 88, pp. 715–730.

Coulthard, G.J. (2009) A Review of the Educational Use and Learning Effectiveness of Simulations and Games [online]. PhD Thesis, Purdue University. Available at: http://www.coulthard.com/library/paper-simulation.html (Accessed 15/12/2015).

Craft, A. (1999) Creative Development in the Early Years: Some Implications of Policy for Practice. *The Curriculum Journal*, 10(1), pp. 135–150.

Craft, A. (2002) *Creativity and Early Years Education*. London: Continuum.

Crawford, C. (1984) *The Art of Computer Game Design*. Berkeley: California, McGraw-Hill/Osborne Media.

Creswell, I.W. (2009) *Research Design: Qualitative, Quantitative, and Mixed Methods Approaches*, 3rd Edition. London: Sage.

Crockett, L. (2018) Bogost's War on Gamification: Irony and Play on Grandma's Sofa. In: *12th European Conference on Games Based Learning (ECGBL 2018)*. Reading: Academic Conferences and Publishing International Ltd.

Cronan, T.P. and Douglas, D.E. (2012) A Student ERP Simulation Game: A Longitudinal Study. *Journal of Computer Information Systems*, 53(1), pp. 3–13.

Csikszentmihalyi, M. (1990) Psychology Today: Finding Flow. Available at: https://www.researchgate.net/publication/200026151_Finding_Flow_The_Psychology_of_Engagement_With_Everyday_Life/link/55ad6c2f08aed9b7dcdae222/download (Accessed 03/05/2016).

Csikszentmihalyi, M. (1997) *Creativity, Flow and the Psychology of Discovery and Invention*. New York: Harper.

Davis, F.D. (1989) Perceived Usefulness, Perceived Ease of Use, and User Acceptance of Information Technology. *MIS Quarterly*, 13(3), pp. 319–340.

De Bono, E. (1992) *Teach Your Child How to Think*. England: McQuaig Group Inc.

Dede, C. (2005) Planning for Neomillennial Learning Styles. *Educause Quarterly*, 28(1), pp. 7–12.

Derryberry, A. (2007) Serious Games: Online Games for Learning. White Paper. [Online]. Available at: http://www.adobe.com/resources/elearning/pdfs/serious_games_wp.pdf (Accessed 10/12/2015).

Desurive, H., Caplan, M. and Toth, J. (2004) Using Heuristics to Evaluate the Playability of Games. *CHI Conference*, 32 (1), pp. 97–110. DOI: 10.1177/104687810103200108

Desurvire, H. and Wiberg, C. (2009) Game Usability Heuristics (PLAY) for Evaluation and Designing Better Games: The Next Iteration. In *Online Communities and Social Computing: Third International Conference, OCSC 2009, Held as Part of HCI International 2009*, San Diego, CA, USA, July 19–24, 2009. Proceedings 3, pp. 557–566. Springer Berlin Heidelberg.

Desurvire, H. and Wiberg, C. (2010) User Experience Design for Inexperienced Gamers: GAP – Game Approachability Principles. In: Regina Bernhaupt (ed.). *Evaluating User Experience in Games*. Springer, pp. 131–148. DOI: 10.1007/978-1-84882-963-3_1

Dondlinger, J.M. (2007) Educational Video Game Design: A Review of the Literature. *Journal of Applied Educational Technology*, 4(1), Spring/Summer, pp. 21–31. Available at: https://www.researchgate.net/publication/238444705_Educational_Video_Game_Design_A_Review_of_the_Literature (Accessed 12/12/2015).

Draper, S. (2013) Social Constructivism. Available at: http://www.psy.gla.ac.uk/~steve/courses/archive/CERE12-13-safari-archive/topic3/webarchive-index.html#social%20con%201 (Accessed 23/12/2019).

Dunne, C. (2011) The Place of The Literature Review in Grounded Theory Research. *International Journal of Social Research Methodology*, 14, pp. 111–124. DOI: 10.1080/13645579.2010.494930

Dweck, C.S. (2006) *Mindset: The New Psychology of Success*. United States: Ballantine Books Trade Paperback Edition.

Dwyer, J. and Slyman, S. (2016) *An Introduction to Quantitative Methods for Business Research*. London: Algana Publishing. 200 pages; ISBN–13: 9781907934094.

Early, R.E. (1992) The Alchemy of Mathematical Experience: A Psychoanalysis of Student Writings, for the Learning of Statistics, 12(1), pp. 15–20.

Fairweather, E. and Cramond, B. (2010) Infusing Creative and Critical Thinking into the Curriculum Together. In: Beghetto, A. and Kaufman, J.C. (ed.). *Nurturing Creativity in the Classroom*. United States: Cambridge University Press.

Falstein, N. and Barwood, H. (2001) The 400 Project. Available at: http://theinspiracy.com/400_project.htm. (Accessed 20/04/2016).

Faria, A.J. (2001) The Changing Nature of Business Simulation/Gaming Research. *Simulation and Gaming*, 32(1), pp. 97–110. DOI: 10.1177/104687810103200108

Faria, A.J., Hutchinson, D. Wellington, W.J. and Gold, S. (2009) Developments in Business Gaming: A Review of the Past 40 Years. *Simulation and Gaming*, 40(4), pp. 464–487. DOI: 10.1177/1046878108327585

Faria, A.J. and Wellington, W.J. (2004) 'A Survey of Simulation Game Users, Former-Users, and Never-Users. *Simulation and Gaming*, 35(2), pp. 178–207. DOI: 10.1177/1046878104263543

Federoff, M.A. (2002) Heuristics and Usability Guidelines for the Creation and Evaluation of Fun in Video Games. Master Thesis, Indiana University, Bloomington.

Fienberg, S.E. (1992) A Brief History of Statistics in Three and One-Half Chapters: A Review Essay. Available at: https://www.jstor.org/stable/2246307?casa_token=Vek3Rmd5Mk0AAAAA%3AJ0zRZC5AxskwvTNY7GlrgsRV6oLrE20DY3FdeaPad0JvNSNrZDav8uZpX-YqY1hXTDtaz6zM87S4vwkQdflDLSzmrLCNlEsaBqWnykepc0jlYXsQXA

Forest, E. (2014) ADDIE Model. Available at: http://educationaltechnology.net/the-addie-model-instructional-design/ (Accessed 20/04/2016).

Fosnot, T.C. (1989) *Enquiring Teachers, Enquiring Learners: A Constructivist Approach for Teaching*. New York: Teachers College Press.

Fosnot, T.C. and Perry, R.S. (2005) Constructivism: A Psychological Theory of Learning. Available at: http://beyondbitsandatoms.org/readings/fosnot_constructivism_1996.pdf (Accessed 20/07/2019).

Garris, R., Ahlers, R. and Driskell, E.J. (2002) Games, Motivation, and Learning: A Research and Practice Model. *Simulation and Gaming*, 33(4), pp. 441–467.

Gee, J.P. (2003a) *What Video Games Have to Teach Us About Learning and Literacy*. New York: Palgrave Macmillan.

Gee, J.P. (2003b) Good Video Games and Good Learning. Available at: https://academiccolab.org/resources/documents/Good_Learning.pdf (Accessed 03/01/2021).

Gee, J.P. (2004) *Situated Language and Learning: A Critique of Traditional Schooling*. London: Routledge.

Gee, J.P. (2005a) Why are Video Games Good for Learning? Retrieved January 1, 2018, from http://academiccolab.org/resources/documents/MacArthur.pdf (Accessed 20/04/2016).

Gee, J.P. (2005b) *Why Video Games Are Good for Your Soul: Pleasure and Learning*. Melbourne: Common Ground.

Gee, J.P. (2007) *Good Video Games and Good Learning: Collected Essays on Video Games, Learning, and Literacy*. New York: Peter Lang.

Gee, J.P. (2008) *Getting over the Slump: Innovation Strategies to Promote Children's Learning*. New York: The Joan Ganz Cooney Center at Sesame Workshop.

Gee, J.P., Hayes, E., Torres, R.J., Games, I.A., Squire, K. and Salen, K. (2008) Playing to Learn Game Design Skills in A Game Context. In: *Proceeding ICLS'08 Proceedings of the 8th International Conference on International Conference for the Learning Sciences*, Vol. 3, pp. 368–374.

Gee, J.P. and Hayes, E.R., 2011. *Language and Learning in the Digital Age*. Routledge.

Gibbons, J. and Gray, M. (2004) Critical Thinking as Integral to Social Work Practice. *Journal of Teaching in Social Work*, 24, pp. 19–37.

Glaser, B.G. and Strauss, A.L. (1967) *The Discovery of Grounded Theory*. Chicago: Aldine.

Gomes, A. and Mendes, A.J. (2007) Learning to Program-Difficulties and Solutions. In *International Conference on Engineering Education–ICEE*, Vol. 7.

Granic, I., Lobel, A. and Engels, R.C. (2014) The Benefits of Playing Video Games. *American Psychological Association*, 69(1), pp. 66–78. DOI: 10.1037/a0034857

Hackett, G. and Betz, N.E. (1989) An Exploration of the Mathematics Self-Efficacy/ Mathematics Performance Correspondence. *Journal for Research in Mathematics Education*, 20(3), pp. 261–273.

Hamari, J., Shernoff, D.J., Rowe, E., Coller, B., Asbell-Clarke, J. and Edwards, T. (2016) Challenging Games Help Students Learn: An Empirical Study on Engagement, Flow and Immersion in Game-Based Learning. *Computers in Human Behavior*, 54, pp. 170–179.

Hamari, J. and Tuunanen, J. (2014) Player Types: A Meta-Synthesis. *Transactions of the Digital Games Research Association*, 1(2), pp. 29–53.

Harel, I. and Papert, S. (1990) *Software Design as a Learning Environment*. Massachusetts Institute of Technology Cambridge, MA: MIT Media Laboratory.

Harel, I. and Papert, S. (1991) *Constructionism*. Norwood, NY: Ablex Publishing Corporation.

Hargreaves, A. (2003) *Teaching in the Knowledge Society: Education in the Age of Insecurity*. New York: Teachers' College Press.

Holbert. N., Penney, L. and Wilensky, U. (2010) *Bringing Constructionism to Action Game-Play, Constructionism*. Paris. Available at: https://www.google.co.uk/search?q=Bringing+Constructionism+to+Action+Game-Playandoq=Bringing+Constructionism+to+Action+Game-Playandaqs=chrome..69i57.7170j0j8andsourceid=chromeandie=UTF-8 (Accessed 20/03/2017).

Hoover, W.A. (1996) The Practice Implications of Constructivism. *SEDL Letter*, 9(3), pp. 1–2.

Horn, B., Hoover, A.K. Barnes, J. Folajimi, Y. Smith, G. and Harteveld, C. (2016) Opening the Black Box of Play: Strategy Analysis of An Educational Game. In *Proceedings of the 2016 Annual Symposium on Computer-Human Interaction in Play*, pp. 142–153.

Hsu, C.-Y., Tsai, C.-C. and Wang, H.-Y. (2012) Facilitating Third Graders' Acquisition of Scientific Concepts Through Digital Game-Based Learning: The Effects of Self-Explanation Principles. *Asia-Pacific Education Researcher*, 21(1), pp. 71–82.

Hunicke, R., LeBlanc M. and Zubek, R. (2004) MDA: A Formal Approach to Game Design and Game Research. Available at: https://www.google.co.uk/search?q=MDA%3A+A+Formal+Approach+to+Game+Design+and+Game+Researchandoq=MDA%3A+A+Formal+Approach+to+Game+Design+and+Game+Researchandaqs=chrome..69i57j69i58.3868j0j8andsourceid=chromeandie=UTF-8 (Accessed 8/10/2016).

Illeris, K. (2007) *How We Learn: Learning and Non-Learning in School and Beyond.* London: Routledge.
ISO (1998) Available at: http://www.standardsglossary.com/iso1.htm (Accessed 8/10/2016).
Jackson, N. and Shaw, M. (2007) Developing Perspectives on Creativity in Higher Education. In: N. Jackson, M. Oliver, M. Shaw, and J. Wisdom (ed.). *Developing Creativity in Higher Education: An imaginative curriculum.* London: Routledge.
Jagger, S., Siala, H. and Sloan, D. (2016) It's All in The Game: A 3D Learning Model for Business Ethics. *Journal of Business Ethics*, 137(2), pp. 383–403.
Jameson, M.M. and Fusco, B.R. (2014) Math Anxiety, Math Self-Concept, and Math Self-Efficacy in Adult Learners Compared to Traditional Undergraduate Students. *Adult Education Quarterly*, 64(4), pp. 306–322.
Jarvis, M. (2005) *'Motivation' in Psychology of Learning and Teaching.* Cheltenham: Nelson Thornes.
Jeffrey, B. and Craft, A. (2004) Teaching Creatively and Teaching for Creativity: Distinctions and Relationships. *Educational Studies*, 30(1), pp. 77–87.
Jenson, J. and Droumeva, M. (2015) Making Games with Game Maker: A Computational Thinking Curriculum Case Study. In *ECGBL 2015–9th European Conference on Games Based Learning: ECGBL2015.* Academic Conferences and Publishing Limited, p. 260.
Jones, M.G. (1998) Creating Engagement in Computer-Based Learning Environments. Available at: https://www.yumpu.com/en/document/view/18776351/creating-engagement-in-computer-based-learning-environments- (Accessed 12/04/2016).
Kafai, Y. (2006) Playing and Making Games for Learning: Instructionist and Constructionist Perspectives for Game Studies. *Games and Culture*, 1(1), pp. 36–40.
Kafai, Y.B. and Resnick, M. eds. (2012) *Constructionism in Practice: Designing, Thinking, and Learning in a Digital World.* Routledge. Available at: https://www.google.co.uk/books/edition/Constructionism_in_Practice/KGkIEQAAQBAJ?hl=en&gbpv=1&dq=Kafai,+Y.B.+and+Resnick,+M.+eds.+(2012)+Constructionism+in+Practice:+Designing,+Thinking,+and+Learning+in+a+Digital+World.+Routledge.&pg=PP1&printsec=frontcover
Kato, P.M., Cole, S. Bradlyn, A.S. and Pollock, B.H. (2008) Video Games Improves Behavioural Outcomes in Adolescents and Young Adults with Cancer: A Randomized Trial. *Pediatrics*, 122(2), pp. e305–e217.
Kessner, M., Wood, J. Dillon, R.F. and West, R.L. (2001) On the Reliability of Usability Testing. In *Proceedings of the Extended Abstracts on Human Factors in Computing Systems (CHI '01)*, p. 97.
Kim, B. (2001) Social Constructivism. *Emerging Perspectives on Learning, Teaching, and Technology*, 1(1), p. 16.
Kirriemuir, J. and Mcfarlane, A. (2004) Literature Review in Games and Learning. Available at: https://telearn.archives-ouvertes.fr/hal-00190453/document (Accessed 12/04/2020).
Knight, P.T. (2001) A Briefing on Key Concepts – Formative and Summative, Criterion and Norm Referenced Assessment. *Learning and Teaching Support Network Generic Centre*, Vol. 30. http://www.heacademy.ac.uk
Kodugamelab (2015) Available at: https://www.kodugamelab.com/ (Accessed 8/11/2018).
Kolb, A.Y. and Kolb, D.A. (2008) The Learning Way: Meta-Cognitive Aspects of Experiential Learning. *Simulation and Gaming*, 40(3), pp. 297–327.
Kolb, D.A. (1984) *Experiential Learning: Experience as the Source of Learning and Development*, Vol. 1. Englewood Cliffs, NJ: Prentice-Hall.

Korhonen, H. and Koivisto, E.M.I. (2007) Playability Heuristics for Mobile Multi-Player Games. *Proceedings of the 2nd International Conference on Digital Interactive Media in Entertainment and Arts*, pp. 28–35.

Korris, J. (2004) Full Spectrum Warrior: How the Institute for Creative Technologies Built a Cognitive Training Tool for The Xbox. Available at: http://scholar.google.co.uk/scholar?hl=enandq=FULL+SPECTRUM+WARRIOR%3A+HOW+THE+INSTITUTE+FOR+CREATIVE+TECHNOLOGIESandbtnG=andas_sdt=1%2C5andas_sdtp= (Accessed 6/10/2015).

Koster, R. (2005) *A Theory of Fun for Game Design, Electronic Resource*, Published Scottsdale, AZ: Paraglyph Press. Available at: http://capitadiscovery.co.uk/roehampton/items/510379?resultsUri=http%3A%2F%2Fcapitadiscovery.co.uk%2Froehampton%2Fitems%3Fquery%3Dtheory%2Bof%2Bfun%26target%3Dcatalogue%26facet%255B0%255D%3Dfulltext%253Ayes#availability (Accessed 29/12/2015).

Kraiger, K., Ford, J.K. and Salas, E. (1993) Application of Cognitive, Skill-based, and Affective Theories of Learning Objectives to New Methods of Training Evaluations, *Journal of Applied Psychology*, 78(2), pp. 311–328.

Kuss, D.J. and Griffiths, M.D. (2012) Internet Gaming Addiction: A Systematic Review of Empirical Research. *International Journal of Mental Health and Addiction*, 10(2), pp. 278–296.

Lave, J. and Wenger E. (1991) *Situated Learning: Legitimate Peripheral Participation*. Cambridge University Press. Available at: https://www.google.co.uk/books/edition/Situated_Learning/CAVIOrW3vYAC?hl=en&gbpv=1&dq=Lave,+J.+and+Wenger+E.+(1991)+Situated+Learning:+Legitimate+Peripheral+Participation.+Cambridge+University+Press&pg=PA11&printsec=frontcover

Lazzaro, N. (2004) Why We Play Games: Four Keys to More Emotion without Story. Available at: http://www.xeodesign.com/xeodesign_whyweplaygames.pdf (Accessed 28/12/2015).

Lee, J.S., Keil, M. and Kasi, V. (2012) The Effects of an Initial Budget and Schedule Goal on Software Project Escalation. *Journal of Management Information System*, 29(1), pp. 53–77. DOI: 10.2753/MIS0742-1222290102

Lenhart, A., Kahne, J., Middaugh, E. Macgill, A.R., Evans, C. and Vitak, J. (2008) Teens, Video Games, and Civics: Teens' Gaming Experiences Are Diverse and Include Significant Social Interaction and Civic Engagement. *Pew Internet & American Life Project*. Available at: https://www.google.co.uk/search?q=Lenhart+et+al.%2C+2008andoq=Lenhart+et+al.%2C+2008andaqs=chrome..69i57j0.1244j0j8andsourceid=chromeandie=UTF-8 (Accessed 20/8/2017).

Levine, D.M., Krehbiel, T.C. and Berenson, M.L. (2003) *Business Statistics: A First Course*, 3rd Edition. New Jersey: Pearson Education Ltd.

Lewis, G. (2013) Emotion and Disaffection with School Mathematics. *Research in Mathematics Education*, 15(1), pp. 70–86.

Li, F.W.B. and Watson, C. (2011) *Game-Based Concept Visualization for Learning Programming*. Scottsdale, Arizona, USA: ACM. 978-1-4503-0994-3/11/12.

Li, Z.Z., Cheng, Y.B. and Liu, C.C. (2013) A Constructionism Framework for Designing Game-Like Learning Systems: Its Effect on Different Learners. *British Journal of Educational Technology*, 44(2), pp. 208–224.

Lister, R., Adams, E.S. Fitzgerald, S. Fone, W. Hamer, J. Lindholm, M. McCartney, R. Moström, J.E. Sanders, K. Seppälä, O. and Simon, B. (2004) A Multi-National Study of Reading and Tracing Skills in Novice Programmers. In *ACM SIGCSE Bulletin*, 36(4), pp. 119–150. ACM.

Litosseliti, L. (2005) *Using Focus Groups in Research*. London: Continuum.

Loftus, C., Thomas, L. and Zander, C. (2011) Can Graduating Students Design: Revisited. In *Proceedings of the 42nd ACM Technical Symposium on Computer Science Education*, pp. 105–110.

Lyons, I.M. and Beilock, S.L. (2012) When Math Hurts: Math Anxiety Predicts Pain Network Activation in Anticipation of Doing Math. *PLoS One*, 7(10), p. e48076.
Majgaard, G. (2016) Creating Games in the Classroom: From Native Gamers to Reflective Designers. In: Munkvold, R. (ed.). *Games Based Learning for Researchers, Teachers and Students*, Vol. 1. Reading: Academic Conferences and Publishing International Ltd. ISBN:978-1-911278-12-890000.
Malaka, R. (2014) How Computer Games Can Improve Your Health and Fitness. In: Gobel, S., and Wiemeyer, J., (ed.). *Games for Training Education, Health and Sports, 4th Conference on Serious Games, GameDays 2014*, Darmstadt, Germany, April 1–5, 2014, Springer.
Malone, T.W. (1980) What Makes Things Fun to Learn? A Study of Intrinsically Motivating Computer Games. Available at: http://cci.mit.edu/malone/Malone%20computer%20games%20bibliography.htm
Malone, T.W. and Lepper, M.R. (1987) 'Making Learning Fun: A Taxonomy of Intrinsic Motivations for Learning. Aptitude, Learning, and Instruction': Volume 3: Cognitive and Affective Process Analyses. R. E. Snow and M. J. Farr. Hillsdale, USA, Erlbaum.
Maloney, E.A., Schaeffer, M.W. and Beilock, S.L. (2013) Mathematics Anxiety and Stereotype Threat: Shared Mechanisms, Negative Consequences and Promising Interventions. *Research in Mathematics Education*, 15(2), pp. 115–128.
Marciano, J.N., Miranda, L.C. and Miranda, E.E.C. (2014) Evaluating Multiple Aspects of Educational Computer Games. Available at: https://www.hindawi.com/journals/ijcgt/2014/139205/ (Accessed 27/08/2017).
Marklund, B.B. and Taylor, A.S.A. (2016) Creating Games in the Classroom: From Native Gamers to Reflective Designers. In: Munkvold, R. (ed.). *Games Based learning for Researchers, Teachers and Students*, Vol. 1. Reading: Academic Conferences and Publishing International Ltd. ISBN: 978-1-911278-12-890000.
Maslow, A.H. (1970) *Motivation and Personality*, 2nd Edition. New York; London: Harper and Row.
Mayer, E.R. (2014) *Computer Games for Learning: An Evidence Based Approach.* London: MIT.
Mayer, I., Warmelink, H. and Bekebrede, G. (2013) Learning in a Game-Based Virtual Environment: A Comparative Evaluation in Higher Education. *European Journal of Engineering Education*, 38(1), pp. 85–106.
Mayo, M. (2009) Video Games: A Route to Large-Scale STEM education? *Science*, 323, pp. 79–82.
McConnell, D. (2000) *Implementing Computer Supported Cooperative Learning*, 2nd Edition. London: Kogan Page.
Mccrindle, M. (2018) What Comes after Generation Z? Introducing Generation Alpha. Available at: https://mccrindle.com.au/insights/blogarchive/what-comes-after-generation-z-introducing-generation-alpha/ (Accessed 15/10/2018).
McFarlane, A., Sparrowhawk, A., and Heald, Y. (2002) *Report on the Educational Use of Games: An Exploration by TEEM on the Contribution Which Games Can Make to the Educational Process.* Cambridge, UK: TEEM. http://www.teem.org.uk/publications/teem_gamesined_full.pdf (Accessed 06/12/2015).
McGonigal, J. (2015) *Superbetter: A Revolutionary Approach to Getting Happier Stronger and Happier*. Braver and More Resilient, New York: Penguin Press.
Milane, I. and Rowe, G. (2002) Difficulties in Learning and Teaching Programming-Views of Students and Tutors. *Education and Information Technologies*, 7(1), pp. 55–66.
Monroy-Hernandez, A. and Resnick, M. (2008) Empowering Kids to Create and Share Programmable Media. *Interactions*, 15(2), pp. 50–53. Retrieved August 30, 2015.
Moreno-Ger, P., Burgos, D. Martínez-Ortiz, I. Sierra, J.L. and Fernández-Manjón, B. (2008) Educational Game Design for Online Education. *Computers in Human Behavior*, 24(6), pp. 2530–2540. DOI: 10.1016/j.chb.2008.03.012

Moreno-Ger, P., Torrente, J. Hsieh, Y.G. and Lester, W.T. (2012) Usability Testing for Serious Games: Making Informed Design Decisions with User Data. *Advances in Human-Computer Interaction*, 2012. DOI: 10.1155/2012/369637

Morgan, D.L. (1988) *Focus Groups as Qualitative Research*. Beverly Hills, CA: Sage.

Morrison, K. (2002) *School Leadership and Complexity Theory*. London: Routledge Falmer.

Muijs, D. (2011) *Doing Quantitative Research in Education with SPSS*. London: Sage Publication Ltd.

Mullins, L. (2013) *Management and Organisational Behaviour*, 10th Edition. Harlow: Pearson Education.

Mullins, L. (2019) *Organisational Behaviour in The Workplace*, 12th Edition. Harlow: Pearson Education.

Multon, K.D., Brown, S.D. and Lent, R.W. (1991) Relation of Self-Efficacy Beliefs to Academic Outcomes: A Meta-Analytic Investigation. *Journal of Counselling Psychology*, 18, pp. 30–38.

Naik, N. (2017) The Use of GBL to Teach Mathematics in Higher Education. *Innovations in Education and Teaching International*, 54(3), pp. 238–246. DOI: 10.1080/14703297.2015.1108857

Nicholls, D., De Charon, L. and Hutkin, R. (2010) Development of Critical Thinking and Creativity: Practical Guidelines for the Postsecondary Classroom. *ATEA Journal*, 38(1), p. 12.

Nielsen, J. (1995) Usability Inspection Methods. In *Conference Companion on Human factors in Computing Systems*, pp. 413–414. Available at: https://dl.acm.org/doi/pdf/10.1145/259963.260531

Nielsen, J. (2000) *Designing Web Usability*. Indianapolis, IN, USA: New Riders Publishing.

Nygård, S., Kolås, L. and Sigurdardottir, H. (2015) Teachers' Experiences Using KODU as a Teaching Tool. In: *European Conference on Games Based Learning*, Academic Conferences International Limited. p. 416.

Otkay, J.S. (2012) *Grounded Theory*. Oxford, England: Oxford University Press.

Paavilainen, J. (2010) Critical Review on Video Game Evaluation Heuristics: Social Games Perspective. *Proceedings of the International Academic Conference on the Future of Game Design and Technology*, pp. 56–65. DOI: 10.1145/1920778.1920787

Pachler, N. (2007) *Mobile Learning: Towards a Research Agenda*. London: WLE Centre, IoE. In: Kress, G. and Pachler, N (ed.) (2007) Thinking about the 'm' in m-learning.

Pajares, F. and Kranzler, J. (1995) Self-Efficacy Beliefs and General Mental Ability in Mathematical Problem-Solving. *Contemporary Educational Psychology*, 20, pp. 426–443.

Palincsar, A.S. (1998) Social Constructivist Perspectives on Teaching and Learning. *Annual Review of Psychology*, 49, pp. 345–375.

Papert, S. (1993) *Mindstorms: Children, Computers and Powerful Ideas*. New York: Basic Books.

Paras, B. and Bizzocchi, J. (2005) Game, Motivation, and Effective Learning: An Integrated Model for Educational Game. In *Proceedings of DiGRA Conference on Changing Views–Worlds of Play*. Available at: https://www.researchgate.net/publication/221217604_Game_Motivation_and_Effective_Learning_An_Integrated_Model_for_Educational_Game_Design (Accessed 27/09/2016).

Park, H.J. and Kim, S.B. (2013) Guidelines of Serious Game Accessibility for the Disabled. In *Proceedings of the International Conference on Information Science and Applications (ICISA '13)*, pp. 1–3, IEEE.

Pelletier, C., 2009. Games and Learning: What's the Connection. *International Journal of Learning and Media*, 1(1), pp. 83–101.

Pinelle, D., Wong, N., Stach, T. and Gutwin, C. (2009) Usability Heuristics for Networked Multiplayer Games. *Proceedings of the ACM 2009 International Conference on Supporting Group Work*, pp. 169–178.

Pollard, A., Anderson, J., Maddock, M. Swaffield, S., Warin, J. and Warwick, P. (2008) *Reflective Teaching*, 3rd Edition. London: Continuum International Publishing Group.
Preconception. (2020) In: Dictionary.com. [Online] Available at: https://www.dictionary.com/browse/preconception (Accessed 01/08/2020).
Preece, Y., Rogers, Y. and Sharp, H. (2015) *Interaction Design: Beyond Human-Computer Interaction*, 4th Edition. Chichester: John Wiley and Sons Ltd.
Prensky, M. (2002) The Motivation of Gameplay: The Real Twenty-First Century Learning Revolution. *On the Horizon*, 10(1), pp. 5–11.
Prensky, M. (2005) Listen to Natives. *Association for Supervision and Curriculum Development Educational Leadership*, 63(December 2005), pp. 9–13.
Prensky, M. (2006) *'Don't Bother Me Mom, I'm Learning!': How Computer and Video Games Are Preparing Your Kids for the Twenty-First Century*. St Paul, MN: Paragon House.
Pritchard, A. and Woollard, J. (2010) *Psychology for the Classroom: Constructivism and Social Learning*. Oxon: Routledge.
Punch, K.F. (2009) *Introduction to Research Methods in Education*. London: Sage Publications Ltd.
Resnick, M. and Robinson, K. (2017) *Lifelong Kindergarten: Cultivating Creativity Through Projects, Passion, Peers, and Play*. USA: MIT Press.
Robins, A., Runtree, J. and Rountree, N. (2003) Learning and Teaching Programming: A Review and a Discussion. *Computer Science Education*, 13(2), pp. 137–172.
Robson, C. (2010) *Real World Research*, 3rd Edition. Chichester: John Wiley and Sons Limited.
Robson, S. (2006) *Developing Thinking and Understanding in Young Children: An Introduction for Students*. London: Routledge Ltd.
Rogoff, B. (1998) Cognition as a Collaborative Process. In W. Damon, D. Kuhn and R.S. Seigler (ed.), *Handbook of Child Psychology*, 5th Edition. New York: Wiley.
Roth, W.M. (2000) Authentic School Science: Intellectual Traditions In R. McCormick and C. Paechter (ed.), *Learning & Knowledge*. London, UK: Paul Chapman Publishing, pp. 6–20.
Russ, T., and Drury-Grogan, M.L. (2013) Assessing the Impacts of a Business Communication Simulation on Students' Self-Perceptions. *Eastern Communication Association*, 61(5), pp. 584–595. DOI: 10.1080/01463373.2013.822404
Salen, K. and Zimmerman, E. (2011) The Game Designer Reader, A Rules of Play Anthology. Roger Caillois: The Definition of Play. *The Classification of Games*. Cambridge, Massachusetts. London, England: The MIT Press.
Salisbury, J.H. and Cole. T. (2016) Grounded Theory in Games Research: Making the Case and Exploring the Options. Available at: https://www.academia.edu/27624170/Grounded_Theory_in_Games_Research_Making_the_Case_and_Exploring_the_Options (Accessed 15/10/2015).
Schell, J. (2015) *The Art of Game Design: A Book of Lenses*, 2nd Edition. USA: CRC Press.
Schell, J. (2020) *The Art of Game Design: A Book of Lenses*, 3rd Edition. USA: CRC Press.
Senge, P. (1990) *The Fifth Discipline: The Art and Practice of Learning Organization*. United Kingdom: Random House Business Books.
Senge, P.M. (2006) *The Fifth Discipline: The Art and Practice of Learning Organisation*. London: Random House Business Books.
Shell, D.F., Murphy, C.C. and Bruning, R.H. (1989) Self-efficacy and Outcome Expectancy Mechanisms in Reading and Writing Achievement. *Journal of Educational Psychology*, 81(1), p. 91.
Shunk, D.H. (2000) *Learning Theories: An Educational Perspective*, 3rd Edition. Upper Saddle River, NJ: Prentice-Hall.

Siegel, R.G., Galassi, J.P. and Ware, W.B. (1985) A Comparison of Two Models for Predicting Mathematics Performance: Social Learning Versus Math Aptitude-Anxiety. *Journal of Counseling Psychology*, 32, pp. 531–538.

Silverman, B.G., Johns, M., Weaver, R. and Mosley, J. (2003) Authoring Edutainment Stories for Online Players (AESOP): Introducing Game play into Interactive Dramas. In: *International Conference on Virtual Storytelling*, Springer.

Silverman, D. (1993) *Interpreting Qualitative Data: Methods for Analysing Talk, Text, and Interaction*. London: Sage Publications Ltd.

Silverman, D. (2014) *Interpreting Qualitative Data*, 5th Edition. London: Sage Publications Ltd.

Sindre, G., Natvig, L. and Jahre, M. (2009) Experimental Validation of the Learning Effect for A Pedagogical Game on Computer Fundamentals. *IEEE Transactions on Education*, 52(1), pp. 10–18.

Skemp, R.R. (1986) *The Psychology of Learning Mathematics Great*. Britain: Penguin Books Ltd.

Skemp, R.R. (1989) *Mathematics in the Primary School*. Bristol: Routledge.

Skilbeck, A. (2017) Dewey on Seriousness, Playfulness and the Role of the Teacher. *Education Sciences*, 7(1), p. 16.

Slyman, S. (2014) I Hate Maths! An Investigation into Students' Attitudes Towards Mathematics. MA Thesis, University of Roehampton, London, United Kingdom.

Slyman, S. (2018) *Games Based Learning in Universities, a Life Learning Experience*. New York: Business Expert Press Expert Insights Publishing. ISBN:978-1-94819-801-1.

Slyman, S. (2022) Acumenous Game-Based Learning in Simulation Games and Applied Statistics. PhD Thesis, Goldsmiths, University of London, UK.

Slyman, S., Gillies, M. and Jessel, J. (2019a) Acumenous or Inquisitional? Towards a New Theoretical Lens within Games Learning. In: *13th European Conference on Games Based Learning*, University of South Denmark, Odense, Denmark. ISBN:978-1-912764-38-9 or E-book: ISSN:2049-100X.

Slyman, S., Gillies, M. and Jessel, J. (2019b) Game-Based Learning to Engage students with Applied Statistics using a Simulation Roleplay Game. In: *9th Irish Conference on Games Based Learning*, Cork, Ireland. Available at: http://igbl-conference.com/2016/programme/generate_book_of_abstracts.php (Accessed 25/09/2019).

Smith Report (2004) Available at: http://www.mathsinquiry.org.uk/report/mathsInquiryFinalReport.pdf (Accessed 20/12/2014).

Smith Report (2017) Available online at: https://www.gov.uk/government/publications/smith-review-of-post-16-maths-report-and-government-response (Accessed 20/12/2018).

Sobke, H., Chan, E. Buttlar, R. Grobe-Wortmann, J. and Londong, J. (2014) Cat King's Metamorphosis. The Reuse of an Educational Game in a Further Technical Domain. In: Gobel, S., and Wiemeyer, J. (ed.), *Games for Training Education, Health and Sports, 4th Conference on Serious Games, GameDays 2014*, Darmstadt, Germany, April 1–5, 2014, Springer.

Song, S. and Lee, J. (2007) Key Factors of Heuristic Evaluation for Game Design: Towards massively multi-player online Roleplaying game. *International Journal of Human-Computer Studies*, 65, pp. 709–723.

Strauss, A. and Corbin, J. (1990) *Basics of Qualitative Research: Grounded Theory Procedures and Techniques*, Newbury Park, CA: Sage.

Sweetser, P., Johnson, D. and Wyeth, P. (2012) Revisiting the Game Flow Model with Detailed Heuristics. *Journal of Creative Technologies*, 2012(3), pp. 1–16.

Sweetser, P. and Wyeth, P. (2005) GameFlow: A Model for Evaluating Player Enjoyment in Games. *Computers in Entertainment*, 3(3), p. 3. DOI: 10.1145/1077246.1077253

Tamura, H. (2007) Foundational Value of Mathematics Education for Management Curriculum. *International Statistical Review*, 75(3), pp. 397–405. DOI: 10.1111/j.1751-5823.2007.00032.x

Tao, Y.H., Cheng, C.J. and Sun, S.Y. (2012) Alignment of Teacher and Students Perceptions on the Continued Use of Business Simulation Games. *Educational Technology and Society*, 15(3), pp. 177–189.

TED (2019) *Souad Slyman: Why Change Attitudes Towards Mathematics [YouTube]*. TedxGul, London: Goldsmiths, University of London. Available at: https://www.youtube.com/watch?v=leNDJMNJoNY (Accessed 26/04/2019).

Timonen, V., Foley. G. and Conlon. C. (2018) Challenges when Using Grounded Theory: A Pragmatic Introduction to Doing GT Research. *International Journal of Qualitative Methods*, 17, pp. 1–10. DOI: 10.1177/1609406918758086

Tondello, G.F., Kappen, D.L. Mekler, E.D. Ganaba, M. and Nacke, L.E. (2016) Heuristic Evaluation for Gameful Design. *CHI PLAY'16*. ACM. 978-1-4503-4458-6/16/10. DOI: 10.1145/2968120.2987729

van der Spek, E.D., van Oostendorp, H. and Meyer, J.-J.C.H. (2013) Introducing Surprising events can stimulate deep learning in a serious game. *British Journal of Educational Technology*, 44(1), 156–169.

Vlachopoulos, D. and Makri, A. (2017) The Effect of Games and Simulations on Higher Education: A Systematic Literature Review. *International Journal of Educational Technology in Higher Education*, 14(1), pp. 22. DOI: 10.1186/s41239-017-0062-1

Vos, L. (2014) Marketing Simulation Games: A Review of Issues in Teaching and Learning. *The Marketing Review*, 14(1), pp. 67–96. DOI: 10.1362/146934714X13948909473220

Vygotsky, L.S. (1980) *Mind in Society. The Development of Higher Psychological Processes*. Edited by Cole, M., John-Steiner, V., Scribner, S. and Souberman, E. London: Harvard University Press.

Walonick, D.S. (2000) *Designing and Using Questionnaire in Survival Statistic*. MN: StatPacInc. Available at Internet WWW page at URL: http://www.statpac.com/surveys/surveys.pdf (Accessed 02/04/2016).

Whitton, N. (2011) Games-Based Experiences for Learning. Futurelab Innovation in Education. Available at: https://www.nfer.ac.uk/media/1775/futl11.pdf (Accessed 02/04/2016).

Wolfe, J. (1997) The Effectiveness of Business Games in Strategic Management Coursework. *Simulation and Gaming*, 28(4), pp. 360–376. DOI: 10.1177/1046878197284003

Women in STEM – 'Facts and Statistics' (Reporting period 2012–2015) IET. Available at: https://communities.theiet.org/files/8042 (Accessed 20/06/2017).

Yin, R.K. (1994) *Case Study Research: Design and Methods*, 2nd Edition. London: Sage.

Young, M. (2008) *"Powerful Knowledge" and "Knowledge of the Powerful": Some Reflections*. London: Presentation at the Institute of Education.

Yusoff, A., Crowder, R. and Gilbert, L. (2010) Validation of Serious Games Attributes Using the Technology Acceptance Model [online]. In: IEEE (ed.), *2010 Second International Conference on Games and Virtual Worlds for Serious Applications*, Braga, Portugal, 25–26 March 2010, pp. 45–51. Available at: http://eprints.soton.ac.uk/270809/ (Accessed 10/05/2015).

Zimmerman, B.J. (2000) Self-Efficacy: An Essential Motive to Learn. *Contemporary Educational Psychology*, 25(1), pp. 82–91. DOI: 10.1006/ceps.1999.1016

Zyda, M. (2005) *From Visual Simulation to Virtual Reality to Games*. USC Information Sciences Institute, IEEE Computer Society Edition September, pp. 25–31. Available at: http://scholar.google.co.uk/scholar?hl=enandq=zyda+2005andbtnG=andas_sdt=1%2C5andas_sdtp= (Accessed 07/10/2015).

3 Exploring the Potential of Game-Based Learning

Theoretical Foundations and Case Studies

Intriguingly, this book focuses on the intersection of simulation games (SGs), self-efficacy, and the domain of Mathematics and Statistics. By utilizing James Paul Gee's esteemed theoretical framework (as introduced in Chapter 2), we delve into the realm of SGs as a powerful learning tool for teaching Mathematics and Statistics. Our empirical investigations, guided by Gee's (2003a, 2003b, 2004, 2005a, 2005b, 2007, 2008, 2011) framework, offer fresh perspectives and novel contributions to the field of GBL. As we journey through the chapters, including Chapters 6–10 where the research data analysis takes place, we deepen our understanding of how SGs can shape educational experiences and ignite meaningful transformations.

Building upon the socio-constructivist perspectives discussed in Chapter 2, this book presents a culmination of research and insights that push the boundaries of GBL. As we venture toward the final chapters, particularly Chapter 10, we unveil exciting new contributions to the field, offering a glimpse into the transformative power of games, SGs, and GBL in the realm of Mathematics and Statistics education.

Prepare to be immersed in a world where theory converges with practice, and where game-based learning becomes a catalyst for unleashing human potential. This chapter marks a pivotal moment in our exploration, paving the way for the innovative discoveries that await in the subsequent chapters of this transformative journey.

Rationale for using GBL

Throughout the educational literature I examined on games and game-based learning (GBL; see Chapter 2), an assumption is made that GBL is intrinsically engaging and motivating (Malone, 1980; Faria, 2001; Gee, 2003a, 2003b, 2004, 2005a, 2005b, 2007, 2008, 2011; Prensky, 2005; Dondlinger, 2007; Gee et al., 2008; Moreno-Ger et al., 2008; Mayo, 2009; Li et al., 2013; Mayer et al., 2013; Vos, 2014; Malaka, 2014; McGonigal, 2015; Jagger et al., 2016; Hamari et al.,

DOI: 10.4324/9781032723426-4

2016; Vlachopoulos and Makri, 2017; Cardinot and Fairfield, 2019; Schell, 2020). However, there is no research book to my knowledge that addresses how we can use GBL to increase self-efficacy in subjects such as Mathematics and Statistics. More research is required to tackle the question of whether and to what extent games can address self-efficacy in Mathematical and Statistical learning, and whether these skills can be generalized to real-world contexts. There is a need therefore for empirical research to determine this. In this book, I look precisely at the intersection between SGs, self-efficacy, Mathematics and Statistics. My theoretical and empirical contribution to the field of GBL will be to examine how we can use SGs (using Gee's theoretical framework, see Gee's Principles Section) as a learning tool to teach Mathematics and Statistics. I will apply Gee's theoretical framework to my data analysis (see Chapter 10 for further details). This book project is building on Gee's (2003a, 2003b, 2004, 2005a, 2005b, 2007, 2008, 2011) work, socio-constructivist perspectives, as highlighted in Chapter 2 and discussed next.

Theoretical Foundations and Empirical Investigations in Game-Based Learning

Evidence from the GBL education literature has shown that there are many reasons behind playing games (Chapter 2). This includes play, fun, confidence-building, problem-solving, challenge, competition, teamwork, and social interaction (Faria, 2001; Gee, 2003a, 2003b, 2004, 2005a, 2005b, 2007, 2011; Lazzaro, 2004; Pelletier, 2009; Granic et al., 2014; McGonigal, 2015; Cardinot and Fairfield, 2019; Schell, 2020; see GBL in Mathematics and Statistics Section for more details). It can create positive experience (Garris et al., 2002; Lazzaro, 2004; Mayer et al., 2013; Jagger et al., 2016; Cardinot and Fairfield, 2019), engaging and empowering learners in problem-based solving and deep understanding (Gee, 2003a, 2003b, 2004, 2005a, 2005b, 2007, 2011). However, games can also elicit negative experience as in negative emotions/preconceptions toward Mathematics and Statistics (see Part III). A view that is missing in James Paul Gee's (2003a, 2003b, 2004, 2005a, 2005b, 2007, 2008, 2011) work and I will be exploring further in this book (see the use of SGs in Mathematics and Statistics Section).

As explained earlier (see Glossary), preconceptions mean an opinion, an attitude, or a concept that someone has prior to learning or experience. It is a social construct linked to emotions and fear. It includes how students are socially constructed in Mathematics and Statistics classrooms (as "good/bad" mathematicians). Throughout this book, preconception is used to mean attitudes, that is, negative preconceptions. To avoid repetition, I refer to "negative preconceptions" as "preconceptions" (see Chapters 7 and 9).

Other pedagogic rationales for considering GBL are the changing profile of modern learners and the way they learn. This game generation, the "digital native" or generation "Z," can adopt and adapt games as a way of learning (Prensky, 2002, 2005; Jagger et al., 2016). They prefer stimulated teaching methods (Anderson and Lawton, 2009, cited in Jagger et al., 2016) that is, "stimulated learning." This led several researchers to use games, as a learning tool, in teaching and learning (e.g., Gee, 2003a, 2003b, 2004, 2005a, 2005b, 2007, 2011; Lazzaro, 2004; Gee et al., 2008; Pelletier, 2009; Whitton, 2011; Mayer et al., 2013; Jagger et al., 2016; Naik, 2017; Cardinot and Fairfield, 2019). However, none of these research studies discuss negative emotions (preconceptions) in Mathematical and Statistical learning in HE teaching, nor how can GBL mitigate these negative emotions (preconceptions) toward Mathematics and Statistics. In other words, there is a gap and the aim of this unique thought-provoking book is to explore this.

Furthermore, the use of SGs in Mathematics and Statistics at the HE level is limited. For instance, one research paper investigates the use of games to teach Mathematics. It argues that students' attitudes toward this subject can be addressed (e.g., Naik, 2017), suggesting a positive experience. However, it does not discuss a negative experience resulting from the use of these games nor how these games can be used to teach Mathematics and Statistics. Another study proposes the use of SGs to teach Statistics to engineering students in HE using storyboard design (e.g., Barbosa et al., 2019). However, it is not clear how Statistics is connected to the game storyboard design nor how it will be implemented since this study is in its initial stages. Furthermore, there is no research book (to my knowledge) that addresses how we can use SGs to address self-efficacy in Mathematics and Statistics (further details can be found in Chapters 7–10).

The next section discusses the positive and negative impacts of GBL (in particular, psychological and social benefits), highlighting empirical evidence of how games can be used as a building tool for increasing self-efficacy and eliciting negative emotions such as preconceptions. It examines a number of case studies of GBL, providing an overview of data collection and analysis of each study, and how games were employed.

Impacts of GBL

Psychological and Social Benefits

A review of the games education literature indicates that GBL has many psychological benefits like increasing motivation and engagement (e.g., Faria, 2001; Granic et al., 2014; McGonigal, 2015). This suggests that GBL has potential in increasing self-efficacy. Self-efficacy as a genre has been embedded in many games as in puzzle games, for instance "Sudoku" and "Bubble shooter." Since most game designers give few instructions about how to play games, players or core game players are left with a range of all possibilities of answers to discover on their own using experience and instincts. In contrast, new evidence is emerging that playing games can improve children's creative abilities and overall, their self-efficacy. For example, among a sample of 500 12-year-old students, video game playing is positively associated with creativity and self-efficacy boost (Granic et al., 2014). However, this study's cross-sectional design has made it unclear whether playing GBL develops self-efficacy or people prefer video games (or both).

Like Granic et al. (2014), McGonigal (2015:29–42) discusses the potential benefits of games on improving emotions. In support of this claim, McGonigal cites a scientific paper by Hoffman and Patterson, the inventors of Snow World, an immersive VR (virtual reality) game designed to help reduce pain and low self-efficacy. Hoffman and Patterson state that games can heal patients better in what psychologists diagnose as "post-traumatic growth" (McGonigal, 2015). This implies that the psychological benefits of games on people's health and emotions are massive as games can promote self-efficacy and motivation as in real-life situations. However, more research is required to thoroughly assess whether games can really promote self-efficacy in real-life situations as in the context of Mathematical and Statistical learning in HE. This is yet to be explored in this book. For instance, some fast-paced action games such as "Call of Duty," "Forza," and "Grand Theft Auto" can be cognitively beneficial to the brain as they can address individuals' intelligence, self-efficacy, and decision-making, and lead to more efficient multitasking and neural processing (McGonigal, 2015). Puzzle games like "Sudoku" or "Tetris" are conducive for improving skills such as memory, Mathematics, and Statistics. Tetris is a Mathematics puzzle game that consists of falling geometrical shapes or blocks and it is widely used, nearly half a billion players. In psychology, Tetris is famously recognized for supporting patients with Post-Traumatic Stress Disorder to recover from flashbacks and undesirable haunting memories that trigger unpleasant cognitive conditions (McGonigal, 2015). This effectively may assist patients' recovery. This also suggests that playing Tetris games regularly may increase patients' self-efficacy to defeat chronic pain or annoying flashbacks. Playing games regularly may occupy patients' brain from thinking about negative experiences. Further research could explore whether there are any correlations or relationships between participants' negative experiences, self-efficacy and games (see Part III).

Nevertheless, games do not just promote positive emotions such as increasing self-efficacy, it can also elicit negative emotions as in preconceptions, fear, violence, and addiction, and in some cases lead to cognitive and pathological behavior/illnesses. For instance, in a systematic literature review, Kuss and Griffiths (2012) review 58 empirical studies and find that excessive gameplay can lead to low self-efficacy, addiction, and health problems such as a decline in verbal memory performance and reduction in the amount of sleep. Reasons for gaming addiction include low self-efficacy, fear, loneliness, and boredom (Kuss and Griffiths, 2012).

I suggest although it is evident that games are fun to play, there exists unresolved questions to answer regarding, for instance, the consequences of negative emotional experiences during gameplay and to what extent these games can improve individuals' emotions such as self-efficacy. Based on my own experience of playing, for instance, "Bubble shot," "Candy Crush," "8 Ball Pool," or "Sudoku" games (see Table 2.1 (Chapter 2)), these games can be more irritating than therapeutic, as the functionalities of these games do not respond (sometimes) as the user would like them to. Likewise, "Assassins' Creed" as a game, for example, I suppose is regarded as a tool for inciting violence. In contrast, some games can be seen as an empowering tool for enhancing self-efficacy. For example, research shows that playing games such as "Angry birds," "Super Mario," or "Bejeweled" can improve emotions such as self-efficacy and curiosity (McGonigal, 2015). On the other hand, some scary challenging games such as "Resident Evil" have been shown to equip individuals with better skills for controlling negative emotions such as anger and fear (McGonigal, 2015:111). However, games do not just prompt positive emotions they can also elicit negative ones such as preconceptions. This is because some of the games are intriguingly challenging and require players to solve problems and make decisions. For instance, "Portal 2" is a puzzle game that consists of shapes and challenges that requires the player to solve problems at each stage of the game, but the rules keep changing dramatically, which can be frustrating (McGonigal, 2015). Similarly, "Call of Duty" (a multiplayer shooting game and based on my own experience playing this game) can be provoking sometimes when the objects are moving too fast, and players cannot shoot them. But, with practice (I suppose) players can find strategies of how to deal with such complexity and therefore increase their self-efficacy. Having said that, whether and to what extent games are influencing players' skills, or any learning, is yet to be explored further in Chapters 7–9.

According to McGonigal (2015), one of the most successful gaming stories that have tangible positive impacts on emotions and have increased self-efficacy, and thus changed behavior is "Re-Mission." "Re-Mission" is a shooter fantasy game, where players control a superhero robot named "Roxxi." Roxxi's job is to fight and shoot the "baddies," in this case cancer cells, using antibiotic weapons. The purpose of the game is to motivate and increase young cancer patients' self-efficacy to perseveringly adhere to taking their long-term medication. This is because, according to McGonigal (2015), most young leukemia patients miss taking their doses for various reasons from becoming frustrated with their enduring treatment intake to the inability to continuously endure its side effects like nausea (Granic et al., 2014). According to Professor Steve Cole, at "Hopelab Research Foundation (HRF)," 30 percent of young people miss 20 percent of doses or more. "Those kids have the risk of having a rebound of the Leukaemia" (McGonigal, 2015). From a neuroscience perspective, games such as this have potential to raise self-efficacy and motivation, reach goals, and improve health and well-being. In medical experiments conducted at a USA clinical trial, with 375 cancer patients (aged 13–29) who have played the Re-Mission game regularly for two hours a day, HRF reports that participants become more aware of the significance of their medication (Granic et al., 2014). This game has currently been distributed to over 20,000 patients, as it is a successful gaming example of positive real impacts on behavior and emotions (Granic et al., 2014). Overall, the game has positive psychological impacts on leukemia patients' self-efficacy, since it empowers young cancer patients, just like in the game, to shoot the cancer cells using antibiotic weapons (Granic et al., 2014). This suggests that GBL can increase self-efficacy, improve positive emotions, and overall change behavior. This fascinating study has inspired me considerably since it demonstrates the huge beneficial impacts of games on cognition, behavior, and social interaction, and this can be used to increase self-efficacy, empower people, motivate them, and initiate change for the better.

Contrary to the stereotypes that state that game players play in isolation (Lenhart et al., 2008), in my view, some games can promote social interaction, creative thinking, and decision-making. For instance, based on my own experience playing these games, "FIFA 14," a multiplayer football sport game, and games such as "Chess" online, are good ways for social interaction, play, and socializing with family and friends. This I propose can also improve self-efficacy, communication, and collaboration skills. Games may have potential to solve social problems, according to Granic et al. (2014), change and tighten relationships or bonds between a family and friends or even strangers. Therefore, this implies that gamers are rapidly learning social skills and positive behavior.

GBL in Mathematics and Statistics Education

Having argued the possible benefits and drawbacks of GBL (a sub-genre of SGs), this section will now discuss its potential use in teaching Mathematics and Statistics unpacking learners' emotions with Mathematics which can affect self-efficacy. It will then move on to examine positive and negative experiences when playing games, focusing on negative experiences which is linked to negative emotions toward Mathematics and Statistics. This can lead subsequently to preconceptions. A concept that is missing from the existing literature and worth exploring further in-depth in this book.

Mathematics and Statistics, Emotions, and Self-Efficacy

In a case study reported at a New Zealand University, Bartholomew et al. (2011) examine participants' emotions with Mathematics using interviews (with 165 third-year undergraduates). Bartholomew et al. find that students whose major is Mathematics tend to see themselves as "naturals" in Mathematics and that they do it for love. Learners' emotions with Mathematics can affect their motivation and self-efficacy (Bartholomew et al., 2011). High self-efficacious students perceive themselves to be naturally good at Mathematics (i.e., this could lead to positive emotions toward these subjects). This suggests that learners' emotions can affect their learning experience. Where they feel that they are naturally good at it (Mathematics), this has led to an increase in their self-efficacy and therefore they become high self-efficacious. Where they feel disaffect, this has led to low self-efficacy and therefore negative emotions toward Mathematical and Statistical learning.

As part of my masters, in a case study conducted at a South London University, I explore factors behind negative emotions (Chapter 1) toward Mathematics. I find that negative emotions can be a barrier to learning. Perceptions of irrelevancy, poor-quality teaching practice, disengagement, and low self-efficacy are factors that can lead to negative emotions toward Mathematics (Slyman, 2014). I conclude that negative emotions are likely caused by negative experiences. This further indicates that negative emotions can affect learning and play a key role in understanding Mathematical and Statistical learning. Accordingly, these factors can have negative impacts on learners' experiences, in particular preconceptions. However, identifying ways to change them was beyond the scope of that study. This is why in this book, GBL is employed as a motivating tool to address or "acumenify" students' Mathematics and Statistics (Chapters 7, 9, and 10).

Games, Emotions, and Self-Efficacy

In a theoretical paper about games and emotions, Garris et al. (2002) discuss previous studies on the positive impacts of GBL and what makes games engaging and motivating. They highlight game features such as user judgement, reaction, interaction, behavior, control, feedback, and motivational processes (e.g., engagement and learning) (Garris et al., 2002). This suggests that GBL has potential impacts on emotions. However, the paper does not discuss any engagement/disengagement within any specific subjects like Mathematics and Statistics. This research book is linked to Garris et al.'s research on games and builds on it through the empirical investigations (Part III).

In an empirical study about games, emotions, and self-efficacy, a group of researchers investigate rationales behind playing games. They find that people played games for different reasons, including challenge, competition, therapy, fun, social interaction, fantasy, excitement, aesthetic appreciation, tension, and self-efficacy. Lazzaro (2004) and her colleagues observe 15 hardcore gamers, 15 casual gamers, and 15 non-players playing their favorite games online or offline on platforms such as PCs, consoles, or handheld devices. The researchers look for players' verbal and non-verbal emotional cues, actions, and interactions during gameplay. Findings indicate that reasons behind playing games are for the positive experiences players had, since a game "creates… [the] adrenaline rush… a mental challenge" and the joy of solving problems. Lazzaro's (2004) findings also suggest four keys to unlock emotion: Hard fun, easy fun, altered states, and the people factor. Hard fun consists of challenge, strategy, and problem-solving, and strategic thinking while easy fun involves curiosity, immersion, adventure, and mystery. Altered states include enjoyment, cognitive, social experience, excitement, and therapy. Finally, the people factor is related to the use of games for social experience, amusement, teamwork, competition, and personal recognition. This suggests that games, as a building tool, have potential for improving emotions and increasing self-efficacy.

Unlike Garris et al. (2002), Lazzaro (2004), and Bartholomew et al. (2011), Pelletier (2009) presents a different perspective on developing games to teach English and media skills, building on Gee's (2003a, 2003b) work and using some of his 36 learning perspectives but not the 13 principles. Pelletier (2009) and her colleagues look at the relationship between games and learning, creating a production tool/3D software for 11 to 14 years old to make their own games, in a variety of schools over a three-year period. The aim is to analyze games as texts and create games, focusing on narrative, rules, and economies. Game-making is designed as a framework or an activity for developing conceptual understanding. Pelletier focuses on the importance of using games across time and on the principles of students' work or strategies that they have developed over the time of the course (five weeks). The process involves creating prototypes of ready-made 3D shapes such as rooms, chairs, etc. This also involves setting up rewards/points if the player selects certain entities, which leads to defining entities and the relationship between them. In particular, the team devises a game production course for 12–13-year-olds media studies class and embed game production into the existing course (50-minute sessions for five weeks). The researcher uses video-recording, microphones and field notes, and copies of students' work at the end of each session. She runs many sessions and experiments with students. They create stories like "Dr Who," and act as game designers. At other times, the researcher gives students some examples of how a story should be conducted. Students create and play each other's games. The game is employed as an object or a narrative tool for creating stories. The process of analysis focuses on how students characterize and define a game. In particular, Pelletier uses Gee's (2003a, 2003b) 36 learning principles: "semiotic signs, spatial composition, provenance, demand and offer structures and modality" but not the 13 principles which came out in 2004 and 2005. For example, objects are symbolized as "semiotic signs" (a sign that it is a textual and can be read). This enables Pelletier to describe the type of interaction the game facilities: Spatial composition (how entities are framed); provenance (where signs come from), guide rhythm; "demand and offer structures" (the demands made on the player, for instance a particular kind of action) or the offers made; and "modality" (signifier of genre). This facilitates a description of how game production is developed and characterized as a game and the type of interaction it facilitates. For instance, between designers and players or between observers of the design and play. Furthermore, students' work, research, data, field notes, and teacher's lesson were analyzed in similar manners using same "social semiotic concepts." This analytic procedure can help produce a "history of game-making." Pelletier (2009) defines a game as an entity whose forms and meanings are both situated and strategic. It is a way of interaction. Findings indicate that principles of design emerged at the beginning as a function/norm, since students design games for themselves, but with the teacher's and researchers' intervention and stir, students create games for different target audience: Teachers. It symbolizes the "significations" of game as a strategic tool, that is, students realize a social purpose/meet a social need.

The findings highlight how games as a social activity create social bonds. Game-making is a means to enact social bonds. Pelletier argues that students realize a social identity by creating different games as a result of teacher's and researchers' intervention. The relationship between games and learning is that learning should not be hidden and covered by games. It can be explained as a process by which games emerge: A method of how game comes to be "playable." Game is a way of social interaction. Implications are that a game is an entity whose "forms" and "meanings" are both situated and strategic as components of a visually organized text.

Similar to Garris et al.'s (2002), Lazzaro's (2004), and Pelletier's (2009) projects, but in a longitudinal research study, Jagger et al. (2016) build a 3D game aimed at teaching business ethics education to HE Business undergraduates. The researchers employ mixed research methods: Online surveys and focus-group interviews. The 3D business game tells a story of how a manager "Anna," in a series of scenarios, faces problems in her day-to-day life and gains her confidence (self-efficacy). The game shows how she makes decisions and solves complex problems using collected clues, quizzes, and puzzles during the gameplay. The simulation game can be played by any player who can take up the responsibilities of the manager. The player must therefore make accurate and sound decisions depending on how ethically educated they are and on their level of understanding of the business ethics and morals. At the end of the game, the player faces the consequences of her/his decision-making and problem-solving skills. The game consists of six levels and includes cartoons and accompanying articles (for reading as assignments). The game also has quizzes and puzzles and some clues to solve problems, and the player must make decisions before proceeding to a higher level. Although the 3D game has succeeded to some extent to portray how final-year undergraduates learn business ethics education, limitations with this study is that morality was a difficult concept to measure and therefore a bigger sample size should be considered for next time. One of the authors of this study, Jagger, is my former colleague at the Business School, University of Roehampton, and her research has influenced my work considerably and inspired me to write this current book on games. This book is linked to Jagger's work on games and builds on it, since future work will be aimed at designing an ESG with successful elements/features such as problem-solving and decision-making activities, scenarios and puzzles, since these can have positive impacts on learners' experiences (see Chapters 7 and 9).

In a comparative study, and unlike Cardinot and Fairfield (2019) and Jagger et al. (2016), Mayer et al. (2013) evaluate 14 games. Mayer et al. use pre-and-post questionnaires and compare 14 different SGs with different topics. Using questionnaires, they assess students' knowledge before and after the game. Next, they develop their own online game called "Cyberdam-2D GBL." Mayer et al. (2013) demonstrate how challenging and fascinating it can be to compare and evaluate a broad assortment of games since timing, topics, institutions, and computing facilities are requirements for the effectiveness of conducting comparative research. The results show that students' emotions, self-efficacy, game design, and facilitators' interaction with students during the game are correlated and connected to students' learning satisfaction.

Like Jagger et al.'s (2016) research, Cardinot and Fairfield (2019) use games in teaching and learning. Cardinot and Fairfield (2019) develop a game to teach physics with astronomy in the Irish Science syllabus. Data collection (n = 119) includes feedback questionnaires, systematic observations, and pre- and post-test surveys. The researchers conclude that GBL can motivate, engage, and increase students' self-efficacy to learn physics. Researchers, designers, and educators are advised to choose a simple easy game to use and to bear in mind that students' emotions are associated with learning, and a key factor to effective learning. In addition, the game is found to be a useful learning tool and an activity to promote social skills.

The next section (and the final section in this chapter) presents the use of SGs in Mathematics and Statistics.

The Use of SGs in Mathematics and Statistics

Following Chapter 2, I indicated that a review of the existing literature suggests that there is no research book (to my knowledge) that addresses how SGs can be used to teach Mathematics and Statistics, as a learning tool in HE. SGs as a pedagogical tool can be utilized for increasing self-efficacy whether one develops or uses SGs, since SGs can be adapted or adopted as a way of learning (Gee, 2003a, 2003b, 2004, 2005a, 2005b, 2007, 2008, 2011; Prensky, 2005; Pelletier, 2009; Chapters 5 and 9). Many "digital native" learners (Prensky, 2005:8), generations "Z" (Dede, 2005; Jagger et al., 2016), and "Alpha" (Mccrindle, 2018) are using SGs. The use of SGs thus may be helpful when solving complex problems and teaching life skills such as increasing focus and creativity (Garris et al., 2002). This is also because SGs provide leisure and entertaining benefits to both parties (service providers and stakeholders) (Kafai, 2006). The purpose of SGs includes to challenge and improve abilities and to increase self-efficacy, accept a goal, make an effort, get feedback on that effort, improve a concrete skill, keep trying, and eventually succeed (McGonigal, 2015:81).

The aim of this subsection thus is to discuss research investigations exploring the use of SGs in Mathematics and Statistics. It is worth noting that in the context of HE, there is limited research discussing Mathematics and Statistics & SGs, as explained next.

In other studies, this time with first-year undergraduates on an engineering course, Majgaard (2016) looks at how game developers can become more confident and reflective through building SGs using "GameMaker" (Habgood, 2006, 2007) (in 2D version). These digital-native students, according to Majgaard, have played games since the age of 5 to 6 years, and have knowledge of game theory and programming from other lectures/modules. The researcher explores how these students can switch roles and develop, from native gamers to become reflective designers. The goal is to promote creativity, experimentation, exploration, trial and error, Mathematical learning and reflection while developing/designing SGs as well as reusing coding/existing games to create new SGs.

In a case study, Naik (2017) uses a SG to teach Mathematics to a Computer Science course offered at the Aberystwyth University. The aim is to address students' attitudes toward Mathematics and easy certain fears about the subject, creating positive experience. Naik (2017) runs an experiment with 69 first-year Computer Science undergraduates, using and adapting a small number of non-digital games to the teaching of selected mathematical principles. He looks at six games. This includes an arithmetic-fractions rummy, a decimal crossword puzzle, a logarithmic and experiential jigsaw, and a binary-to-decimal conversion magic box. He then tests and creates a deck of 54 playing cards with range of topics such as multiplication, fractions, and equality operators. These SGs are accompanied by some instructions/rules of playing the game and played as single-or-multiple players while others, depending on the games, as a group in class matching the lecture requirement. He measures the learning outcomes through assessment and students' feedback through evaluation outcomes (questionnaires). Findings indicate that SGs support student learning and motivation in line with the previous literature (see Perrotta et al., 2013). It also suggests that SGs have potential in encouraging a more positive attitude to mathematical learning (Fengfeng, 2008; Naik, 2017). Furthermore, Naik advises that SGs should be an integral part of everyday HE pedagogical practice besides well-established lectures, seminars, and workshops. Findings also emphasize the importance of incentives used in SGs in encouraging more positive attitudes toward Mathematics since the high level of social interactivity in games are favored by students and therefore can be motivational.

Another study proposes the use of SGs to teach Statistics to engineering students in HE using storyboard design (e.g., Barbosa et al., 2019). The researchers build an overview of the structure of Statistics game and divide it into four main stages: Exposition, rising action, climax, and falling action. They then discuss how the game is related to these concepts. However, it is not clear how Statistics is connected to the future proposed Statistics game nor how it will be implemented since this study is in its initial stages.

Overall, as I explained above, this book project builds on socio-constructivist perspectives following the footsteps of Gee's work where games have been used as a learning tool. I will be using and applying Gee's (2003a, 2003b, 2004, 2005a, 2005b, 2007, 2011) 13 principles, as a theoretical framework, to my empirical work (Part III). I will discuss this in Chapter 10 (Outline of the Main Findings and My Theoretical Contribution to avoid repetition; see also Chapter 2: Gee's Principles Section).

Having reviewed a wide range of research literature studies (as afore-indicated) from Faria (2001), Garris et al. (2002), Gee (2003a, 2003b, 2004, 2005a, 2005b, 2007, 2011), for instance, to Lenhart et al. (2008), Pelletier (2009), Bartholomew et al. (2011), Kuss and Griffiths (2012), Mayer et al.

(2013), Granic et al. (2014), Slyman (2014), McGonigal (2015), Jagger et al. (2016), Jenson and Droumeva (2015), Naik (2017), Barbosa et al. (2019) and Cardinot and Fairfield (2019), there appears to be further research questions to be explored (i.e., there is a gap) that the current book seeks to address. These include exploring the connections between Mathematics and GBL and participants' experience (Chapter 7). Research questions/objectives will investigate ways of addressing negative emotions (preconceptions) toward Mathematics and Statistics using games, for instance. It will be interesting to examine participants' experiences in learning Statistics using SGs (Chapter 9). This will be investigated in this book (see Chapters 7–9). What game evaluation heuristics, game design, developments, and usability principles are more significant than others and which ones are less important (see Chapters 4 and 6).

In this book, I will look at how games can be integrated into the curriculum as an activity to teach Mathematics and Statistics, I will look at how we can use SGs creatively and playfully, namely "SimCity" as a learning tool for improving participants' Mathematical and Statistical learning (Chapter 9). I have taught "correlation and regression" (Chapter 5) in a scenario-based learning environment (see Simulation GBL Approach to Teaching Section in Chapter 5). My aim is to test whether there are any effects on participants' self-efficacy, with pre- and post-test surveys (Chapter 6). Computing/business undergraduates will not create SGs but develop Mathematical and Statistical skills in the series of gameplay workshops. They will creatively investigate a new approach of learning Statistics using a SG in a narrative style, with Excel calculations (Chapter 5).

In all, this Chapter 3 has presented an overview of the previous literature on GBL, and its impacts, looking at learners' emotions, self-efficacy, and Mathematics. I then moved on to examine positive and negative experiences when playing games. I have identified that negative experiences can lead to negative emotions and subsequently to preconceptions toward Mathematics and Statistics. This research focus is missing from the existing literature and worth exploring further in this book. I then discussed the use of SGs in Mathematics and Statistics and how there is limited existing literature in HE discussing these subjects. This chapter has shown the theoretical potential for SGs to be an effective tool for learning if it is accompanied by effective pedagogy.

The next Chapter, **Chapter 4**, will explore the Power of Game Design: Mechanics, Interactions, and Evaluation Heuristics.

Bibliography

Acumen (2017) In: Cambridge University Press. [Online] Available at: http://dictionary.cambridge.org/dictionary/english/acumen (Accessed 30/05/2017).

Adams, A., Lunt, P. and Cairns, P. (2008) A Qualitative Approach to HCI Research. In: Cairns, P. and Cox, A. (ed.). *Research Methods for Human-Computer Interaction*. Cambridge, UK: Cambridge University Press, pp. 138–157. Available at: http://www.cambridge.org/catalogue/catalogue.asp?isbn=9780521870122andss=toc (Accessed 20/11/2016).

ADDIE Model (2016) Available at: http://www.learning-theories.com/addie-model.html (Accessed 20/04/2016).

Anderson, S. (2020) Quantitative Analysis. Available at: https://www.investopedia.com/terms/q/quantitativeanalysis.asp (Accessed 20/02/2021).

Anon (2018) Efficacy In: Cambridge University Press. [Online] Available at: https://dictionary.cambridge.org/dictionary/english/efficacy (Accessed 30/09/2018).

Applefield, J.M., Huber, R. and Moallem, M. (2000) Constructivism in Theory and Practice: Toward a Better Understanding. *The High School Journal*, 84(2) pp. 35–53.

Arthur, J., Waring, M. Coe, R. and Hedges, L.V. (2012) *Research Methods and Methodologies in Education*. California: Sage Publications Inc.

Bandura, A. (1986) *Social Foundations of Thought and Action: A Social Cognitive Theory*. Englewood Cliffs, NJ: Prentice-Hall.

Bandura, A. (1994) Self-efficacy In: Ramachaudra, V.S. (ed.), *Encyclopedia of Human Behaviour*, 4, pp. 71–81.

Bandura, A. (1997) *Self-efficacy: The Exercise of Control*. New York: W.H. Freeman and Company.

Bandura, A. and Schunk, D. (1981) Cultivating Competence, Self-Efficacy, and Intrinsic Interest Through Proximal Self-Motivation. *Journal of Personality and Social Psychology*, 41(3), pp. 586–598.

Barbosa, T., Lopes, S.P. Leao. C. Soares, F. and Carvalho, V. (2019) Serious Game for Teaching Statistics in Higher Education: Storyboard Design. In: *7th EAI International Conference*, Braga, Portugal, on 'Interactivity, Game Creation, Design, Learning, and Innovation' (ed.) Springer, pp. 169–175. DOI: 10.1007/978-3-030-06134-0_18

Bartholomew, H., Darragh, L., Ell, F. and Saunders, J. (2011) "I am Natural, and I Do It for a Love!": Exploring Students' Accounts of Studying Mathematics. *International Journal of Mathematical Education in Science and Technology*, 42(7), pp. 915–924.

Bassey, M. (1999) *Case Study Research in Educational Settings*. Buckingham: Open University Press.

Bellotti, F., Kapralos, B. Lee, K. Moreno-Ger, P. and Berta, R. (2013) Assessment in and of Serious Games: An Overview. *Advances in Human-Computer Interaction*, 2013, p. 1.

Belluigi, D.Z. (2009) Exploring the Discourses Around 'Creativity' and 'Critical Thinking' in a South African Creative Arts Curriculum. *Studies in Higher Education*, 34(6), pp. 699–717.

Bennedesen, J. and Caspersen, M.E. (2004) Programming in Context: A Model-First Approach to CS1. In *Proceedings of the 35th SIGCSE Technical Symposium on the Computer Science Education*, pp. 477–481. ACM Press.

Bevan, N. (2001) International standards for HCl and usability. *International Journal of Human Computer Studies*, 55(4), pp. 533–552.

Boaler, J. (2009) *The Elephant in the Classroom*. London: Souvenir Press.

Brennan, R. and Vos, L. (2010) Marketing Simulation Games: Student and Lecturer Perspectives. *Marketing Intelligence and Planning*.

Brown, J.S., Collins, A. and Duguid, P. (1989) Situated Cognition and the Culture of Learning. *Educational Researcher*, 18(1), 32–42.

Bruner, J. (1973) *Going Beyond the Information Given*. New York: Norton.

Bryant, A. (2017) *Grounded Theory and Grounded Theorizing: Pragmatism in Research Practice*. New York, NY: Oxford University Press.

Bryman, A. (2008) *Social Research Methods*, 3rd Edition. Oxford: Oxford University Press (first published: 2001).

Bryman, A. (2012) *Social Research Methods*, 4th Edition. United States: Oxford University Press.

Bryman, A. (2016) *Social Research Methods*, 5th Edition. United States: Oxford University Press.

Caillois, R. (1961) *Man, Play, and Games*. New York: Free Press.

Cardinot, A. and Fairfield, J.A. (2019) Game-Based Learning to Engage Students with Physics and Astronomy Using a Board Game. *International Journal of Game-Based Learning (IJGBL)*, 9(1), pp. 42–57.

Charmaz. K. (2006) *Constructing Grounded Theory*, 1st Edition. London: Sage Publications Ltd.

Charmaz. K. (2014) *Constructing Grounded Theory*, 2nd Edition. London: Sage Publications Ltd.

Chen, M.P., Wong, Y.T. and Wang, L.C. (2014) Effects of Type of Exploratory Strategy and Prior Knowledge on Middle School Students' Learning of Chemical Formulas from a 3D Roleplaying Game. *Educational Technology Research and Development*, 62, pp. 163–185. DOI: 10.1007/s11423-013-9324-3

Clark, D.B., Tanner-Smith, E.E. and Killingsworth, S.S. (2015) Digital Games, Design, and Learning a Systematic Review and Meta-Analysis. *Review of Educational Research*, 86(1), pp. 79–122.

Claxton, G. (2002) *Building Learning Power*, 1st Edition. Bristol: TLO Limited.

Cohen, L., Manion, L. and Morrison, K. (2011) *Research Methods in Education*. London: Routledge.

Cohen, L., Manion, L. and Morrison, K. (2018) *Research Methods in Education*. London: Routledge.

Conlon, C., Carney, G., Timonen, V. and Scharf, T. (2015) Emergent Reconstruction in Grounded Theory: Learning from Team-Based Interview Research. *Qualitative Research*, 15, pp. 39–56. DOI: 10.1177/1468794113495038

Cook, J. (1992) *Negotiating the Curriculum: Programming for Learning. Negotiating the Curriculum: Educating for the 21st Century*, pp. 15–31.

Cooper, P.A. (1993) Paradigm Shifts in Designed Instruction: From Behaviorism to Cognitivism to Constructivism. *Educational Technology*, 33(5), pp. 12–19.

Corbin, J. and Strauss, A. (2015) *Basics of Qualitative Research: Techniques and Procedures for Developing Grounded Theory*, 4th Edition. Thousand Oaks, CA: Sage.

Cordova, D.I. and Lepper, M.R. (1996) Intrinsic Motivation and the Process of Learning: Beneficial Effects of Contextualization, Personalization, and Choice. *Journal of Educational Psychology*, 88, pp. 715–730.

Coulthard, G.J. (2009) A Review of the Educational Use and Learning Effectiveness of Simulations and Games [online]. PhD Thesis, Purdue University. Available at: http://www.coulthard.com/library/paper-simulation.html (Accessed 15/12/2015).

Craft, A. (1999) Creative Development in The Early Years: Some Implications of Policy for Practice. *The Curriculum Journal*, 10(1), pp. 135–150.

Craft, A. (2002) *Creativity and Early Years Education*. London: Continuum.

Crawford, C. (1984) *The Art of Computer Game Design*. Berkeley, California, McGraw-Hill/Osborne Media.

Creswell, I.W. (2009) *Research Design: Qualitative, Quantitative, and Mixed Methods Approaches*, 3rd Edition. London: Sage.

Crockett, L. (2018) Bogost's War on Gamification: Irony and Play on Grandma's Sofa. In: *12th European Conference on Games Based Learning (ECGBL 2018)*. Reading: Academic Conferences and Publishing International Ltd.

Cronan, T.P. and Douglas, D.E. (2012) A Student ERP Simulation Game: A Longitudinal Study. *Journal of Computer Information Systems*, 53(1), pp. 3–13.

Csikszentmihalyi, M. (1990) Psychology Today: Finding Flow. Available at: https://www.researchgate.net/publication/200026151_Finding_Flow_The_Psychology_of_Engagement_With_Everyday_Life/link/55ad6c2f08aed9b7dcdae222/download (Accessed 03/05/2016).

Csikszentmihalyi, M. (1997) *Creativity, Flow and the Psychology of Discovery and Invention*. New York: Harper.

Data Protection Act (1998) Available at: http://www.legislation.gov.uk/ukpga/1998/29/pdfs/ukpga_19980029_en.pdf (Accessed 10/12/2016).
Data Protection Act (2018) Available at: http://www.legislation.gov.uk/ukpga/2018/12/pdfs/ukpga_20180012_en.pdf (Accessed 10/09/2019).
Davis, F.D. (1989) Perceived Usefulness, Perceived Ease of Use, and User Acceptance of Information Technology. *MIS Quarterly*, 13(3), pp. 319–340.
De Bono, E. (1992) *Teach Your Child How to Think*. England: McQuaig Group Inc.
Dede, C. (2005) Planning for Neomillennial Learning Styles. *Educause Quarterly*, 28(1), pp. 7–12.
Derryberry, A. (2007) Serious Games: Online Games for Learning. White Paper. [Online]. Available at: http://www.adobe.com/resources/elearning/pdfs/serious_games_wp.pdf (Accessed 10/12/2015).
Desurive, H., Caplan, M. and Toth, J. (2004) Using Heuristics to Evaluate the Playability of Games. *CHI Conference*, 32 (1), pp. 97–110. DOI: 10.1177/104687810103200108
Desurvire. H. and Wiberg, C. (2009) Game Usability Heuristics (PLAY) for Evaluation and Designing Better Games: The Next Iteration. In *Online Communities and Social Computing: Third International Conference, OCSC 2009, Held as Part of HCI International 2009*, San Diego, CA, USA, July 19–24, 2009. Proceedings 3, pp. 557–566. Springer Berlin Heidelberg.
Desurvire, H. and Wiberg, C. (2010) User Experience Design for Inexperienced Gamers: GAP – Game Approachability Principles. In: Regina Bernhaupt (ed.), *Evaluating User Experience in Games*, Springer, pp. 131–148. DOI: 10.1007/978-1-84882-963-3_1
Dondlinger, J.M. (2007) Educational Video Game Design: A Review of the Literature. *Journal of Applied Educational Technology*, 4(1), Spring/Summer, pp. 21–31. Available at: http://reserachgate.net/publication/238444705 (Accessed 12/12/2015).
Draper, S. (2013) Social Constructivism. Available at: http://www.psy.gla.ac.uk/~steve/courses/archive/CERE12-13-safari-archive/topic3/webarchive-index.html#social%20con%201 (Accessed 23/12/2019).
Dunne, C., (2011) The Place of The Literature Review in Grounded Theory Research. *International Journal of Social Research Methodology*, 14, pp. 111–124. DOI: 10.1080/13645579.2010.494930
Dweck, C.S. (2006) *Mindset: The New Psychology of Success*. United States: Ballantine Books Trade Paperback Edition.
Dwyer, J. and Slyman, S. (2016) *An Introduction to Quantitative Methods for Business Research*. London: Algana Publishing, 200 pages, ISBN−13: 9781907934094.
Early, R.E. (1992) The Alchemy of Mathematical Experience: A Psychoanalysis of Student Writings. *For the Learning of Statistics*, 12(1), pp. 15–20.
Fairweather, E. Cramond, B. (2010) Infusing Creative and Critical Thinking into the Curriculum Together. In: Beghetto, A. and Kaufman, J.C. (ed.). *Nurturing Creativity in the Classroom*. United States: Cambridge University Press.
Falstein, N. and Barwood, H. (2001) The 400 Project. Available at http://theinspiracy.com/400_project.htm. (Accessed 20/04/2016).
Faria, A.J. (2001) The Changing Nature of Business Simulation/Gaming Research. *Simulation and Gaming*, 32(1), pp. 97–110. DOI: 10.1177/104687810103200108
Faria, A.J., Hutchinson, D. Wellington, W.J. and Gold, S. (2009) Developments in Business Gaming: A Review of the Past 40 Years. *Simulation and Gaming*, 40(4), pp. 464–487. DOI: 10.1177/1046878108327585
Faria, A.J. and Wellington, W.J. (2004) 'A Survey of Simulation Game Users, Former-Users, and Never-Users. *Simulation and Gaming*, 35(2), pp. 178–207. DOI: 10.1177/1046878104263543
Federoff. M.A. (2002) Heuristics and Usability Guidelines for the Creation and Evaluation of Fun in Video Games. Master Thesis, Indiana University, Bloomington.

Fienberg, S.E. (1992) A Brief History of Statistics in Three and One-Half Chapters: A Review Essay. Available at: https://www.jstor.org/stable/2246307?casa_token=Vek3Rmd5Mk0AAAAA%3AJ0zRZC5AxskwvTNY7GlrgsRV6oLrE20DY3FdeaPad0JvNSNrZDav8uZpX-YqY1hXTDtaz6zM87S4vwkQdflDLSzmrLCNlEsaBqWnykepc0jlYXsQXA

Forest, E. (2014) ADDIE Model. Available at: http://educationaltechnology.net/the-addie-model-instructional-design/ (Accessed 20/04/2016).

Fosnot, T.C. (1989) *Enquiring Teachers, Enquiring Learners: A Constructivist Approach for Teaching*. New York: Teachers College Press.

Fosnot, T.C. and Perry, R.S. (2005) Constructivism: A Psychological Theory of Learning. Available at: http://beyondbitsandatoms.org/readings/fosnot_constructivism_1996.pdf (Accessed 20/07/2019).

Garris, R., Ahlers, R. and Driskell, E.J. (2002) Games, Motivation, and Learning: A Research and Practice Model. *Simulation and Gaming*, 33(4), pp. 441–467.

Gee, J.P. (2003a) *What Video Games Have to Teach Us About Learning and Literacy*. New York: Palgrave Macmillan.

Gee, J.P. (2003b) Good Video Games and Good Learning. Available at: https://academiccolab.org/resources/documents/Good_Learning.pdf (Accessed 03/01/2021).

Gee, J.P. (2004) *Situated Language and Learning: A Critique of Traditional Schooling*. London: Routledge.

Gee, J.P. (2005a) Why are Video Games Good for Learning? Retrieved January 1, 2018, from http://academiccolab.org/resources/documents/MacArthur.pdf (Accessed 20/04/2016).

Gee, J.P. (2005b) *Why Video Games Are Good for Your Soul: Pleasure and Learning*. Melbourne: Common Ground.

Gee, J.P. (2007) *Good Video Games and Good Learning: Collected Essays on Video Games, Learning, and Literacy*. New York: Peter Lang.

Gee, J.P. (2008) *Getting over the Slump: Innovation Strategies to Promote Children's Learning*. New York: The Joan Ganz Cooney Center at Sesame Workshop.

Gee, J.P., Hayes, E., Torres, R.J., Games, I.A., Squire, K. and Salen, K. (2008) Playing to Learn Game Design Skills in A Game Context. In *Proceeding ICLS'08 Proceedings of the 8th International Conference on International Conference for the Learning Sciences*, Vol. 3, pp. 368–374.

Gee, J.P. and Hayes, E.R., 2011. *Language and Learning in the Digital Age*. Routledge.

Gibbons, J. and Gray, M. (2004) Critical Thinking as Integral to Social Work Practice. *Journal of Teaching in Social Work*, 24, pp. 19–37.

Glaser, B.G. and Strauss, A.L. (1967) *The Discovery of Grounded Theory*. Chicago: Aldine.

Gomes, A. and Mendes, A.J. (2007) Learning to Program-Difficulties and Solutions. In *International Conference on Engineering Education–ICEE*, Vol. 7.

Granic, I., Lobel, A. and Engels, R.C. (2014) The Benefits of Playing Video Games. *American Psychological Association*, 69(1), pp. 66–78. DOI: 10.1037/a0034857

Hackett, G. and Betz, N.E. (1989) An Exploration of the Mathematics Self-Efficacy/Mathematics Performance Correspondence. *Journal for Research in Mathematics Education*, 20(3), pp. 261–273.

Hamari, J., Shernoff, D.J., Rowe, E., Coller, B., Asbell-Clarke, J. and Edwards, T. (2016) Challenging Games Help Students Learn: An Empirical Study on Engagement, Flow and Immersion in Game-Based Learning. *Computers in Human Behavior*, 54, pp. 170–179.

Hamari, J. and Tuunanen, J. (2014) Player Types: A Meta-Synthesis. *Transactions of the Digital Games Research Association*, 1(2), pp. 29–53.

Harel, I. and Papert, S. (1990) *Software Design as a Learning Environment*. Massachusetts Institute of Technology Cambridge, MA: MIT Media Laboratory.

Harel, I. and Papert, S. (1991) *Constructionism.* Norwood, NY: Ablex Publishing Corporation.

Hargreaves, A. (2003) *Teaching in the Knowledge Society: Education in the Age of Insecurity*, New York: Teachers' College Press.

Holbert. N., Penney, L. and Wilensky, U. (2010) *Bringing Constructionism to Action Game-Play, Constructionism.* Paris. Available at: https://www.google.co.uk/search?q=Bringing+Constructionism+to+Action+Game-Playandoq=Bringing+Constructionism+to+Action+Game-Playandaqs=chrome..69i57.7170j0j8andsourceid=chromeandie=UTF-8 (Accessed 20/03/2017).

Hoover, W.A. (1996) The Practice Implications of Constructivism. *SEDL Letter*, 9(3), pp. 1–2.

Horn, B., Hoover, A.K. Barnes, J. Folajimi, Y. Smith, G. and Harteveld, C. (2016) Opening the Black Box of Play: Strategy Analysis of An Educational Game. In *Proceedings of the 2016 Annual Symposium on Computer-Human Interaction in Play*, pp. 142–153.

Hsu, C.-Y., Tsai, C.-C. and Wang, H.-Y. (2012) Facilitating Third Graders' Acquisition of Scientific Concepts Through Digital Game-Based Learning: The Effects of Self-Explanation Principles. *Asia-Pacific Education Researcher*, 21(1), pp. 71–82.

Hunicke, R., LeBlanc M. and Zubek, R. (2004) MDA: A Formal Approach to Game Design and Game Research. Available at: https://www.google.co.uk/search?q=MDA%3A+A+Formal+Approach+to+Game+Design+and+Game+Researchandoq=MDA%3A+A+Formal+Approach+to+Game+Design+and+Game+Researchandaqs=chrome..69i57j69i58.3868j0j8andsourceid=chromeandie=UTF-8 (Accessed 8/10/2016).

Illeris, K. (2007) *How We Learn: Learning and Non-Learning in School and Beyond.* London: Routledge.

ISO (1998) Available at: http://www.standardsglossary.com/iso1.htm (Accessed 8/10/2016).

Jackson, N. and Shaw, M. (2007) Developing Perspectives on Creativity in Higher Education. In: Jackson, N., Oliver, M., Shaw M., and Wisdom, J. (ed.). *Developing Creativity in Higher Education: An Imaginative Curriculum.* London: Routledge.

Jagger, S., Siala, H. and Sloan, D. (2016) It's All in The Game: A 3D Learning Model for Business Ethics. *Journal of Business Ethics*, 137(2), pp. 383–403.

Jameson, M.M. and Fusco, B.R. (2014) Math Anxiety, Math Self-Concept, and Math Self-Efficacy in Adult Learners Compared to Traditional Undergraduate Students. *Adult Education Quarterly*, 64(4), pp. 306–322.

Jarvis, M. (2005) *'Motivation' in Psychology of Learning and Teaching*. Cheltenham: Nelson Thornes.

Jeffrey, B. and Craft, A. (2004) Teaching Creatively and Teaching for Creativity: Distinctions and Relationships. *Educational Studies*, 30(1), pp. 77–87.

Jenson, J. and Droumeva, M. (2015) Making Games with Game Maker: A Computational Thinking Curriculum Case Study. In *ECGBL 2015-9th European Conference on Games Based Learning: ECGBL2015*, p. 260. Academic Conferences and Publishing Limited.

Jones, M.G. (1998) Creating Engagement in Computer-Based Learning Environments. Available at: http://itech1.coe.uga.edu/itforum/paper30/paper30.htmI (Accessed 12/04/2016).

Kafai, Y. (2006) Playing and Making Games for Learning: Instructionist and Constructionist Perspectives for Game Studies. *Games and Culture*, 1(1), pp. 36–40.

Kafai, Y.B. and Resnick, M. eds. (2012) *Constructionism in Practice: Designing, Thinking, and Learning in a Digital World.* Routledge. Available at: https://www.google.co.uk/books/edition/Constructionism_in_Practice/KGkIEQAAQBAJ?hl=en&gbpv=1&dq=Kafai,+Y.B.+and+Resnick,+M.+eds.+(2012)+Constructionism+in+Practice:+Designing,+Thinking,+and+Learning+in+a+Digital+World.+Routledge.&pg=PP1&printsec=frontcover

Kato, P.M., Cole, S. Bradlyn, A.S. and Pollock, B.H. (2008) Video Games Improves Behavioural Outcomes in Adolescents and Young Adults with Cancer: A Randomized Trial. *Pediatrics*, 122(2), pp. e305–e217.

Kessner, M., Wood, J. Dillon, R.F. and West, R.L. (2001) On the Reliability of Usability Testing. In *Proceedings of the Extended Abstracts on Human Factors in Computing Systems (CHI '01)*, p. 97.

Kim, B. (2001) Social Constructivism. *Emerging Perspectives on Learning, Teaching, and Technology*, 1(1), pp. 16.

Kirriemuir, J. and Mcfarlane, A. (2004) Literature Review in Games and Learning. Available at: https://telearn.archives-ouvertes.fr/hal-00190453/document (Accessed 12/04/2020).

Knight, P.T. (2001) A Briefing on Key Concepts – Formative and Summative, Criterion and Norm Referenced Assessment. *Learning and Teaching Support Network Generic Centre*, 30. http://www.heacademy.ac.uk

Kodugamelab (2015) Available at: https://www.kodugamelab.com/ (Accessed 8/11/2018).

Kolb, A.Y. and Kolb. D.A. (2008) The Learning Way: Meta-Cognitive Aspects of Experiential Learning. *Simulation and Gaming*, 40(3), pp. 297–327.

Kolb, D.A. (1984) *Experiential Learning: Experience as the Source of Learning and Development*, Vol. 1. Englewood Cliffs, NJ: Prentice-Hall.

Korhonen, H. and Koivisto, E.M.I. (2007) Playability Heuristics for Mobile Multi-Player Games. In *Proceedings of the 2nd International Conference on Digital Interactive Media in Entertainment and Arts*, pp. 28–35.

Korris, J. (2004) Full Spectrum Warrior: How the Institute for Creative Technologies Built a Cognitive Training Tool for The Xbox. Available at: http://scholar.google.co.uk/scholar?hl=enandq=FULL+SPECTRUM+WARRIOR%3A+HOW+THE+INSTITUTE+FOR+CREATIVE+TECHNOLOGIESandbtnG=andas_sdt=1%2C5andas_sdtp= (Accessed 6/10/2015).

Koster, R. (2005) *A Theory of Fun for Game Design, Electronic Resource*. Published Scottsdale, AZ: Paraglyph Press. Available at: http://capitadiscovery.co.uk/roehampton/items/510379?resultsUri=http%3A%2F%2Fcapitadiscovery.co.uk%2Froehampton%2Fitems%3Fquery%3Dtheory%2Bof%2Bfun%26target%3Dcatalogue%26facet%255B0%255D%3Dfulltext%253Ayes#availability (Accessed 29/12/2015).

Kraiger, K., Ford, J.K. and Salas, E. (1993) Application of Cognitive, Skill-based, and Affective Theories of Learning Objectives to New Methods of Training Evaluations, *Journal of Applied Psychology*, 78(2), pp. 311–328.

Kuss, D.J. and Griffiths, M.D. (2012) Internet Gaming Addiction: A Systematic Review of Empirical Research. *International Journal of Mental Health and Addiction*, 10(2), pp. 278–296.

Lave, J. and Wenger E. (1991) *Situated learning: Legitimate Peripheral Participation*. Cambridge University Press. Available at: https://www.google.co.uk/books/edition/Situated_Learning/CAVIOrW3vYAC?hl=en&gbpv=1&dq=Lave,+J.+and+Wenger+E.+(1991)+Situated+Learning:+Legitimate+Peripheral+Participation.+Cambridge+University+Press&pg=PA11&printsec=frontcover

Lazzaro, N. (2004) Why We Play Games: Four Keys to More Emotion without Story. Available at: http://www.xeodesign.com/xeodesign_whyweplaygames.pdf (Accessed 28/12/2015).

Lee, J.S., Keil, M. and Kasi, V. (2012) The Effects of an Initial Budget and Schedule Goal on Software Project Escalation. *Journal of Management Information System*, 29(1), pp. 53–77. DOI: 10.2753/MIS0742-1222290102

Lenhart, A., Kahne, J. Middaugh, E. Macgill, A.R. Evans, C. and Vitak, J. (2008) Teens, Video Games, and Civics: Teens' Gaming Experiences Are Diverse and Include Significant Social Interaction and Civic Engagement. *Pew Internet & American Life*

Project. Available at: https://www.google.co.uk/search?q=Lenhart+et+al.%2C+2008andoq=Lenhart+et+al.%2C+2008andaqs=chrome..69i57j0.1244j0j8andsourceid=chromeandie=UTF-8 (Accessed 20/8/2017).

Levine, D.M., Krehbiel, T.C. and Berenson, M.L. (2003) *Business Statistics: A First Course*, 3rd Edition. New Jersey: Pearson Education Ltd.

Lewis, G. (2013) Emotion and Disaffection with School Mathematics. *Research in Mathematics Education*, 15(1), pp. 70–86.

Li, F.W.B. and Watson, C. (2011) *Game-Based Concept Visualization for Learning Programming*. Scottsdale, Arizona, USA: ACM. 978-1-4503-0994-3/11/12.

Li, Z.Z., Cheng, Y.B. and Liu, C.C. (2013) A Constructionism Framework for Designing Game-Like Learning Systems: Its Effect on Different Learners. *British Journal of Educational Technology*, 44(2), pp. 208–224.

Lister, R., Adams, E.S. Fitzgerald, S. Fone, W. Hamer, J. Lindholm, M. McCartney, R. Moström, J.E. Sanders, K. Seppälä, O. and Simon, B. (2004) A Multi-National Study of Reading and Tracing Skills in Novice Programmers. In *ACM SIGCSE Bulletin*, 36(4), pp. 119–150. ACM.

Litosseliti, L. (2005) *Using Focus Groups in Research*. London: Continuum.

Loftus, C., Thomas, L. and Zander, C. (2011) Can Graduating Students Design: Revisited. In *Proceedings of the 42nd ACM Technical Symposium on Computer Science Education*, pp. 105–110.

Lyons, I.M. and Beilock, S.L. (2012) When Math Hurts: Math Anxiety Predicts Pain Network Activation in Anticipation of Doing Math. *PLoS One*, 7(10), p. e48076.

Majgaard, G. (2016) Creating Games in the Classroom: From Native Gamers to Reflective Designers. In: Munkvold, R. (ed.). *Games Based learning for Researchers, Teachers and Students*, Vol. 1. Reading: Academic Conferences and Publishing International Ltd. ISBN: 978-1-911278-12-890000.

Malaka, R. (2014) How Computer Games Can Improve Your Health and Fitness. Gobel, S. and Wiemeyer, J. (ed.). *Games for Training Education, Health and Sports, 4th Conference on Serious Games, GameDays 2014*, Darmstadt, Germany, April 1–5, 2014, Springer.

Malone. T.W. (1980) What Makes Things Fun to Learn? A Study of Intrinsically Motivating Computer Games. Available at: http://cci.mit.edu/malone/Malone%20computer%20games%20bibliography.htm

Malone, T.W. and Lepper, M.R. (1987) Making Learning Fun: A Taxonomy of Intrinsic Motivations for Learning. Aptitude, Learning, and Instruction. In: Snow, R.E. and Farr, M.J. (ed.). *Volume 3: Cognitive and Affective Process Analyses*. Hillsdale, USA: Erlbaum.

Maloney, E.A., Schaeffer, M.W. and Beilock, S.L. (2013) Mathematics Anxiety and Stereotype Threat: Shared Mechanisms, Negative Consequences and Promising Interventions. *Research in Mathematics Education*, 15(2), pp. 115–128.

Marciano, J.N., Miranda, L.C. and Miranda, E.E.C. (2014) Evaluating Multiple Aspects of Educational Computer Games. Available at: https://www.hindawi.com/journals/ijcgt/2014/139205/ (Accessed 27/08/2017).

Marklund, B.B. and Taylor, A.S.A. (2016) Creating Games in the Classroom: From Native Gamers to Reflective Designers. In: Munkvold, R. (ed.). *Games Based learning for Researchers, Teachers and Students*, Vol. 1. Reading: Academic Conferences and Publishing International Ltd. ISBN: 978-1-911278-12-890000.

Maslow, A.H. (1970) *Motivation and Personality*, 2nd Edition. New York; London: Harper and Row.

Mayer, E.R. (2014) *Computer Games for Learning: An Evidence Based Approach*. London: MIT.

Mayer, I., Warmelink, H. and Bekebrede, G. (2013) Learning in a Game-Based Virtual Environment: A Comparative Evaluation in Higher Education. *European Journal of Engineering Education*, 38(1), pp. 85–106.

Mayo, M. (2009) Video Games: A Route to Large-Scale STEM Education? *Science*, 323, pp. 79–82.
McConnell, D. (2000) *Implementing Computer Supported Cooperative Learning*, 2nd Edition. London: Kogan Page.
Mccrindle, M. (2018) What Comes after Generation Z? Introducing Generation Alpha. Available at: https://mccrindle.com.au/insights/blogarchive/what-comes-after-generation-z-introducing-generation-alpha/ (Accessed 15/10/2018).
McFarlane, A., Sparrowhawk, A. and Heald, Y. 2002. *Report on the Educational Use of Games: An Exploration by TEEM on the Contribution Which Games Can Make to the Educational Process*. Cambridge, UK: TEEM. http://www.teem.org.uk/publications/teem_gamesined_full.pdf (Accessed 06/12/2015).
McGonigal, J. (2015) *Superbetter: A Revolutionary Approach to Getting Happier Stronger and Happier*. Braver and More Resilient, New York: Penguin Press.
Milane, I. and Rowe, G. (2002) Difficulties in Learning and Teaching Programming- Views of Students and Tutors. *Education and Information Technologies*, 7(1), pp. 55–66.
Monroy-Hernandez, A. and Resnick, M. (2008) Empowering kids to create and share programmable media. *Interactions*. Retrieved August 30, 2015.
Moreno-Ger, P., Burgos, D. Martínez-Ortiz, I. Sierra, J.L. and Fernández-Manjón, B. (2008) Educational Game Design for Online Education. *Computers in Human Behavior*, 24(6), pp. 2530–2540. DOI: 10.1016/j.chb.2008.03.012
Moreno-Ger, P., Torrente, J. Hsieh, Y.G. and Lester, W.T. (2012) Usability Testing for Serious Games: Making Informed Design Decisions with User Data. *Advances in Human-Computer Interaction*, 2012. DOI: 10.1155/2012/369637
Morgan, D.L. (1988) *Focus Groups as Qualitative Research*, Beverly Hills, CA: Sage.
Morrison, K. (2002) *School Leadership and Complexity Theory*. London: Routledge Falmer.
Muijs, D. (2011) *Doing Quantitative Research in Education with SPSS*. London: Sage Publication Ltd.
Mullins, L. (2013) *Management and Organisational Behaviour*, 10th Edition. Harlow: Pearson Education.
Mullins, L. (2019) *Organisational Behaviour in the Workplace*, 12th Edition. Harlow: Pearson Education.
Multon, K.D., Brown, S.D. and Lent, R.W. (1991) Relation of Self-Efficacy Beliefs to Academic Outcomes: A Meta-Analytic Investigation. *Journal of Counselling Psychology*, 18, pp. 30–38.
Naik, N. (2017) The use of GBL to Teach Mathematics in Higher Education. *Innovations in Education and Teaching International*, 54(3), pp. 238–246. DOI: 10.1080/1470 3297.2015.1108857
Newby, P. (2010) *Research Methods for Education*, Harlow: Pearson Education Ltd.
Nicholls, D., De Charon, L. and Hutkin, R. (2010) Development of Critical Thinking and Creativity: Practical Guidelines for the Postsecondary Classroom. *ATEA Journal*, 38(1), p. 12.
Nielsen, J. (1995) Usability Inspection Methods. *CHI'95 Mosaic of Creativity*, pp. 7–11.
Nielsen, J. (2000) *Designing Web Usability*. New Riders Publishing.
Otkay, J.S. (2012) *Grounded Theory*. Oxford, England: Oxford University Press.
Paavilainen, J. (2010) Critical Review on Video Game Evaluation Heuristics: Social Games Perspective. *Proceedings of the International Academic Conference on the Future of Game Design and Technology*, pp. 56–65. DOI: 10.1145/1920778.1920787
Pachler, N. (2007) *Mobile Learning: Towards a Research Agenda*. London: WLE Centre, IoE. In: Kress, G. and Pachler, N (ed.) (2007) Thinking about the 'm' in m-learning.
Pajares, F. and Kranzler, J. (1995) Self-Efficacy Beliefs and General Mental Ability in Mathematical Problem-Solving. *Contemporary Educational Psychology*, 20, pp. 426–443.

Palincsar, A.S. (1998) Social Constructivist Perspectives on Teaching and Learning. *Annual Review of Psychology*, 49, pp. 345–375.

Papert, S. (1993) *Mindstorms: Children, Computers and Powerful Ideas*. New York: Basic Books.

Paras, B. and Bizzocchi, J. (2005) Game, Motivation, and Effective Learning: An Integrated Model for Educational Game. In *Proceedings of DiGRA Conference on Changing Views–Worlds of Play*. Available at: https://www.researchgate.net/publication/221217604_Game_Motivation_and_Effective_Learning_An_Integrated_Model_for_Educational_Game_Design (Accessed 27/09/2016).

Park, H.J. and Kim, S.B. (2013) Guidelines of Serious Game Accessibility for the Disabled. In *Proceedings of the International Conference on Information Science and Applications (ICISA '13)*, pp. 1–3, IEEE.

Pelletier, C., 2009. Games and Learning: What's the Connection. *International Journal of Learning and Media*, *1*(1), pp. 83–101.

Pinelle, D., Wong, N., Stach, T. and Gutwin, C. (2009) Usability Heuristics for Networked Multiplayer Games. *Proceedings of the ACM 2009 International Conference on Supporting Group Work*, pp. 169–178.

Pollard, A., Anderson, J., Maddock, M., Swaffield, S., Warin, J. and Warwick, P. (2008) *Reflective Teaching*, 3rd Edition. London: Continuum International Publishing Group.

Preconception (2020) In: Dictionary.com. [Online] Available at: https://www.dictionary.com/browse/preconception (Accessed 01/08/2020).

Preece, Y., Rogers, Y. and Sharp, H. (2015) *Interaction Design: Beyond Human-Computer Interaction*, 4th Edition. Chichester: John Wiley and Sons Ltd.

Prensky, M. (2002) The Motivation of Gameplay: The Real Twenty-First Century Learning Revolution. *On the Horizon*, 10(1), pp. 5–11.

Prensky, M. (2005) Listen to Natives. *Association for Supervision and Curriculum Development Educational Leadership*, 63(December 2005), pp. 9–13.

Prensky, M. (2006) *'Don't Bother Me Mom, I'm Learning!': How Computer and Video Games Are Preparing Your Kids for the Twenty-First Century*. St Paul, MN: Paragon House.

Pritchard, A. and Woollard, J. (2010) *Psychology for the Classroom: Constructivism and Social Learning*. Oxon: Routledge.

Punch. K.F. (2009) *Introduction to Research Methods in Education*, London: Sage Publications Ltd.

Resnick, M. and Robinson, K. (2017) *Lifelong Kindergarten: Cultivating Creativity Through Projects, Passion, Peers, and Play*. MIT Press.

Robson. C. (2010) *Real World Research*, 3rd Edition. Chichester: John Wiley and Sons Limited.

Robson, S. (2006) *Developing Thinking and Understanding in Young Children: An Introduction for Students*. London: Routledge Ltd.

Rogoff, B. (1998) Cognition as a Collaborative Process. In: Damon, W., Kuhn, D. and Seigler, R.S. (ed.). *Handbook of Child Psychology*, 5th Edition, p. 2. New York: Wiley.

Roth, W.M. (2000) Authentic School Science: Intellectual Traditions. In: McCormick, R. and Paechter, C. (ed.). *Learning & Knowledge*. London, UK: Paul Chapman Publishing, pp. 6–20.

Russ, T., and Drury-Grogan, M.L. (2013) Assessing the Impacts of a Business Communication Simulation on Students' Self-Perceptions. *Eastern Communication Association*, 61(5), pp. 584–595. DOI: 10.1080/01463373.2013.822404

Salen, K. and Zimmerman, E. (2011) The Game Designer Reader, A Rules of Play Anthology. Roger Caillois: The Definition of Play. *The Classification of Games*. Cambridge, Massachusetts. London, England: The MIT Press.

Salisbury, J.H. and Cole. T. (2016) Grounded Theory in Games Research: Making the Case and Exploring the Options. Available at: https://www.academia.edu/27624170/Grounded_Theory_in_Games_Research_Making_the_Case_and_Exploring_the_Options (Accessed 15/10/2015).

Schell, J. (2015) *The Art of Game Design: A Book of Lenses*, 2nd Edition. USA: CRC Press.

Schell, J. (2020) *The Art of Game Design: A Book of Lenses*, 3rd Edition. USA: CRC Press.

Senge, P. (1990) *The Fifth Discipline: The Art and Practice of Learning Organization*. United Kingdom: Random House Business Books.

Senge, P.M. (2006) *The Fifth Discipline: The Art and Practice of Learning Organisation*. London: Random House Business Books.

Shell, D.F., Murphy, C.C. and Bruning, R.H. (1989) Self-efficacy and Outcome Expectancy Mechanisms in Reading and Writing Achievement. *Journal of Educational Psychology*, 81(1), p. 91.

Shunk, D.H. (2000) *Learning Theories: An Educational Perspective*, 3rd Edition. Upper Saddle River, NJ: Prentice-Hall.

Siegel, R.G., Galassi, J.P. and Ware, W.B. (1985) A Comparison of Two Models for Predicting Mathematics Performance: Social Learning Versus Math Aptitude-Anxiety. *Journal of Counseling Psychology*, 32, pp. 531–538.

Silverman, B.G., Johns, M. Weaver, R. Mosley, J. (2003) Authoring Edutainment Stories for Online Players (AESOP): Introducing Game play into Interactive Dramas. In: *International Conference on Virtual Storytelling*. Springer.

Silverman, D. (1993) *Interpreting Qualitative Data: Methods for Analysing Talk, Text, and Interaction*. London: Sage Publications Ltd.

Silverman, D. (2014) *Interpreting Qualitative Data*, 5th Edition. London: Sage Publications Ltd.

Sindre, G., Natvig, L. and Jahre, M. (2009) Experimental Validation of the Learning Effect for A Pedagogical Game on Computer Fundamentals. *IEEE Transactions on Education*, 52 (1), pp. 10–18.

Skemp, R.R. (1986) *The Psychology of Learning Mathematics Great*. Britain: Penguin Books Ltd.

Skemp, R.R. (1989) *Mathematics in the Primary School*. Bristol: Rutledge.

Skilbeck, A. (2017) Dewey on Seriousness, Playfulness and the Role of the Teacher. *Education Sciences*, 7(1), p. 16.

Slyman, S. (2014) I Hate Maths! An Investigation into Students' Attitudes Towards Mathematics. MA Thesis, University of Roehampton, London, United Kingdom.

Slyman, S. (2018) *Games Based Learning in Universities, A Life Learning Experience*. New York: Business Expert Press Expert Insights Publishing. ISBN:978-1-94819-801-1.

Slyman, S. (2022) Acumenous Game-Based Learning in Simulation Games and Applied Statistics. PhD Thesis, Goldsmiths, University of London, UK.

Slyman, S., Gillies, M. and Jessel, J. (2019a) Acumenous or Inquisitional? Towards a New Theoretical Lens within Games Learning. In: *13th European Conference on Games Based Learning*. Odense, Denmark: University of South Denmark. ISBN:978-1-912764-38-9 or E-book: ISSN:2049-100X.

Slyman, S., Gillies, M. and Jessel, J. (2019b) Game-Based Learning to Engage students with Applied Statistics using a Simulation Roleplay Game. In: *9th Irish Conference on Games Based Learning*, Cork, Ireland. Available at: http://igbl-conference.com/2016/programme/generate_book_of_abstracts.php (Accessed 25/09/2019).

Smith Report, (2004) Available at: http://www.mathsinquiry.org.uk/report/mathsInquiryFinalReport.pdf (Accessed 20/12/2014).

Smith Report, (2017) Available online at: https://www.gov.uk/government/publications/smith-review-of-post-16-maths-report-and-government-response (Accessed 20/12/2018).

Sobke, H., Chan, E., Buttlar, R., Grobe-Wortmann, J. and Londong, J. (2014) Cat King's Metamorphosis. The Reuse of an Educational Game in a Further Technical Domain. Gobel, S., and Wiemeyer, J. (ed.) *Games for Training Education, Health and Sports, 4th Conference on Serious Games, GameDays 2014* Darmstadt, Germany, April 1–5, 2014, Springer.

Song, S. and Lee, J. (2007) Key Factors of Heuristic Evaluation for Game Design: Towards massively multi-player online Roleplaying game. *International Journal of Human-Computer Studies*, 65, pp. 709–723.

Sweetser, P., Johnson, D. and Wyeth, P. (2012) Revisiting the GameFlow Model with Detailed Heuristics. *Journal of Creative Technologies*, 2012(3), pp. 1–16.

Sweetser. P. and Wyeth, P. (2005) GameFlow: A Model for Evaluating Player Enjoyment in Games. *Computers in Entertainment*, 3(3), pp. 3. DOI: 10.1145/1077246.1077253

Tamura, H. (2007) Foundational Value of Mathematics Education for Management Curriculum. *International Statistical Review*, 75(3), pp. 397–405. DOI: 10.1111/j.1751-5823.2007.00032.x

Tao, Y.H., Cheng, C.J. and Sun, S.Y. (2012) Alignment of Teacher and Students Perceptions on the Continued Use of Business Simulation Games. *Educational Technology and Society*, 15(3), pp. 177–189.

TED (2019) *Souad Slyman: Why Change Attitudes Towards Mathematics [YouTube].* TedxGul, London: Goldsmiths, University of London. Available at: https://www.youtube.com/watch?v=leNDJMNJoNY (Accessed 26/04/2019).

Timonen, V., Foley, G. and Conlon, C. (2018) Challenges when Using Grounded Theory: A pragmatic introduction to Doing GT Research. *International Journal of Qualitative Methods*, 17, pp. 1–10. DOI: 10.1177/1609406918758086

Tondello, G.F., Kappen, D.L. Mekler, E.D. Ganaba, M. and Nacke, L.E. (2016) Heuristic Evaluation for Gameful Design. *CHI PLAY'16*. ACM. 978-1-4503-4458-6/16/10. DOI: 10.1145/2968120.2987729

van der Spek, E.D., van Oostendorp, H. and Meyer, J.-J.C.H. (2013) Introducing Surprising events can stimulate deep learning in a serious game. *British Journal of Educational Technology*, 44(1), pp. 156–169.

Vlachopoulos, D. and Makri, A. (2017) The Effect of Games and Simulations on Higher Education: A Systematic Literature Review. *International Journal of Educational Technology in Higher Education*, 14(1), p. 22. DOI: 10.1186/s41239-017-0062-1

Vos, L. (2014) Marketing Simulation Games: A Review of Issues in Teaching and Learning. *The Marketing Review*, 14(1), pp. 67–96. DOI: 10.1362/146934714X13948909473220

Vygotsky, L.S. (1980) *Mind in Society. The Development of Higher Psychological Processes*. Edited by Cole, M., John-Steiner, V., Scribner, S. and Souberman, E. London: Harvard University Press.

Walonick, D.S. (2000) *Designing and Using Questionnaire in Survival Statistic*. MN: StatPacInc. Available at Internet WWW page at URL: http://www.statpac.com/surveys/surveys.pdf (Accessed 02/04/2016).

Whitton, N. (2011) Games-Based Experiences for Learning. *Futurelab Innovation in Education*. Available at: https://www.nfer.ac.uk/media/1775/futl11.pdf (Accessed 02/04/2016).

Wolfe, J. (1997) The Effectiveness of Business Games in Strategic Management Coursework. *Simulation and Gaming*, 28(4), pp. 360–376. DOI: 10.1177/1046878197284003

Women in STEM – 'Facts and Statistics' (Reporting period 2012–2015) IET. Available at: https://communities.theiet.org/files/8042 (Accessed 20/06/2017).

Yin, R.K. (1994) *Case Study Research: Design and Methods*, 2nd Edition. London: Sage.

Young, M. 2008. *"Powerful Knowledge" and "Knowledge of the Powerful": Some Reflections*. London: Presentation at the Institute of Education.

Yusoff, A., Crowder, R. and Gilbert, L. (2010) Validation of Serious Games Attributes Using the Technology Acceptance Model [online]. In: IEEE (ed.), *2010 Second International Conference on Games and Virtual Worlds for Serious Applications*, Braga, Portugal, 25–26 March 2010, pp. 45–51. Available from: http://eprints.soton.ac.uk/270809/ (Accessed 10/05/2015).

Zimmerman, B.J. (2000) Self-Efficacy: An Essential Motive to Learn. *Contemporary Educational Psychology*, 25(1), pp. 82–91. DOI: 10.1006/ceps.1999.1016

Zyda, M. (2005) *From Visual Simulation to Virtual Reality to Games*. USC Information Sciences Institute, IEEE Computer Society Edition September, pp. 25–31. Available at: http://scholar.google.co.uk/scholar?hl=enandq=zyda+2005andbtnG=andas_sdt=1%2C5andas_sdtp= (Accessed 07/10/2015).

Part II

Game Design in Simulation Games and the Transformation of Learning in Statistics

Having delved into the realm of game-based learning and its potential as a tool for enhancing the teaching and learning of Mathematics and Statistics, Part II, and in particular Chapter 4 will shift the focus towards aspects of game design and development, evaluation heuristics, Human-Computer Interaction (HCI), and usability guidelines. Next, Chapter 5 will embark on an exploration of teaching Statistics in higher education, employing a creative approach through gameplay investigation, utilizing games such as 'SimCity' as a part of the core curriculum.

DOI: 10.4324/9781032723426-5

4 Exploring the Power of Game Design

Mechanics, Interactions, and Evaluation Heuristics

Game design emerges as a rich tapestry of knowledge and skills, including system-based thinking, iterative critical problem-solving, art and aesthetics, writing, storytelling, interactive design, game logic, rules, and computer skills (Gee et al., 2008:2).

As we navigate this chapter, we delve into the multidimensional nature of game design, recognizing its pivotal role in crafting computer systems that facilitate human engagement and foster both individual and collective social interactions. The chapter unravels the socio-technical engineering skills, exploring how games shape and are shaped by the dynamic interplay between players and the game environment.

Our exploration then turns toward the crucial realm of game evaluation. We embark on an in-depth examination of evaluation heuristics, usability testing, and human-computer interaction (HCI) guidelines. With a particular focus on both entertainment and educational games, we present compelling examples of evaluation heuristics, honing our attention on the realm of simulation games (SGs) and the exciting game mechanics and designs within game-based learning (GBL). Through this exploration, we shed light on the contributions and limitations of these evaluation approaches, paving the way for informed decision-making in game design and development.

As we reach the conclusion of this chapter, we are equipped with a profound appreciation for the intricate nuances of game design. We stand at the precipice of understanding how game mechanics, interactions, and evaluation form the foundation for captivating and transformative game experiences. This chapter serves as a vital bridge, connecting theory with practice, and offering a springboard for the chapters that lie ahead. Prepare to unlock the potential of game design as we embark on a remarkable expedition into the world of game-based learning and its transformative power.

DOI: 10.4324/9781032723426-6

Game Design and Development: Evaluation Heuristics, Game Design, and HCI Principles

Game evaluation heuristics are a set of guidelines for reviewing, analyzing, and evaluating games, and they are important for designing games. There are many evaluation heuristics available for both entertainment and educational games, but they are inconsistent, invalid, and some do not adhere to HCI principles (e.g., easy-to-use, etc.). This is because entertainment games focus on engagement, empowerment, embodiment, leadership, identity, and enjoyment but lack educational perspectives. In contrast, educational games are complex to design and focus mainly on education, and they ignore engagement and enjoyment aspects. Given that there may not exist any evaluation games that have a balance (both educational and entertainment perspectives simultaneously), the aim of this section is to briefly review current literature on game design and development, and evaluation heuristics for both educational and entertainment games (with a focus on educational games) and derive the best evaluation heuristics for educational games (in particular SGs).

Definitions and Categories: Game Design and Evaluation Heuristics

Game design is the art of creating games. According to Schell (2020) and Gee et al. (2008:2), game design entails building, assessing, and transforming systems that require "game mechanics and human interactions with technology and other humans." This suggests that game design plays a key role in creating computer systems which entails mechanics and human interactions with technology and other humans. Game design entails a rich array of knowledge and skills, including system-based thinking, iterative critical problem-solving, art and aesthetics, writing, storytelling, interactive design, game logic, rules and computer skills (Gee et al., 2008:2).

Game design also involves thinking about socio-technical engineering skills: How people will interact with the game and how the game will shape both individual and competitive or collaborative social interaction. Game design is but a start in learning to think of complex interactions among variables, people and technology, but it is also a domain where eventually, through SGs, designers can benefit others by reflecting on issues of social change and complex interaction systems.

In recent years, the role of game design has become increasingly significant, especially in the field of user-centered design, as more professionals are developing new methodologies for analyzing and evaluating games using usability testing and evaluation heuristics.

Evaluation heuristics are generally regarded as a set of design principles, and they are closely linked with more general design principles relating to the usability of interfaces (Malone, 1980). It is a field which originates from HCI and aims at designing easy-to-use interfaces and activities (Nielsen, 1995, 2000; Desurive et al., 2004; Preece et al., 2015). While usability testing is a key process in designing innovative technologies in industry and particularly in game design, analyzing and evaluating games necessitate evaluation heuristics and guidelines.

Evaluation heuristics are important because they play a major role in the software development life cycle where the aim is to develop a new software design after several iterative stages of implementation.

Many game evaluation heuristics models exist for entertainment and educational games (e.g., Nielsen, 1995, 2000; Falstein and Barwood, 2001; Garris et al., 2002; Desurive et al., 2004; Hunicke et al., 2004; Paras and Bizzocchi, 2005; Sweetser and Wyeth, 2005; Song and Lee, 2007; Gee et al., 2008; Desurvire and Wiberg, 2009; 2010; Paavilainen, 2010; Whitton, 2011; Sweetser et al., 2012; Schell, 2015, 2020; Tondello et al., 2016; for further clarifications see Evaluation Heuristics and HCI Principles Sections).

Examples include

- HEP (Heuristics Evaluation for Playability) (Desurive et al., 2004).
- MDA (Mechanics, Dynamics and Aesthetics) (Hunicke et al., 2004).
- PLAY (Desurvire and Wiberg, 2009).
- GAP (Game Approachability Playability) (Desurvire and Wiberg, 2010).
- Heuristics for Social games (Paavilainen, 2010).
- GameFlow (Garris et al., 2002; Sweetser and Wyeth, 2005; Sweetser et al., 2012).
- Gamification Heuristics (Tondello et al., 2016).
- A taxonomy of learning elements of digital games-based learning experiences (Whitton, 2011).
- The 13 learning principles (Gee, 2003a, 2003b, 2004, 2005a, 2005b, 2007, 2008, 2011; see Chapter 2 Gee's 36 Principles Section).

However, most of these heuristics tend not to combine educational with entertainment, or HCI and usability principles, and I could not find a game evaluation heuristics model that is educational and still entertaining/engaging. While some heuristics are not based on any educational perspectives, others seem not to adhere to any HCI principles, game design or usability testing guidelines, and game evaluation heuristics models. In other words, there appears to be no balance between educational and entertainment games. These are discussed in detail next.

In the next section, existing research literature on game design, evaluation heuristics, and usability testing from HCI principles are examined. The differences between game evaluation heuristics models for entertainment and educational games are discussed, in order to understand how and why they are used, and backed up with examples of both evaluation heuristics (entertainment and educational). I will start with evaluation heuristics for entertainment games and then move on to educational ones.

Evaluation Heuristics and HCI Principles: Entertainment Games

In 1980, Malone compiles a list of heuristics for designing enjoyable programs followed by usability frameworks in 1990. This is employed as a way of assessing user interfaces (UI). It consists of four basic evaluation principles, usability measures, usability testing, formal usability models, and rules of thumb (ISO (International Standards Organizations), see Bevan, 2001).

In other studies, Falstein and Barwood (2001) write a list of 400 proven rules of game design. Subsequently, Desurive et al. (2004) employ user studies to indicate the value of these heuristics and design a new prototype game that consists of screenshots using Macromedia Flash. However, this study has not demonstrated how players socially interact with the game since the gameplay and game mechanics are not fully developed, because this game is only a prototype. Additionally, over the next few years, Desurvire and Wiberg (2010) compile a set of heuristics based on the HEP (Desurive et al., 2004) and the online website games "metacritic.com" popularity scores, Nielsen, and the literature. For example, HEP is a list of heuristics based on the literature and usability testing for evaluating the usefulness of computer, videos, and board games, and it is developed by Desurive et al. (2004). However, this method is broad and unsuited for evaluating specific issues in software development and HCI game usability and playability (Desurvire and Wiberg, 2009).

Furthermore, Nielsen (1995, 2000) develops a set of HCI designs, while Federoff (2002) compiles a record of game heuristics from a case study at a game development company and compares them with current game industry guidelines and with Nielsen's heuristics. However, there is no information regarding participants' gaming experiences (positive or negative). Further questions could examine how people do perceive gameplay, game mechanics, and design (see Chapters 6–9).

In addition, Clanton (1998) develops game interface, gameplay, and game mechanics to which Song and Lee (2007) add game narrative. Hunicke et al. (2004) come up with MDA (Mechanics, Dynamics, and Aesthetics). MDA is a design methodology for guiding the creative process of thinking. It is for building prototypes and designing software "interfaces and constructing an argument or implementing a series of...experiments" (Hunicke et al., 2004:1–2). Hunicke and the team (2004) design MDA as a framework/lens for game design from HCI perspectives, that is, designer➔creates➔ a game➔ is consumed by➔ the player.

Schell (2015, 2020), however, summarizes the key ingredients/heuristics for game design under four categories:

- Mechanics: Goals, objectives, and rules of the game.
- Story: Characteristics of the story, driving instincts and events.
- Aesthetics: Appearance of the game, sounds, animation.
- Technology: Tools and materials that facilitate the creation of the game.

When it comes to Human Computer Interaction (HCI), the user interface design is how the user interacts with the computer and the information technology in general, which includes the display screen (interface), keyboard, and the mouse, etc.

Having a well-designed professional interface that is accessibly consistent, usable, and easy-to-use is one of the golden rules in Nielsen's (1995, 2000) usability heuristics. Although Nielsen's heuristics are more practical for website design rather than game design, some researchers, such as Desurive et al. (2004), Desurvire and Wiberg (2009), and Moreno-Ger et al. (2008, 2012), have gone as far as to adapt these heuristics, resulting in game evaluation models. This suggests that game evaluation heuristics are necessary for game design and creating accessible, consistent, usable, and easy-to-use systems. Further research questions/objectives could explore how to use these in educational context (see Chapters 6 and 10).

In Tondello et al.'s (2016) research about evaluation heuristics for a gameful design, the researchers present an inclusive review of existing literature on gameful design methods such as HEP (Heuristics Evaluation for Playability) (Desurive et al., 2004), Playability Heuristics for Mobile Games (Korhonen and Koivisto, 2007), Networked Game Heuristics (NGH) (Pinelle et al., 2009), PLAY (Desurvire and Wiberg, 2009), GAP (Game Approachability Playability) (Desurvire and Wiberg, 2010), Heuristics for Social Video Games (Paavilainen, 2010), GameFlow (Sweetser and Wyeth, 2005; Sweetser et al., 2012), and how these are currently utilized. They next relate this to motivational affordances theory and propose 28 new sets of gameful design heuristics. However, the heuristics have not been used or validated, according to researchers, and therefore designers are urged to establish how these heuristics can be used for their future research. Furthermore, these heuristics are in general (i.e., not specific to any subjects such as Mathematics and Statistics). In this book, specific game evaluations for Mathematics and Statistics in HE will be designed (see Chapter 6).

To this end, this research book builds on the above heuristics by proposing a new game evaluation heuristics model for educational simulation games (ESGs) called a ESGAE, since as I explained earlier there is a gap and there appears to be no balance between educational and entertainment games. The ESGAE Model is guided by this chapter and is based on the current research findings from Chapter 6: Game Design and Development.

Evaluation Heuristics and HCI Principles: Educational Games

Furthermore, with regard to existing game evaluation heuristics and additional features that make educational games motivating and engaging, there are many different educational features (see Chapter 2). These educational game features/elements or characteristics include: Gameplay, challenge, fantasy, and curiosity, rewards, feedback, learning objectives/goals, immersion and storytelling; game design and mechanics; game graphics, sounds, and music (e.g., Malone, 1980; Malone and Lepper, 1987; Jones, 1998; Garris et al., 2002; Prensky, 2002, 2005; Gee, 2003a, 2003b, 2004, 2005a, 2005b, 2007, 2011; Schell, 2015, 2020; McGonigal, 2015; Arnab and Clarke, 2017). However, there is no balance as I highlighted earlier (Game Design and Development Section): They do not adhere to HCI principles and game design and mechanics, making learning hard and disengaging.

Similarly, Whitton's (2011) report concludes that common engaging elements include learning, creativity, challenge, control, competition, fantasy, feedback, goals, sensory stimuli, and social interaction (see Whitton, 2011; see q2l.org).

Sweetser and Wyeth (2005) establish a GameFlow Model and identify eight core game characteristics/elements for sustaining enjoyment in educational games. These are: Concentration, challenge, skills, control, clear objectives, feedback, immersion, and social interaction.

Moreover, Sweetser et al. (2012) use the GameFlow Model on real-time strategy games, and they identify a number of game characteristics (165 heuristics). These heuristics are crucial for player engagement such as GameFlow, immersion, social interaction, challenge, confidence building, competition, and group vs. individual experience. They then test it using fantasy-based strategy games. They also provide details on how to achieve GameFlow for games, including immersion, graphics, sound, and gameplay. Further research questions could focus on HCI principles in the context of HE and Mathematics and Statistics (see Chapters 6 and 10).

In another study, and contrary to previous research but like Whitton (2011), Moreno-Ger et al. (2012) develop a step-by-step methodology with explicit annotations/categories for a SG called "MasterMed." The aim is to educate patients in hospital about their medication intake, match it to the right health conditions, and to increase their self-efficacy. The play time lasts for 30 minutes, and this helps researchers to understand how to improve the game and solve any usability and learning issues. The researchers also conduct an in-depth evaluation of the "MasterMed" game to gain significant insights into how to improve the usability, user interface (UI) design, and the learning outcomes.

Like Moreno-Ger et al. (2012), Heintz and Law (2015) create a web-based questionnaire to collect data on how people play and perceive games. They then launch web questionnaires on game classification to refine existing game genres. They base the genre definitions on a comprehensive model called Game Element Attribute Model (GEAM). Their aim is to research how they can use existing games to inform research on educational games (EGs). Unlike previous research (mentioned above), Heintz and Law analyze the web questionnaires using hierarchical methods and a mapping technique that computes "the degree of relevance of individual game attributes to game types," and "visualises them with hues of grey." By creatively deploying a "mapping technique, they visualise the relations between game genres" (Heintz and Law, 2015:1). Overall, the researchers argue that educational games should address both the pedagogical and game design aspects, which should draw on the same mechanisms of entertainment games. Since a set of attributes or game genres are diverse and inconsistent, and when deciding which games to use for teaching, there is no "Silver Bullet." There is thus a need for a well-defined classification. This book project builds on Heintz's and Law's study and will look at developing game heuristics for educational games, in particular SGs. The game evaluation heuristics and the ESGAE (Educational Simulation Games Analysis and Evaluation) Model will be looked at in Chapter 6.

In all, many researchers such as Prensky (2002, 2005), Squire (2005), Gee et al. (2008), Whitton (2011) and Park and Kim (2013) argue that educational games can be a motivating and engaging learning experience if games have the embedded features discussed above (e.g., Gee, 2003a, 2003b, 2004, 2005a, 2005b, 2007, 2008, 2011; Whitton, 2011; see also the First and Second Sections in Chapters 2 and 3). It is argued that educational games can be beneficial to students over classroom teaching because games have many benefits (Chapters 2 and 3). It can, for instance, give learners control, immediate feedback, and support (Malone, 1980; Jones, 1998; Garris et al., 2002; Prensky, 2005; Squire, 2005; Mayo, 2009; Schell, 2015, 2020; Jagger et al., 2016). However, when entertainment fails in educational games (such as Mathematics and Statistics games), the learning experience suffers too (e.g., Gee, 2003a, 2003b; Kirriemuir and McFarlane, 2004; Moreno-Ger et al., 2008; Whitton, 2011; Malaka, 2014).

Evaluation Heuristics for Simulation Games

To this end, there are many and various game evaluation heuristics (e.g., Desurive et al., 2004; Desurvire and Wiberg, 2010; Heintz and Law, 2015; Tondello et al., 2016; Chapter 3 and as I will explain later on). However, none of these heuristics appear to combine educational content, with HCI, usability, and game evaluation heuristics as well as entertainment. For example, some heuristics are based on HCI principles (accessible and usable) such as HEP, MDA, and PLAY; others are solely educational (e.g., GAP, GameFlow, Gee's (2003a, 2003b, 2004, 2005a, 2005b, 2007, 2008, 2011) learning principles) or mainly focus on gameful design (e.g., Gee et al.'s research (2008); Tondello et al.'s research (2016)). For this purpose, educational games are considered in regard to their intended learning purposes as well as their engaging properties as games. They must have learning objectives to succeed educationally, if they are not engaging or enjoyable as games they will fail. In contrast, entertainment games like "SimCity," "Minecraft" or even "MMORP" games are intended to concentrate more on entertainment rather than learning. This suggests that having a comprehensive framework for evaluating educational games, from various perspectives such as game design principles (Hunicke et al., 2004; Desurive et al., 2004; Desurvire and Wiberg, 2009; Schell, 2015, 2020), HCI, usability testing guidelines (Nielsen, 1995, 2000; Falstein and Barwood, 2001), and educational purposes is complex and challenging. While some researchers evaluate and generate a set of guidelines that game researchers should consider when designing games (such as Park and Kim, 2013), others reuse existing ones (such as Heintz and Law, 2015). However, some evaluation heuristics are not used, inconsistent or invalidated, and there is no best recipe. In other words, most evaluation heuristics tend to be designed for entertainment rather than educational purposes, and those with learning intentions usually do not meet HCI or game design principles. Furthermore, there is no research for analyzing and evaluating Mathematics and Statistics games in the context of HE or for increasing self-efficacy in those subjects and thus creating a better game experience. In other words, there is a need for game evaluation heuristics models that detect negative emotions (i.e., preconceptions) toward Mathematics and Statistics and customize learning (see Chapter 6), because there is no game evaluation heuristics model that addresses these issues. In this book, I feel that there is a necessity to develop new game evaluation heuristics models based on both computing and educational perspectives, that is, HCI principles, usability testing guidelines, and educational perspectives for the reasons discussed above. Furthermore, these evaluation heuristics are based on the empirical data outcomes and thus offer insights into a new way of employing these evaluation heuristics to investigate the effects of games on players' gaming experiences (a transformation/a journey: For the better/worse) (Schell, 2015, 2020; Hamari et al., 2016). This is because for educational games, additional features must be accounted for such as gameplay, game story, mechanics, competition, engagement, learning, feedback, challenge, entertainment, and assessment (Nielsen, 1995; Hunicke et al., 2004; Whitton, 2011; Schell, 2015, 2020; Arnab and Clarke 2017); empowering learners, problem-based learning, and deep understanding (Gee, 2003a, 2003b, 2004, 2005a, 2005b, 2007, 2011). It has been argued that these elements should be balanced in order to create a game worth playing (Desurvire and Wiberg, 2010; Schell, 2015, 2020). Due to differences in the above approaches, a new set of design principles is necessary. To address this, I will present a new set of game evaluation heuristics from multifaceted perspectives for educational games such as game design principles, HCI, usability testing guidelines, and educational/theoretical perspectives, which I call an ESGAE.

The ESGAE (Educational Simulation Games Analysis and Evaluation) Model was designed in order to analyze and evaluate simulation games (SGs). The ESGAE consists of *Educational Simulation Games Analysis* (analyzing users' needs, eliciting requirements) and *Evaluation* (testing and reviewing). The proposed ESGAE Model is the outcome of Game Design and Development Chapter 6; see this chapter for more details.

In all, this chapter has presented an overview of game design and development and its potential use in educational context, focusing on game design, evaluation heuristics, usability guidelines, and HCI principles. It has also given examples of the existing gap in the evaluation heuristics literature for both entertainment and educational games, with highlights on contributions and limitations, and how this resulted in the development of the ESGAE Model, before concluding.

The next Chapter, **Chapter 5**, will unleash the Power of Mathematics and Statistics: Integrating Simulation Games for Transformative Learning.

Bibliography

Acumen (2017) In: Cambridge University Press. [Online] Available at: http://dictionary.cambridge.org/dictionary/english/acumen (Accessed 30/05/2017).

Adams, A., Lunt, P. and Cairns, P. (2008) A Qualitative Approach to HCI Research. In: Cairns, P. and Cox, A. (ed.). *Research Methods for Human-Computer Interaction*. Cambridge, UK: Cambridge University Press, pp. 138–157. Available at: http://www.cambridge.org/catalogue/catalogue.asp?isbn=9780521870122andss=toc (Accessed 20/11/2016).

Anderson, S., (2020) Quantitative Analysis. Available at: https://www.investopedia.com/terms/q/quantitativeanalysis.asp (Accessed 20/02/2021).

Anon (2018) Efficacy In: Cambridge University Press. [Online] Available at: https://dictionary.cambridge.org/dictionary/english/efficacy (Accessed 30/09/2018).

Applefield, J.M., Huber, R. and Moallem, M. (2000) Constructivism in Theory and Practice: Toward a Better Understanding. *The High School Journal*, pp. 35–53.

Arnab, S. and Clarke, S. (2017) Towards a Trans-disciplinary Methodology for a Game-based Intervention Development Process. *British Journal of Educational Technology*, 48(2), pp. 279–312.

Arthur, J., Waring, M. Coe, R. and Hedges, L.V. (2012) *Research Methods and Methodologies in Education*. California: Sage Publications Inc.

Barbosa, T., Lopes, S.P. Leao. C. Soares, F. and Carvalho, V. (2019) Serious Game for Teaching Statistics in Higher Education: Storyboard Design. In: *7th EAI International Conference*, Braga, Portugal, on 'Interactivity, Game Creation, Design, Learning, and Innovation' (ed.) Springer, pp. 169–175. DOI: 10.1007/978-3-030-06134-0_18

Bartholomew, H., Darragh, L., Ell, F. and Saunders, J. et al. (2011) "I am Natural, and I Do It for a Love!": Exploring Students' Accounts of Studying Mathematics. *International Journal of Mathematical Education in Science and Technology*, 42(7), pp. 915–924.

Bassey, M. (1999) *Case Study Research in Educational Settings*. Buckingham: Open University Press.

Bellotti, F., Kapralos, B. Lee, K. Moreno-Ger, P. and Berta, R. (2013) Assessment in and of Serious Games: An Overview. *Advances in Human-Computer Interaction*, 2013, p. 1.

Belluigi, D.Z. (2009) Exploring the Discourses Around 'Creativity' and 'Critical Thinking' in a South African Creative Arts Curriculum. *Studies in Higher Education*, 34(6), pp. 699–717.

Bennedesen, J. and Caspersen, M.E. (2004) Programming in Context: A Model-First Approach to CS1. In *Proceedings of the 35th SIGCSE Technical Symposium on the Computer Science Education*, pp. 477–481. ACM Press.

Bevan, N. (2001) International standards for HCl and usability. *International Journal of Human Computer Studies*. 55(4), pp. 533–552.

Boaler, J. (2009) *The Elephant in the Classroom*. London: Souvenir Press.

Brown. J.S., Collins. A. and Duguid P. (1989) Situated Cognition and the Culture of Learning. *Educational Researcher*, 18(1), 32–42.

Bruner, J. (1973) *Going Beyond the Information Given*. New York: Norton.

Bryant, A. (2017). *Grounded Theory and Grounded Theorizing: Pragmatism in Research Practice*. New York, NY: Oxford University Press.

Bryman, A. (2008) *Social Research Methods*, 3rd Edition, Oxford: Oxford University Press (first published: 2001).

Bryman, A. (2012) *Social Research Methods*, 4th Edition, United States: Oxford University Press.

Bryman, A. (2016) *Social Research Methods*, 5th Edition, United States: Oxford University Press.

Burton, R.R., Brown, J.S. and Fischer, G. (1984) Skiing as a Model of Instruction. In: Rogoff, B. and Lave, J. (ed.). *Everyday Cognition: Its Development in Social Context*, pp. 139–150.

Caillois, R. (1961) *Man, Play, and Games*. New York: Free Press.
Cardinot, A. and Fairfield, J.A. (2019). Game-Based Learning to Engage Students with Physics and Astronomy Using a Board Game. *International Journal of Game-Based Learning (IJGBL)*, 9(1), pp. 42–57.
Charmaz, K. (2006) *Constructing Grounded Theory*. 1st Edition. London: Sage Publications Ltd.
Charmaz, K. (2014) *Constructing Grounded Theory*, 2nd Edition. London: Sage Publications Ltd.
Chen, M.P., Wong, Y.T. and Wang, L.C. (2014) Effects of Type of Exploratory Strategy and Prior Knowledge on Middle School Students' Learning of Chemical Formulas from a 3D Roleplaying Game. *Educational Technology Research and Development*, 62, pp. 163–185. DOI: 10.1007/s11423-013-9324-3
Clark, D.B., Tanner-Smith, E.E. and Killingsworth, S.S. (2015) Digital Games, Design, and Learning a Systematic Review and Meta-Analysis. *Review of Educational Research*, 86(1), pp. 79–122.
Claxton, G. (2002) *Building Learning Power*, 1st Edition, Bristol: TLO Limited.
Cohen, L., Manion, L. and Morrison, K. (2011) *Research Methods in Education*. London: Routledge.
Cohen, L., Manion, L. and Morrison, K. (2018) *Research Methods in Education*. London: Routledge.
Conlon, C., Carney, G., Timonen, V. and Scharf, T. (2015) Emergent Reconstruction in Grounded Theory: Learning from Team-Based Interview Research. *Qualitative Research*, 15, pp. 39–56. DOI: 10.1177/1468794113495038
Cook, J. (1992) *Negotiating the Curriculum: Programming for Learning. Negotiating the Curriculum: Educating for the 21st Century*, pp. 15–31.
Cooper, P.A. (1993) Paradigm Shifts in Designed Instruction: From Behaviorism to Cognitivism to Constructivism. *Educational Technology*, 33(5), pp. 12–19.
Corbin, J. and Strauss, A. (2015) *Basics of Qualitative Research: Techniques and Procedures for Developing Grounded Theory*, 4th Edition. Thousand Oaks, CA: Sage.
Cordova, D.I. and Lepper, M.R. (1996) Intrinsic Motivation and the Process of Learning: Beneficial Effects of Contextualization, Personalization, and Choice. *Journal of Educational Psychology*, 88, pp. 715–730.
Coulthard, G.J. (2009) A Review of the Educational Use and Learning Effectiveness of Simulations and Games [online]. PhD Thesis, Purdue University. Available at: http://www.coulthard.com/library/paper-simulation.html (Accessed 15/12/2015).
Craft, A. (1999) Creative Development in The Early Years: Some Implications of Policy for Practice. *The Curriculum Journal*, 10(1), pp. 135–150.
Craft, A. (2002) *Creativity and Early Years Education*. London: Continuum.
Crawford, C. (1984) *The Art of Computer Game Design*. Berkeley, California, McGraw-Hill/Osborne Media.
Davis, F.D. (1989) Perceived Usefulness, Perceived Ease of Use, and User Acceptance of Information Technology. *MIS Quarterly*, 13(3), pp. 319–340.
De Bono, E. (1992) *Teach Your Child How to Think*. England: McQuaig Group Inc.
Dede, C. (2005) Planning for Neomillennial Learning Styles. *Educause Quarterly*, 28(1), pp. 7–12.
Derryberry, A. (2007) Serious Games: Online Games for Learning. White Paper. [Online]. Available at: http://www.adobe.com/resources/elearning/pdfs/serious_games_wp.pdf (Accessed 10/12/2015).
Desurive, H., Caplan, M. and Toth, J. (2004) Using Heuristics to Evaluate the Playability of Games. *CHI Conference*, 32 (1), pp. 97–110. DOI: 10.1177/104687810103200108
Desurvire. H. and Wiberg, C. (2009) Game Usability Heuristics (PLAY) for Evaluation and Designing Better Games: The Next Iteration. In *Online Communities and Social Computing: Third International Conference, OCSC 2009, Held as Part of HCI International 2009*, San Diego, CA, USA, July 19–24, 2009. Proceedings 3, pp. 557–566. Springer Berlin Heidelberg.

Desurvire, H. and Wiberg, C. (2010) User Experience Design for Inexperienced Gamers: GAP – Game Approachability Principles. In: Regina Bernhaupt (ed.), *Evaluating User Experience in Games*, Springer, pp. 131–148. DOI: 10.1007/978-1-84882-963-3_1

Dondlinger, J.M. (2007) Educational Video Game Design: A Review of the Literature. *Journal of Applied Educational Technology*, 4(1), Spring/Summer, pp. 21–31. Available at: http://reserachgate.net/publication/238444705 (Accessed 12/12/2015).

Draper, S. (2013) Social Constructivism. Available at: http://www.psy.gla.ac.uk/~steve/courses/archive/CERE12-13-safari-archive/topic3/webarchive-index.html#social%20con%201 (Accessed 23/12/2019).

Dunne, C., (2011) The Place of the Literature Review in Grounded Theory Research. *International Journal of Social Research Methodology*, 14, pp. 111–124. DOI: 10.1080/13645579.2010.494930

Dweck, C.S. (2006) *Mindset: The New Psychology of Success*. United States: Ballantine Books Trade Paperback Edition.

Dwyer, J. and Slyman, S. (2016) *An Introduction to Quantitative Methods for Business Research*. London: Algana Publishing. 200 pages; ISBN−13: 9781907934094.

Early, R.E. (1992) The Alchemy of Mathematical Experience: A Psychoanalysis of Student Writings. *For the Learning of Statistics*, 12(1), pp. 15–20.

Fairweather, E. Cramond, B. (2010) Infusing Creative and Critical Thinking into the Curriculum Together. In: Beghetto, A. and Kaufman, J.C. (ed.). *Nurturing Creativity in the Classroom*. United States: Cambridge University Press.

Falstein, N. and Barwood, H. (2001) The 400 Project. Available at http://theinspiracy.com/400_project.htm. (Accessed 20/04/2016).

Faria, A.J. (2001) The Changing Nature of Business Simulation/Gaming Research. *Simulation and Gaming*, 32(1), pp. 97–110. DOI: 10.1177/104687810103200108

Faria, A.J., Hutchinson, D. Wellington, W.J. and Gold, S. (2009) Developments in Business Gaming: A Review of the Past 40 Years. *Simulation and Gaming*, 40(4), pp. 464–487. DOI: 10.1177/1046878108327585

Faria, A.J. and Wellington, W.J. (2004) 'A Survey of Simulation Game Users, Former-Users, and Never-Users. *Simulation and Gaming*, 35(2), pp. 178–207. DOI: 10.1177/1046878104263543

Federoff, M.A. (2002) Heuristics and Usability Guidelines for the Creation and Evaluation of Fun in Video Games. Master Thesis. Indiana University, Bloomington.

Fienberg, S.E. (1992) A Brief History of Statistics in Three and One-Half Chapters: A Review Essay. Available at: https://www.jstor.org/stable/2246307?casa_token=Vek3Rmd5Mk0AAAAA%3AJ0zRZC5AxskwvTNY7GlrgsRV6oLrE20DY3FdeaPad0JvNSNrZDav8uZpX-YqY1hXTDtaz6zM87S4vwkQdflDLSzmrLCNlEsaBqWnykepc0jlYXsQXA

Forest, E. (2014) ADDIE Model. Available at: http://educationaltechnology.net/the-addie-model-instructional-design/ (Accessed 20/04/2016).

Fosnot, T.C. (1989) *Enquiring Teachers, Enquiring Learners: A Constructivist Approach for Teaching*. New York: Teachers College Press.

Fosnot, T.C. and Perry, R.S. (2005) Constructivism: A Psychological Theory of Learning. Available at: http://beyondbitsandatoms.org/readings/fosnot_constructivism_1996.pdf (Accessed 20/07/2019).

Garris, R., Ahlers, R. and Driskell, E.J. (2002) Games, Motivation, and Learning: A Research and Practice Model. *Simulation and Gaming*, 33(4), pp. 441–467.

Gee, J.P. (2003a) *What Video Games Have to Teach Us About Learning and Literacy*. New York: Palgrave Macmillan.

Gee, J.P. (2003b) Good Video Games and Good Learning. Available at: https://academiccolab.org/resources/documents/Good_Learning.pdf (Accessed 03/01/2021).

Gee, J.P. (2004) *Situated Language and Learning: A Critique of Traditional Schooling*. London: Routledge.

Gee, J.P. (2005a) Why are Video Games Good for Learning? Retrieved January 1, 2018, from http://academiccolab.org/resources/documents/MacArthur.pdf (Accessed 20/04/2016).

Gee, J.P. (2005b) *Why Video Games Are Good for Your Soul: Pleasure and Learning*. Melbourne: Common Ground.

Gee, J.P. (2007) *Good Video Games and Good Learning: Collected Essays on Video Games, Learning, and Literacy*. New York: Peter Lang.

Gee, J.P. (2008). *Getting over the Slump: Innovation Strategies to Promote Children's Learning*. New York: The Joan Ganz Cooney Center at Sesame Workshop.

Gee, J.P., Hayes, E., Torres, R.J., Games, I.A., Squire, K. and Salen, K. (2008) Playing to Learn Game Design Skills in A Game Context. In *Proceeding ICLS'08 Proceedings of the 8th International Conference on International Conference for the Learning Sciences*, Vol. 3, pp. 368–374.

Gibbons, J. and Gray, M. (2004) Critical Thinking as Integral to Social Work Practice. *Journal of Teaching in Social Work*, 24, pp. 19–37.

Glaser, B.G. and Strauss, A.L. (1967) *The Discovery of Grounded Theory*. Chicago: Aldine.

Gomes, A. and Mendes, A.J. (2007) Learning to Program-Difficulties and Solutions. In *International Conference on Engineering Education–ICEE*, Vol. 7.

Granic, I., Lobel, A. and Engels, R.C. (2014) The Benefits of Playing Video Games. *American Psychological Association*, 69(1), pp. 66–78. DOI: 10.1037/a0034857

Hackett, G. and Betz, N.E. (1989) An Exploration of the Mathematics Self-Efficacy/ Mathematics Performance Correspondence. *Journal for Research in Mathematics Education*, 20(3), pp. 261–273.

Hamari, J., Shernoff, D.J., Rowe, E., Coller, B., Asbell-Clarke, J. and Edwards, T. (2016) Challenging Games Help Students Learn: An Empirical Study on Engagement, Flow and Immersion in Game-Based Learning. *Computers in Human Behavior*, 54, pp. 170–179.

Hamari, J. and Tuunanen, J. (2014) Player Types: A Meta-Synthesis. *Transactions of the Digital Games Research Association*, 1(2), pp. 29–53.

Hsu, C.-Y., Tsai, C.-C. and Wang, H.-Y. (2012) Facilitating Third Graders' Acquisition of Scientific Concepts Through Digital Game-Based Learning: The Effects of Self-Explanation Principles. *Asia-Pacific Education Researcher*, 21(1), pp. 71–82.

Hunicke, R., LeBlanc M. and Zubek, R. (2004) MDA: A Formal Approach to Game Design and Game Research. Available at: https://www.google.co.uk/search?q=MDA%3A+A+Formal+Approach+to+Game+Design+and+Game+Researchandoq=MDA%3A+A+Formal+Approach+to+Game+Design+and+Game+Researchandaqs=chrome..69i57j69i58.3868j0j8andsourceid=chromeandie=UTF-8 (Accessed 8/10/2016).

Illeris, K. (2007) *How We Learn: Learning and Non-Learning in School and Beyond*. London: Routledge.

ISO (1998) Available at: http://www.standardsglossary.com/iso1.htm (Accessed 8/10/2016).

Jackson, N. and Shaw, M. (2007) Developing Perspectives on Creativity in Higher Education. In: Jackson, N., Oliver, M., Shaw M., and Wisdom, J. (ed.). *Developing Creativity in Higher Education: An Imaginative Curriculum*. London: Routledge.

Jagger, S., Siala, H. and Sloan, D. (2016) It's All in the Game: A 3D Learning Model for Business Ethics. *Journal of Business Ethics*, 137(2), pp. 383–403.

Jameson, M.M. and Fusco, B.R. (2014) Math Anxiety, Math Self-Concept, and Math Self-Efficacy in Adult Learners Compared to Traditional Undergraduate Students. *Adult Education Quarterly*, 64(4), pp. 306–322.

Jeffrey, B. and Craft, A. (2004) Teaching Creatively and Teaching for Creativity: Distinctions and Relationships. *Educational Studies*, 30(1), pp. 77–87.

Jenson, J. and Droumeva, M. (2015) Making Games with Game Maker: A Computational Thinking Curriculum Case Study. In *ECGBL 2015-9th European Conference on Games Based Learning: ECGBL2015*, p. 260. Academic Conferences and Publishing Limited.

Jones, M.G. (1998) Creating Engagement in Computer-Based Learning Environments. Available at: http://itech1.coe.uga.edu/itforum/paper30/paper30.htmI (Accessed 12/04/2016).
Kafai, Y. (2006) *Playing and Making Games for Learning: Instructionist and Constructionist Perspectives for Game Studies. Games and Culture*, 1(1), pp. 36–40.
Kafai, Y.B. and Resnick, M. eds. (2012) *Constructionism in Practice: Designing, Thinking, and Learning in a Digital World.* Routledge. Available at: https://www.google.co.uk/books/edition/Constructionism_in_Practice/KGkIEQAAQBAJ?hl=en&gbpv=1&dq=Kafai,+Y.B.+and+Resnick,+M.+eds.+(2012)+Constructionism+in+Practice:+Designing,+Thinking,+and+Learning+in+a+Digital+World.+Routledge.&pg=PP1&printsec=frontcover
Kato, P.M., Cole, S. Bradlyn, A.S. and Pollock, B.H. (2008) Video Games Improves Behavioural Outcomes in Adolescents and Young Adults with Cancer: A Randomized Trial, *Pediatrics*, 122(2), pp. e305–e217.
Kessner, M., Wood, J. Dillon, R.F. and West, R.L. (2001) On the Reliability of Usability Testing. In *Proceedings of the Extended Abstracts on Human Factors in Computing Systems (CHI '01)*, p. 97.
Kim, B. (2001) Social Constructivism. *Emerging Perspectives on Learning, Teaching, and Technology*, 1(1), pp. 16.
Kirriemuir, J. and Mcfarlane, A. (2004) Literature Review in Games and Learning. Available at: https://telearn.archives-ouvertes.fr/hal-00190453/document (Accessed 12/04/2020).
Knight, P.T. (2001) A Briefing on Key Concepts – Formative and Summative, Criterion and Norm Referenced Assessment. *Learning and Teaching Support Network Generic Centre*, 30. http://www.heacademy.ac.uk
Kodugamelab (2015) Available at: https://www.kodugamelab.com/ (Accessed 8/11/2018).
Kolb, A.Y. and Kolb. D.A. (2008) The Learning Way: Meta-Cognitive Aspects of Experiential Learning. *Simulation and Gaming*, 40(3), pp. 297–327.
Kolb, D.A. (1984) *Experiential Learning: Experience as the Source of Learning and Development* (Vol. 1). Englewood Cliffs, NJ: Prentice-Hall.
Korhonen, H. and Koivisto, E.M.I. (2007) Playability Heuristics for Mobile Multi-Player Games. *Proceedings of the 2nd International Conference on Digital Interactive Media in Entertainment and Arts*, pp. 28–35.
Kuss, D.J. and Griffiths, M.D. (2012) Internet Gaming Addiction: A Systematic Review of Empirical Research. *International Journal of Mental Health and Addiction*, 10(2), pp. 278–296.
Lave, J. and Wenger E. (1991) *Situated learning: Legitimate Peripheral Participation*. Cambridge University Press. Available at: https://www.google.co.uk/books/edition/Situated_Learning/CAVIOrW3vYAC?hl=en&gbpv=1&dq=Lave,+J.+and+Wenger+E.+(1991)+Situated+Learning:+Legitimate+Peripheral+Participation.+Cambridge+University+Press&pg=PA11&printsec=frontcover
Lazzaro, N. (2004) Why We Play Games: Four Keys to More Emotion without Story. Available at: http://www.xeodesign.com/xeodesign_whyweplaygames.pdf (Accessed 28/12/2015).
Lee, J.S., Keil, M. and Kasi, V. (2012) The Effects of an Initial Budget and Schedule Goal on Software Project Escalation. *Journal of Management Information System*, 29(1), pp. 53–77. DOI: 10.2753/MIS0742-1222290102
Levine, D.M., Krehbiel, T.C. and Berenson, M.L. (2003) *Business Statistics: A First Course*, 3rd Edition. New Jersey: Pearson Education Ltd.
Lewis, G. (2013) Emotion and Disaffection with School Mathematics, *Research in Mathematics Education*, 15(1), pp. 70–86.

Li, F.W.B. and Watson, C. (2011) *Game-Based Concept Visualization for Learning Programming*. Scottsdale, Arizona, USA: ACM. 978-1-4503-0994-3/11/12.

Li, Z.Z., Cheng, Y.B. and Liu, C.C. (2013) A Constructionism Framework for Designing Game-Like Learning Systems: Its Effect on Different Learners, *British Journal of Educational Technology*, 44(2), pp. 208–224.

Lister, R., Adams, E.S. Fitzgerald, S. Fone, W. Hamer, J. Lindholm, M. McCartney, R. Moström, J.E. Sanders, K. Seppälä, O. and Simon, B. (2004) A Multi-National Study of Reading and Tracing Skills in Novice Programmers. In *ACM SIGCSE Bulletin*, 36(4), pp. 119–150. ACM.

Litosseliti, L. (2005) *Using Focus Groups in Research*. London: Continuum.

Loftus, C., Thomas, L. and Zander, C. (2011) Can Graduating Students Design: Revisited. In *Proceedings of the 42nd ACM Technical Symposium on Computer Science Education*, pp. 105–110.

Lyons, I.M. and Beilock, S.L. (2012) When Math Hurts: Math Anxiety Predicts Pain Network Activation in Anticipation of Doing Math. *PLoS One*, 7(10), p. e48076.

Majgaard, G. (2016) Creating Games in the Classroom: From Native Gamers to Reflective Designers. In: Munkvold, R. (ed.). *Games Based learning for Researchers, Teachers and Students*, Vol. 1. Reading: Academic Conferences and Publishing International Ltd. ISBN: 978-1-911278-12-890000.

Malaka, R. (2014) How Computer Games Can Improve Your Health and Fitness. Gobel, S. and Wiemeyer, J. (ed.). *Games for Training Education, Health and Sports, 4th Conference on Serious Games, GameDays 2014*, Darmstadt, Germany, April 1–5, 2014, Springer.

Malone. T.W. (1980) What Makes Things Fun to Learn? A Study of Intrinsically Motivating Computer Games. Available at: http://cci.mit.edu/malone/Malone%20computer%20games%20bibliography.htm

Malone, T.W. and Lepper, M.R. (1987) Making Learning Fun: A Taxonomy of Intrinsic Motivations for Learning. Aptitude, Learning, and Instruction. In: Snow, R.E. and Farr, M.J. (ed.). *Volume 3: Cognitive and Affective Process Analyses*. Hillsdale, USA: Erlbaum.

Maloney, E.A., Schaeffer, M.W. and Beilock, S.L. (2013) Mathematics Anxiety and Stereotype Threat: Shared Mechanisms, Negative Consequences and Promising Interventions. *Research in Mathematics Education*, 15(2), pp. 115–128.

Marciano, J.N., Miranda, L.C. and Miranda, E.E.C. (2014) Evaluating Multiple Aspects of Educational Computer Games. Available at: https://www.hindawi.com/journals/ijcgt/2014/139205/ (Accessed 27/08/2017).

Marklund, B.B. and Taylor, A.S.A. (2016) Creating Games in the Classroom: From Native Gamers to Reflective Designers. In: Munkvold, R. (ed.). *Games Based learning for Researchers, Teachers and Students*, Vol. 1. Reading: Academic Conferences and Publishing International Ltd. ISBN: 978-1-911278-12-890000.

Maslow, A.H. (1970) *Motivation and Personality*, 2nd Edition. New York; London: Harper and Row.

Mayer, E.R. (2014) *Computer Games for Learning: An Evidence Based Approach*. London: MIT.

Mayer, I., Warmelink, H. and Bekebrede, G. (2013) Learning in a Game-Based Virtual Environment: A Comparative Evaluation in Higher Education. *European Journal of Engineering Education*, 38(1), pp. 85–106.

Mayo, M. (2009) Video Games: A Route to Large-Scale STEM Education? *Science*, 323, pp. 79–82.

McConnell, D. (2000) *Implementing Computer Supported Cooperative Learning*, 2nd Edition. London: Kogan Page.

Mccrindle, M. (2018) What Comes after Generation Z? Introducing Generation Alpha. Available at: https://mccrindle.com.au/insights/blogarchive/what-comes-after-generation-z-introducing-generation-alpha/ (Accessed 15/10/2018).

McFarlane, A., Sparrowhawk, A. and Heald, Y. 2002. *Report on the Educational Use of Games: An Exploration by TEEM on the Contribution Which Games Can Make to the Educational Process*. Cambridge, UK: TEEM. http://www.teem.org.uk/publications/teem_gamesined_full.pdf (Accessed 06/12/2015).
McGonigal, J. (2015) *Superbetter: A Revolutionary Approach to Getting Happier Stronger and Happier*. Braver and More Resilient, New York: Penguin Press.
Milane, I. and Rowe, G. (2002) Difficulties in Learning and Teaching Programming- Views of Students and Tutors. *Education and Information Technologies*, 7(1), pp. 55–66.
Monroy-Hernandez, A. and Resnick, M. (2008) Empowering kids to create and share programmable media. *Interactions*, 15(2), pp. 50–53. Retrieved August 30, 2015.
Moreno-Ger, P., Burgos, D. Martínez-Ortiz, I. Sierra, J.L. and Fernández-Manjón, B. (2008) Educational Game Design for Online Education. *Computers in Human Behavior*, 24(6), pp. 2530–2540. DOI: 10.1016/j.chb.2008.03.012
Moreno-Ger, P., Torrente, J. Hsieh, Y.G. and Lester, W.T. (2012) Usability Testing for Serious Games: Making Informed Design Decisions with User Data. *Advances in Human-Computer Interaction*, 2012. DOI: 10.1155/2012/369637
Morgan, D.L. (1988) *Focus Groups as Qualitative Research*. Beverly Hills, CA: Sage.
Morrison, K. (2002) *School Leadership and Complexity Theory*. London: Routledge Falmer.
Muijs, D. (2011) *Doing Quantitative Research in Education with SPSS*. London: Sage Publication Ltd.
Mullins, L. (2013) *Management and Organisational Behaviour*, 10th Edition. Harlow: Pearson Education.
Mullins, L. (2019) *Organisational Behaviour in the Workplace*, 12th Edition. Harlow: Pearson Education.
Multon, K.D., Brown, S.D. and Lent, R.W. (1991) Relation of Self-Efficacy Beliefs to Academic Outcomes: A Meta-Analytic Investigation. *Journal of Counselling Psychology*, 18, pp. 30–38.
Naik, N. (2017) The use of GBL to Teach Mathematics in Higher Education. *Innovations in Education and Teaching International*, 54(3), pp. 238–246. DOI: 10.1080/14703297.2015.1108857
Newby, P. (2010) *Research Methods for Education*, Harlow: Pearson Education Ltd.
Nicholls, D., De Charon, L. and Hutkin, R. (2010) Development of Critical Thinking and Creativity: Practical Guidelines for the Postsecondary Classroom. *ATEA Journal*, 38(1), pp. 12.
Nielsen, J., 1994, April. Usability Inspection Methods. In *Conference Companion on Human Factors in Computing Systems*, pp. 413–414. Available at: https://dl.acm.org/doi/pdf/10.1145/259963.260531
Nielsen, J. (1995) Usability Inspection Methods. *CHI'95 Mosaic of Creativity*, pp. 7–11.
Nielsen, J. (2000) *Designing Web Usability*. New Riders Publishing.
Paavilainen, J. (2010) Critical Review on Video Game Evaluation Heuristics: Social Games Perspective. *Proceedings of the International Academic Conference on the Future of Game Design and Technology*, pp. 56–65. DOI: 10.1145/1920778.1920787
Pajares, F. and Kranzler, J. (1995) Self-Efficacy Beliefs and General Mental Ability in Mathematical Problem-Solving. *Contemporary Educational Psychology*, 20, pp. 426–443.
Palincsar, A.S. (1998) Social Constructivist Perspectives on Teaching and Learning. *Annual Review of Psychology*, 49, pp. 345–375.
Paras, B. and Bizzocchi, J. (2005) Game, Motivation, and Effective Learning: An Integrated Model for Educational Game. In *Proceedings of DiGRA Conference on Changing Views–Worlds of Play*. Available at: https://www.researchgate.net/publication/221217604_Game_Motivation_and_Effective_Learning_An_Integrated_Model_for_Educational_Game_Design (Accessed 27/09/2016).

Park, H.J. and Kim, S.B. (2013) Guidelines of Serious Game Accessibility for the Disabled. In *Proceedings of the International Conference on Information Science and Applications (ICISA '13)*, pp. 1–3, IEEE.

Piaget, J. and Inhelder B. (1973) *Memory and Intelligence*. London: Routledge and Kegan Paul.

Pinelle, D., Wong, N., Stach, T. and Gutwin, C. (2009) Usability Heuristics for Networked Multiplayer Games. *Proceedings of the ACM 2009 International Conference on Supporting Group Work*, pp. 169–178.

Pollard, A., Anderson, J., Maddock, M., Swaffield, S., Warin, J. and Warwick, P. (2008) *Reflective Teaching*, 3rd Edition. London: Continuum International Publishing Group.

Preconception. (2020) In: Dictionary.com. [Online] Available at: https://www.dictionary.com/browse/preconception (Accessed 01/08/2020).

Preece, Y., Rogers, Y. and Sharp, H. (2015) *Interaction Design: Beyond Human-Computer Interaction*, 4th Edition. Chichester: John Wiley and Sons Ltd.

Prensky, M. (2002) The Motivation of Gameplay: The Real Twenty-First Century Learning Revolution. *On the Horizon*, 10(1), pp. 5–11.

Prensky, M. (2005) Listen to Natives. *Association for Supervision and Curriculum Development Educational Leadership*, 63(4), pp. 9–13.

Prensky, M. (2006) *'Don't Bother Me Mom, I'm Learning!': How Computer and Video Games Are Preparing Your Kids for the Twenty-First Century*. St Paul, MN: Paragon House.

Pritchard, A. and Woollard, J. (2010) *Psychology for the Classroom: Constructivism and Social Learning*. Oxon: Routledge.

Punch. K.F. (2009) *Introduction to Research Methods in Education*. London: Sage Publications Ltd.

Resnick, M. and Robinson, K. (2017) *Lifelong Kindergarten: Cultivating Creativity Through Projects, Passion, Peers, and Play*. MIT Press. Available at: https://www.google.co.uk/books/edition/Lifelong_Kindergarten/nGYyDwAAQBAJ?hl=en&gbpv=1&dq=Resnick,+M.+and+Robinson,+K.+(2017)+Lifelong+Kindergarten:+Cultivating+Creativity+Through+Projects,+Passion,+Peers,+and+Play.+MIT+Press.&pg=PR5&printsec=frontcover

Robins, A., Runtree, J. and Rountree, N. (2003) Learning and Teaching Programming: A Review and a Discussion. *Computer Science Education*, 13(2), pp. 137–172.

Robson. C. (2010) *Real World Research*, 3rd Edition. Chichester: John Wiley and Sons Limited.

Robson, S. (2006) *Developing Thinking and Understanding in Young Children: An Introduction for Students*. London: Routledge Ltd.

Rogoff, B. (1998) Cognition as a Collaborative Process. In: Damon, W., Kuhn, D. and Seigler, R.S. (ed.). *Handbook of Child Psychology*, 5th Edition, p. 2. New York: Wiley.

Roth, W.M. (2000) Authentic School Science: Intellectual Traditions. In: McCormick, R. and Paechter, C. (ed.). *Learning & Knowledge*. London, UK: Paul Chapman Publishing, pp. 6–20.

Russ, T., and Drury-Grogan, M.L. (2013) Assessing the Impacts of a Business Communication Simulation on Students' Self-Perceptions. *Eastern Communication Association*, 61(5), pp. 584–595. DOI: 10.1080/01463373.2013.822404

Salen, K. and Zimmerman, E. (2011) The Game Designer Reader, A Rules of Play Anthology. Roger Caillois: The Definition of Play. *The Classification of Games*. Cambridge, Massachusetts. London, England: The MIT Press.

Salisbury, J.H. and Cole. T. (2016) Grounded Theory in Games Research: Making the Case and Exploring the Options. Available at: https://www.academia.edu/27624170/Grounded_Theory_in_Games_Research_Making_the_Case_and_Exploring_the_Options (Accessed 15/10/2015).

Schell, J. (2015) *The Art of Game Design: A Book of Lenses*, 2nd Edition. USA: CRC Press.
Schell, J. (2020) *The Art of Game Design: A Book of Lenses*, 3rd Edition. USA: CRC Press.
Senge, P. (1990) *The Fifth Discipline: The Art and Practice of Learning Organization*. United Kingdom: Random House Business Books.
Senge, P.M. (2006) *The Fifth Discipline: The Art and Practice of Learning Organisation*. London: Random House Business Books.
Shell, D.F., Murphy, C.C. and Bruning, R.H. (1989) Self-efficacy and Outcome Expectancy Mechanisms in Reading and Writing Achievement. *Journal of Educational Psychology*, 81(1), pp. 91.
Shunk, D.H. (2000) *Learning Theories: An Educational Perspective*, 3rd Edition. Upper Saddle River, NJ: Prentice-Hall.
Siegel, R.G., Galassi, J.P. and Ware, W.B. (1985) A Comparison of Two Models for Predicting Mathematics Performance: Social Learning Versus Math Aptitude-Anxiety. *Journal of Counseling Psychology*, 32, pp. 531–538.
Silverman, B.G., Johns, M. Weaver, R. Mosley, J. (2003) Authoring Edutainment Stories for Online Players (AESOP): Introducing Game play into Interactive Dramas. In: *International Conference on Virtual Storytelling*, Springer.
Silverman, D. (1993) *Interpreting Qualitative Data: Methods for Analysing Talk, Text, and Interaction*. London: Sage Publications Ltd.
Silverman, D. (2014) *Interpreting Qualitative Data*, 5th Edition. London: Sage Publications Ltd.
Sindre, G., Natvig, L. and Jahre, M. (2009) Experimental Validation of the Learning Effect for A Pedagogical Game on Computer Fundamentals. *IEEE Transactions on Education*, 52 (1), pp. 10–18.
Skemp, R.R. (1986) *The Psychology of Learning Mathematics Great* Britain: Penguin Books Ltd.
Skemp, R.R. (1989) *Mathematics in the Primary School*, Bristol: Routledge.
Skilbeck, A. (2017) Dewey on Seriousness, Playfulness and the Role of the Teacher. *Education Sciences*, 7(1), p. 16.
Slyman, S. (2014) I Hate Maths! An Investigation into Students' Attitudes Towards Mathematics. MA Thesis, University of Roehampton, London, United Kingdom.
Slyman, S. (2018) *Games Based Learning in Universities, A Life Learning Experience*. New York: Business Expert Press Expert Insights Publishing. ISBN:978-1-94819-801-1.
Slyman, S., (2022) Acumenous Game-Based Learning in Simulation Games and Applied Statistics. PhD Thesis, Goldsmiths, University of London, UK.
Slyman, S., Gillies, M. and Jessel, J. (2019a) Acumenous or Inquisitional? Towards a New Theoretical Lens within Games Learning. In: *13th European Conference on Games Based Learning*. Odense, Denmark: University of South Denmark. ISBN:978-1-912764-38-9 or E-book: ISSN:2049-100X.
Slyman, S., Gillies, M. and Jessel, J. (2019b) Game-Based Learning to Engage students with Applied Statistics using a Simulation Roleplay Game. In: *9th Irish Conference on Games Based Learning*, Cork, Ireland. Available at: http://igbl-conference.com/2016/programme/generate_book_of_abstracts.php (Accessed 25/09/2019).
Smith Report (2004). Available at: http://www.mathsinquiry.org.uk/report/mathsInquiryFinalReport.pdf (Accessed 20/12/2014).
Smith Report (2017). Available online at: https://www.gov.uk/government/publications/smith-review-of-post-16-maths -report-and-government-response (Accessed 20/12/2018).
Sobke, H., Chan, E., Buttlar, R., Grobe-Wortmann, J. and Londong, J. (2014) Cat King's Metamorphosis. The Reuse of an Educational Game in a Further Technical Domain. Gobel, S., and Wiemeyer, J. (ed.) *Games for Training Education, Health and Sports, 4th Conference on Serious Games, GameDays 2014* Darmstadt, Germany, April 1–5, 2014, Springer.

Song, S. and Lee, J. (2007) Key Factors of Heuristic Evaluation for Game Design: Towards Massively Multi-Player Online Roleplaying Game. *International Journal of Human-Computer Studies*, 65, pp. 709–723.

Strauss, A. and Corbin, J. (1990) *Basics of Qualitative Research: Grounded Theory Procedures and Techniques*. Newbury Park, CA: Sage.

Sweetser, P., Johnson, D. and Wyeth, P. (2012) Revisiting the GameFlow Model with Detailed Heuristics. *Journal of Creative Technologies*, 2012(3), pp. 1–16.

Sweetser. P. and Wyeth, P. (2005) GameFlow: A Model for Evaluating Player Enjoyment in Games. *Computers in Entertainment* 3, 3: 3. DOI: 10.1145/1077246.1077253

Tamura, H. (2007) Foundational Value of Mathematics Education for Management Curriculum, *International Statistical Review* (2007), 75(3), pp. 397–405. DOI: 10.1111/j.1751-5823.2007.00032.x

Tao, Y.H., Cheng, C.J. and Sun, S.Y. (2012) Alignment of Teacher and Students Perceptions on the Continued Use of Business Simulation Games. *Educational Technology and Society*, 15(3), pp. 177–189.

TED (2019) *Souad Slyman: Why Change Attitudes Towards Mathematics [YouTube]*. TedxGul, London: Goldsmiths, University of London. Available at: https://www.youtube.com/watch?v=leNDJMNJoNY (Accessed 26/04/2019).

Tondello, G.F., Kappen, D.L. Mekler, E.D. Ganaba, M. and Nacke, L.E. (2016) Heuristic Evaluation for Gameful Design. *CHI PLAY'16*. ACM. 978-1-4503-4458-6/16/10. DOI: 10.1145/2968120.2987729

van der Spek, E.D., van Oostendorp, H. and Meyer, J.-J.C.H. (2013) Introducing Surprising events can stimulate deep learning in a serious game. *British Journal of Educational Technology*, 44(1), pp. 156–169.

Vlachopoulos, D. and Makri, A. (2017) The Effect of Games and Simulations on Higher Education: A Systematic Literature Review. *International Journal of Educational Technology in Higher Education*, 14(1), pp. 22. DOI: 10.1186/s41239-017-0062-1

Vos, L. (2014) Marketing Simulation Games: A Review of Issues in Teaching and Learning. *The Marketing Review*, 14(1), pp. 6796. DOI: 10.1362/146934714X13948909473220

Vygotsky, L.S. (1980) *Mind in Society. The Development of Higher Psychological Processes*. Edited by Cole, M., John-Steiner, V., Scribner, S. and Souberman, E. London: Harvard University Press.

Walonick, D.S. (2000) *Designing and Using Questionnaire in Survival Statistic*. MN: StatPacInc. Available at Internet WWW page at URL: http://www.statpac.com/surveys/surveys.pdf (Accessed 02/04/2016).

Whitton, N. (2011) Games-Based Experiences for Learning. *Futurelab Innovation in Education*. Available at: https://www.nfer.ac.uk/media/1775/futl11.pdf (Accessed 02/04/2016).

Wolfe, J. (1997) The Effectiveness of Business Games in Strategic Management Coursework. *Simulation and Gaming*, 28(4), pp. 360–376. Doi: 10.1177/1046878197284003

Women in STEM – 'Facts and Statistics' (Reporting period 2012–2015), IET. Available at: https://communities.theiet.org/files/8042 (Accessed 20/06/2017).

Yin, R.K. (1994) *Case Study Research: Design and Methods*, 2nd Edition. London: Sage.

Young, M. 2008. *"Powerful Knowledge" and "Knowledge of the Powerful": Some Reflections*. London: Presentation at the Institute of Education.

Yusoff, A., Crowder, R. and Gilbert, L. (2010) Validation of Serious Games Attributes Using the Technology Acceptance Model [online]. In: IEEE (ed.), *2010 Second International Conference on Games and Virtual Worlds for Serious Applications*, Braga, Portugal, 25–26 March 2010, pp. 45–51. Available from: http://eprints.soton.ac.uk/270809/ (Accessed 10/05/2015).

Zimmerman, B.J. (2000) Self-Efficacy: An Essential Motive to Learn. *Contemporary Educational Psychology*, 25(1), pp. 82–91. DOI: 10.1006/ceps.1999.1016

Zyda, M. (2005) *From Visual Simulation to Virtual Reality to Games.* USC Information Sciences Institute, IEEE Computer Society Edition September, pp. 25–31. Available at: http://scholar.google.co.uk/scholar?hl=enandq=zyda+2005andbtnG=andas_sdt=1%2C5andas_sdtp= (Accessed 07/10/2015).

5 Understanding the Power of Statistics

Integrating Simulation Games for Transformative Learning

Our journey begins by demystifying the essence of Statistics, unravelling its complexities, and providing a firm foundation for the chapters to come. As we navigate this chapter, we delve into my innovative approach to teaching correlation and regression, showcasing the creative methodologies and learning objectives that underpin gameplay investigation. We delve into the integration of gameplay as a powerful tool within the curriculum, fostering engagement, enhancing learning outcomes, and cultivating essential skills.

At the heart of this chapter lies the exploration of the transformative potential of GBL in Statistical learning. Drawing upon the captivating world of "SimCity," an entertainment game that possesses remarkable potential for meeting learning objectives within the domain of Statistics, we delve into the profound question of whether GBL can effectively address the negative emotions and preconceptions that hinder the mastery of this complex subject. With the omnipresence of technology in students' lives and their inherent affinity for gaming, we envision a future where every student can unlock the power of Statistics through the captivating realm of GBL.

Furthermore, the chapter unveils a second game, accompanied by an immersive gameplay scenario, serving as a compelling case study within the teaching structure. With practical exercises utilizing Excel, we bridge theory and application, empowering students to actively participate in their own learning journey.

As we approach the culmination of this chapter, we are left with a profound appreciation for the transformative potential of GBL in Statistical learning. We are poised to witness the remarkable impacts that this immersive and engaging approach can yield, revolutionizing the way students' approach and comprehend this critical subject. This chapter paves the way for the chapters that lie ahead, laying the groundwork for a paradigm shift in Statistical education, where simulation games become catalysts for empowering learners and reshaping the landscape of Statistics education.

DOI: 10.4324/9781032723426-7

Teaching Statistics using GBL

This chapter explores the teaching of Statistics to first-year undergraduates, as a part of the HE core curriculum, and particularly my approach to teaching correlation and regression (As I will explain later on) and the learning objectives/skills behind the gameplay investigation (see the following section for further clarifications). As previously outlined, as part of the GBL approach to teaching, the first game I am using is "SimCity," which is an entertainment rather than a so-called educational game, but it can meet some learning objectives in the area I am teaching (see Simulation GBL Approach to Teaching Section for extra details). The concept behind this is to investigate whether GBL approach to teaching can have impacts on Statistical learning (due to negative emotions, i.e., preconceptions; see Chapter 1). Can GBL affect self-efficacy? Given that many students play games every day and most tend to perceive technology as useful and fun, my vision is that every student can learn Statistics, a complex subject, through GBL.

The second game that I will present is "Numbas." This is a Mathematics shooting game and it is utilized for comparison reasons (see Quiz GBL Approach to Teaching Section for supplementary insights). Furthermore, I am aware from my experience as a HE senior lecturer that this type of learning (quiz GBL) is widely used as a part of curriculum teaching. It is employed sometimes in lecture/lab/seminar work to complement learning. I will return to examine this further with the "SimCity" gameplay (the teaching pedagogy), but first I will introduce the curriculum in Statistics to which I taught to first-year undergraduates, and how "SimCity" has been creatively embedded in the lecture.

This chapter focuses on teaching Statistics (correlation and regression) to first-year undergraduates and how I employed gameplay as a learning tool in teaching and learning. In this book, I have focused on two games, "Numbas" and "SimCity." I will start, however, by defining Statistics and Applied Statistics discussing reasons behind using the term "Applied Statistics" and how I have taught it using SGs, moving on to a gameplay scenario before concluding.

Simulation GBL Approach to Teaching Statistics in HE

Statistics is a field of Mathematics that is associated with data analysis (Dwyer and Slyman, 2016). Statistics is the science of collecting, organizing, interpreting, and reporting data where the data consist of observations taken in/from the real world (Fienberg, 1992:209). Applied Statistics is a term I will be referring to throughout this book and it can mean applying Statistics in different disciplines/situations such as improving quality, production and making better decisions, as in business, education, computing or marketing and therefore it becomes applicable, i.e., applied. Applied Statistics is used in "almost all functional areas of management, marketing, human resources… [and] finance, and there is a fundamental reason for this." The success of every business organization depends on its managers or leaders in these areas making quality decisions (Tamura, 2007:403; Dwyer and Slyman, 2016). Decision-making skills require managers to make quality decisions based on analyzing, interpreting, and forecasting data.

Data can be defined as information required to make more informed decisions. Some examples of where data are needed, according to Levine et al. (2003:6) and Dwyer and Slyman (2016), are as follows:

- A potential investor would like to determine what firms within which industries are likely to have accelerated growth in a period of economic recovery.
- A researcher needs to assess product characteristics to distinguish one product from another.
- An operation manager wants to monitor an assembly process on a regular basis to find whether the quality of the product that is being produced conforms to business company standards.

Reasons why we analyze data include measuring performance and production, evaluating business quality and standards (Levine et al., 2003), designing IT systems, improving business operations, staying competitive, increasing profits, and sustaining growth, and overall making informed decisions.

Statistical methods and equations can be applied to a data set in order to analyze and interpret results, explain variations in the data, find associations between variables, or predict future data. Statistics is widely applicable in many disciplines such as business, social sciences, computing, education, and engineering. There are two types of Statistics, descriptive or inferential. Descriptive statistics is where we describe/report what is happening in the data. Examples of Statistical measures include average (arithmetic mean), mode (most frequent value), variance (how spread data is from the mean), and standard deviations (square root of the variance). Inferential Statistics, on the other hand, is employed to assist in drawing conclusions from data. Examples of inferential Statistics may include correlation, linear or multiple regression, and probability. In this book, I will focus on correlation and regression, since I taught these two statistical measures as part of this empirical project.

Correlation and Regression

Correlation is when two or more variables are associated in some way. An example is the relationship between the rise of temperature (variable x) in summer in relation to the increase in sales of ice-creams (variable y), see Figure 5.2. We say that the two variables (x and y) are correlated or have a strong positive correlation, since as temperature increases, the sales of ice creams increase. This can be denoted by the coefficient r, which is calculated using Pearson's Product Moment and Spearman's Rank correlation coefficient (r). Once calculated, r is between 1 and −1. If it is close to:

- 1 or −1 there is a strong positive/negative correlation.
- 0.5 or −0.5 moderate correlation.
- 0 no correlation.

Furthermore, this can be graphically represented in a scatter diagram (as in Figure 5.1) where we can draw a line of best fit (by eye), but this is not always accurate, consistent, and practical especially for future predications. Moreover, we do not know which factors are the most important (Bryman, 2016; Dwyer and Slyman, 2016). Hence the significance of regression analysis becomes apparent, discussed next.

In any business organization, regression is the "most useful statistical method" and "the standard way" for analyzing data, identifying, combining, and predicting future forecasts (Tamura, 2007:402). Regression analysis is a way of finding the straight line of best fit algebraically using regression equations ($Y = a + bX$) to represent the connection between variables where b is the slope and a is the intercept where the line cuts the Y (see Figure 5.2). If we are looking at two variables as in temperature (x) and sales of ice-creams (y), we call this type of relationship simple linear regression, but if there are more than two variables (three or more), we call it multiple regression. Finding a regression equation algebraically is accurate since it enables us to make predications and draw more conclusions (unlike drawing a line by eye as aforementioned), as shown below.

Figure 5.2 illustrates the relationship between the temperature (x) and the sales (y). As x increases, y increases and $R^2 = 0.922$ (the goodness of fit) is close to 1 and it is positive. We say that there is a strong positive correlation. Using the regression equation ($y = 0.6328x + 12.568$), we can now predict any future in/decrease of temperature (x) and obtain the value of ice creams sales (y).

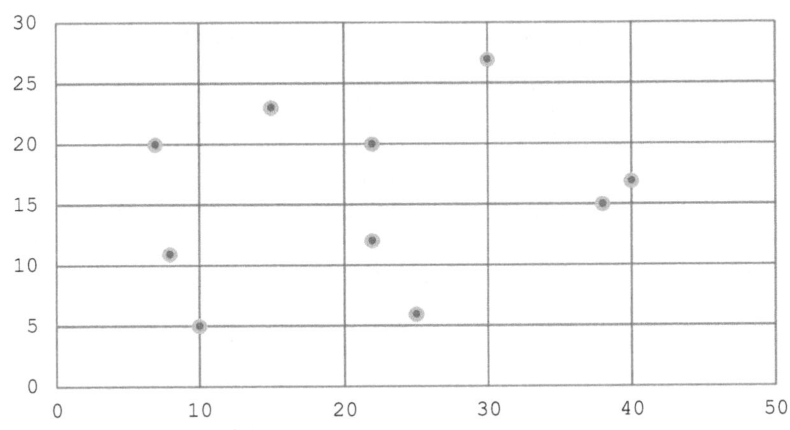

Figure 5.1 Scatter diagrams with no correlation (Slyman, 2022:68).

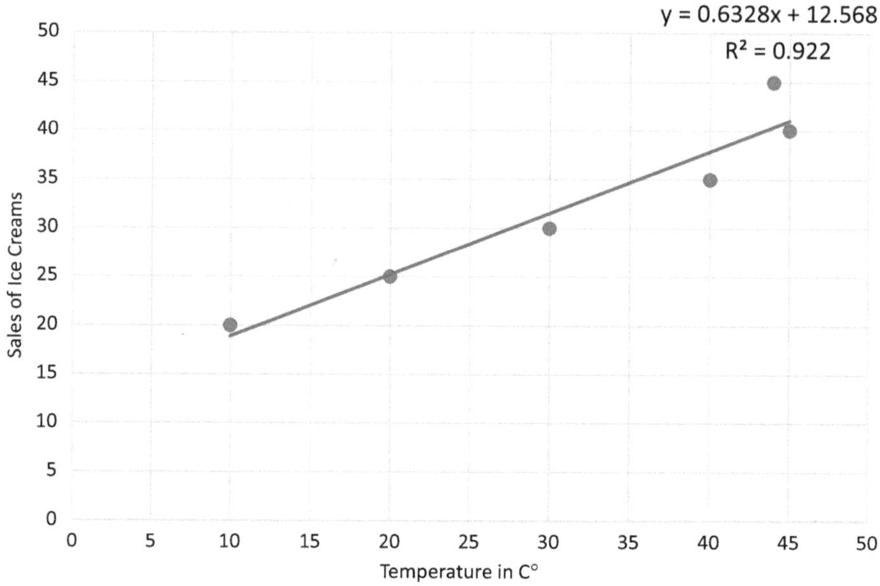

Figure 5.2 Sales of ice creams.

There are many statistical software tools available for analyzing data such as Microsoft Office Excel, SPSS, MATLAB, R, and SAS (Dwyer and Slyman, 2016). In this book, Excel will be employed to analyze data (correlation and regression). Excel is used to identify and interpret quantitative data and solve a range of business problems like business modelling, investment, profits, and growth (as in mayor's gameplay scenario, discussed next, see also Chapter 9).

Simulation GBL Approach to Teaching: SimCity Gameplay

"SimCity" is a 3D entertainment simulation game (SG) that was developed by Will Wright. The first "SimCity" appeared in 1989, and several series have been introduced since then. The latest version was released in 2014 (according to "SimCity" games 2017 website (http://www2.ea.com/simulation)). The purpose of choosing this game is because "SimCity" is a 3D simulation environment where players must plan, expand, and manage a city. Such learning environments may achieve educational objectives, because they will enable participants to test their problem-solving and decision-making skills. The idea is to give students autonomy to create, build, and design a city of their choice, i.e., a scenario using "SimCity." In this 3D SG, the player is given the role of a mayor who has an empty map consisting of a green space. The mission of the mayor is to build the city and provide residents with public services including hospitals, parks, factories, schools, banks, and government buildings, for instance.

Gameplay Scenario and Excel Calculations: Structure/Plan

Within SimCity, the only sources for generating revenues are taxation, prisons, military bases, and investment within neighboring cities. As a mayor of the city, the player's role is to look after the city's infrastructure, public services, and all necessities. In this scenario, participants were asked to investigate how they can generate capital and so consider investing in some companies, i.e., investing in buying/selling shares. Can players help the mayor decide which company (of the five companies) is worth investing in?"

Using the game, Excel, and Statistics, what patterns and future predications can they make?

The learning objectives in using Excel are to teach students:

- The meaning of Statistics, and how it can be applied to real-life scenarios.
- How to calculate and compute correlation and regression in Excel.
- How to solve problems (real-life ones) using correlation and regression (Statistical analysis).

The gameplay teaching experiment outline is as follows:

- Introduction (5 min) (introducing background information about Statistics and Applied Statistics Education).
- Pre-test/Self-efficacy questionnaires (5 min testing participants' confidence before attempting the experiment using Zimmerman's (2000) questionnaire).
- Correlation and regression (teach it for 10 min).
- Playing the "SimCity" game (10 min).
- Excel Calculations (5 min).
- Post-test/Self-efficacy questionnaires (5 min using Zimmerman's (2000) questionnaire).
- Focus-group interviews (20 min, see further details in Chapter 9).
- Summary (5 min).

I began the lecture introducing my gameplay experiment, the BERA ethical guidelines (i.e., their rights to withdraw at any time, anonymity, and confidentiality, etc.); the gameplay scenario and the lecture outlines which included introduction to Statistics, as a discipline and to Applied Statistics in particular with its meaning, moving on to correlation and regression as subtopics, with explanations about what they are, why we require such knowledge, and how it can be applied to real-life scenarios (as mentioned above).

The information was presented sequentially, breaking down complex concepts into manageable sections to ensure students grasped key topics before proceeding to the gameplay segment. Prior to gameplay initiation, participants were briefed on their upcoming activity: a ten-minute session of "SimCity." Each participant had a Mac computer with the latest "SimCity" version installed and was given a blank game map. Most participants were unfamiliar with the game's mechanics and were intentionally not instructed on how to play, encouraging spontaneous exploration and learning. Circulating in the lecture room/lab, I found that participants were creatively planning and building their cities and managing their resources. They were engaged in buying and selling shares, paying taxes, and making money. They were involved in a process of understanding, reasoning, and deciding. After ten minutes, I halted the gameplay and directed participants to refocus on the lecture, introducing a scenario involving investment decisions within their simulated cities, which included this information: "Mayors, you are in a mission, some companies are coming to invest in your city, and you are required to employ Statistics and Excel to assist you with your understanding, reasoning and decision-making. Which company is worth investing in?" I did it this way because users' perception of IT as useful and fun can have some impacts on their preconceptions and subsequently their self-efficacy (see Chapter 8), thus changing their behavior and how they interact with the game system. Once interacting with the game, this will produce some desirable outcomes/skills, which can be **Acumenous or Inquisitional**. After circulating and observing the participants in the Statistics lecture, it was apparent that participants were creating, interacting, and building their cities. They were **"acumenifying"** their learning (see Chapters 7–10 for supplementary details). It was evident from observing their screens that they were employing Excel for Statistical analysis, choosing the best company for their city based on Table 5.1.

The goal of the game with Excel is for participants to practice skills like problem-solving, decision-making, and numerical skills in Excel while playing the "SimCity" gameplay. From Table 5.1, participants were asked to compute whether there were any relationships between companies' years in the market vs. their shares. What is the best advice can they offer to the mayor? Which company is a valuable opportunity for investment? Do companies' experiences (how many years there are in the market) reflect the market shares (they have)? The data in Excel are shown in Table 5.1:

Table 5.1 Statistics and Excel exercise

Companies/supermarket chain	*Years in the market*	*Market share in (%)*
Mresco	23	47
Talisbury	9	12
Ftsa	22	45
Horizone	25	40
Co-Federasion	27	46

As I will articulate later, in my empirical investigation (Chapter 9), I have taught students Statistics lecture (Correlation and Regression) using "SimCity" game scenario as a learning tool. In this gameplay experiment, I used "SimCity Buildit," 2013, version 4. As aforementioned participants were tested before and after the empirical investigation using a self-efficacy questionnaire adopted from Zimmerman (2000) with the aim of testing their confidence in Statistics before and after the game scenario (Chapters 2, 4, and 9). These participants conducted the gameplay: "SimCity," which lasted for 40 minutes, and used Excel to calculate Statistics within the game scenario. The quantitative results are presented below for convenience. For more detailed analysis of the qualitative findings and its nascent themes, see Chapters 9 and 10.

Results

As aforesaid, the SimCity gameplay scenario with Statistics lecture looked at how we can enhance students' self-efficacy in Mathematics and Statistics education using SGs. There were three focus group interviews (FGIs) with 20 participants, 11 males, and 9 females. Of these, 16 did the pre- and post-test surveys. The results (both quantitative and qualitative) presented in this book focus on the usefulness of the gameplay for teaching and learning Mathematics and Statistics education (Chapter 9). I will present quantitative results first then move on to discuss the qualitative nascent findings in Chapter 9, see this chapter for further details.

Quantitative Results

As previously explained, the pre- and post-test surveys are the self-efficacy questionnaires that were adopted from Zimmerman (2000). These tests were analyzed using a pairwise t-test.

The pairwise t-test (Table 5.2, Figure 5.3) was conducted to compare participants' self-efficacy in learning Statistics before and after playing the game. With no game, participants' mean score was 2.7773 (SD = .41581), and with the game, the mean score was 3.1758 (SD = .54640). This difference was statistically significant, $t(15) = 3.292$, $p < .005$. This indicates that participants' confidence significantly improved after playing the game.

Figure 5.3 shows that with the game (post-test), participants' self-efficacy in learning Statistics significantly increased, and it is higher than before the pre-test. This difference is statistically significant, see also Table 5.2. For more detailed qualitative findings (i.e., FGIs) and nascent themes, see Chapters 9 and 10.

As previously explained, in this book, I have focused on two games "SimCity" and "Numbas." The purpose of including "Numbas," a quiz game (discussed next), is for comparison reasons with "SimCity" (Simulation GBL Approach to Teaching Section). "Numbas" is discussed next.

Table 5.2 Pairwise t-test results of participants' confidence in learning statistics before and after playing the game (Slyman, 2022:170).

	M	*N*	*SD*	*t*	*Df*	*Sig. (two-tailed)*
Pre-Test – Post-Test	−.39844	16	.48406	−3.292	15	<.005

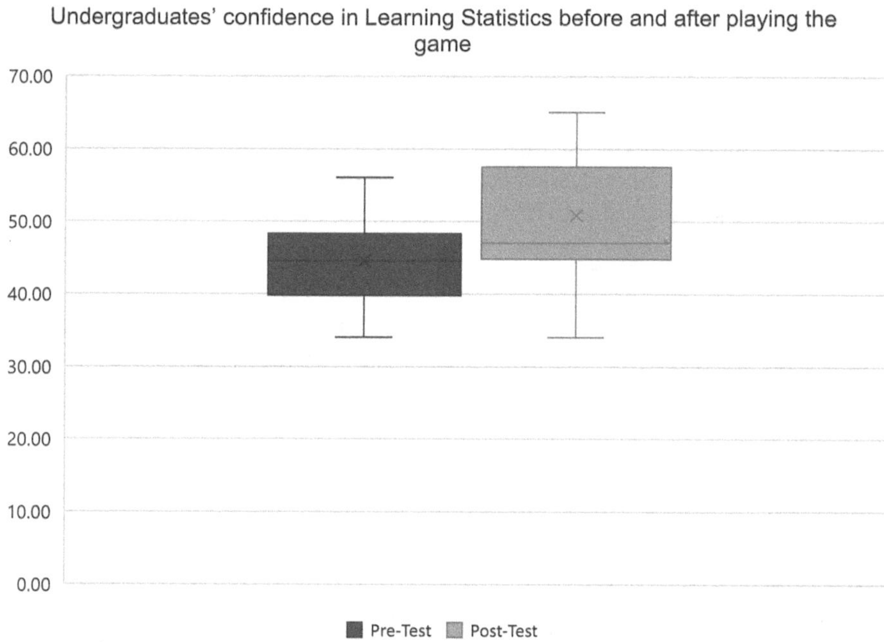

Figure 5.3 Pairwise t-test results of participants' confidence in learning Statistics before and after playing the game (Slyman, 2022:170).

Quiz GBL Approach to Teaching: Numbas

"Numbas" is a Mathematics educational game. This shooting game is set in space. The players take up the role of a defender whose mission is to keep the planet safe from space threats. The player's weapons are Mathematics and speed. Their mission is to destroy asteroids using addition, subtraction, multiplication, and division. The user sees the numbers and types in the correct answers using an on-screen calculator or a keyboard in order to shoot the asteroids and succeed. "Numbas" has a simple user interface consisting of buttons such as a calculator, levels, and scores. It includes operations like addition, subtraction, multiplication, and division. "Numbas" is based on the Mathematical and Statistical knowledge of players, who are required to select correct answers using a tab of numbers. It starts at basic levels and moves to more difficult levels. The game is available both online and offline on PCs and mobile phones. It is not called quiz as such, but if we consider this game from game evaluation perspectives (see Chapters 4 and 6 for more details) like game design and the level of complexity etc., compared to "SimCity," "Assassins" Creed' games or "Sea of Thieves," this 2D game follows basic and easy rules. In this aspect, GBL is encouraged through question-answer formats (Chapter 3). This game requires knowledge and memory, and there is no sophistication or mission involved. "Numbas" is about mental arithmetic, speed, and concentration. This type of GBL can be employed for testing knowledge and assessments but not for consolidating learning or as part of curriculum teaching. It is more like any traditional lecture/lesson. As in real life, knowledge is power: Users can use Mathematical skills to kill asteroids and thus succeed in keeping the planet safe, but this knowledge should be meaningful for GBL to take place. I will return to elucidate this theory further in Chapters 7 and 9.

In all, this chapter has focused on the possibility of teaching Statistics to first-year undergraduates, as a part of the HE core curriculum, and particularly, my GBL approach to teaching correlation and regression, and the learning objectives behind the gameplay scenario, and Excel, moving on to introduce "Numbas" and "SimCity" games.

The next section (a summary) will mark the end of Part I and move on to Part II: Empirical investigations.

Part I: Summary

In Part I, I presented the Research Literature Review (Chapters 1–5) as follows:

Chapter 1 presented an overview of this research project focusing on research background and questions, and why I set out on this book journey. I explained that this empirical research is grounded in my experience as a business computing senior lecturer and researcher at a South London University; it is also a continuation of the empirical research that began during my Master of Arts in Education in 2013–2014. I also explained that as a university senior lecturer, I have convened many programs and modules like Mathematics and Statistics, and I have met many undergraduates. However, I could not envision how to mitigate what appeared to be negative emotions and attitudes that appeared to be exhibited by undergraduates toward Mathematics (Slyman, 2014). My empirical research in 2014 indicates that there are many factors behind such attitudes (see Chapter 1 for full details), but identifying ways to change them was beyond the scope of that work. As I explained in Chapter 1, this question led me to embark on this book journey looking for new ways (i.e., GBL) to address Mathematics and Statistics. I then moved on to the research questions of this empirical work, my contributions to knowledge, and I concluded the Section by explaining how this book is structured.

Chapter 2 examined previous research on learning, self-efficacy, and games. Key discussions of learning theories that were relevant to teaching and learning from socio-constructivist perspectives to Gee's principles were debated. Self-efficacy theory was also defined and critically discussed (along with its implications). This chapter also discussed games, with its different types, looking at the characteristics of games, considering how they can be defined in relation to GBL, and examined theories relating to gaming and engagement, providing a pedagogical rationale for its use in relation to GBL, and overall SGs. I also stated that I am focusing on GBL or games for learning.

To this end, and as indicated earlier, a game can be defined as a fun playful dynamic learning activity with a fixed set of rules and limited time that can simulate real-world problems. GBL is a type of game that has learning outcomes where players can investigate real-life examples in a risk-free environment. Games can teach many skills and thus have many benefits from cognitive, emotional to social ones.

Chapter 3 presented an overview of the gaming research literature, followed by a discussion of GBL and the characteristics of games (for game definition see Chapter 2, the Game Section). A rationale for using GBL was provided. The psychological and social benefits of GBL were also discussed; gaps, contributions, and limitations in the research literature were identified (Chapter 3) I explained that I am focusing on successful examples of GBL building on Gee's 13 principles (Chapter 2). I also indicated that following discussions in Chapter 2, this project will look at how we can use games/SGs as a learning tool to address Mathematics and Statistics.

In Chapter 4, an overview of game design and development literature was presented, focusing on game evaluation heuristics, usability testing and HCI guidelines, and I provided examples of evaluation heuristics for both entertainment and educational games, with highlights on contributions and limitations, before concluding.

Chapter 5 defined what Statistics is. It also explored teaching Statistics (correlation and regression) to first-year undergraduates, as part of the HE core curriculum, using the "SimCity" in a GBL approach to teaching. I explained that "SimCity" was not designed as an educational game, but it can meet some learning objectives. I furthermore highlighted the gameplay scenario with Excel, and how participants would be asked if they could generate capital and so consider investing in some companies, i.e., investing in stock markets: buying/selling shares so that they can problem solve, reason, and decide from the data about the best strategy they should follow. What patterns and future predications can they make? The concept behind this, as I explained, is to investigate and see if this game can have some effects on students' preconceptions and self-efficacy toward Mathematics and Statistics. Could it affect their self-efficacy? My vision is that every student can learn Statistics, a complex subject, using GBL.

I also mentioned that the SimCity gameplay scenario with Statistics lecture was a success and it looked at how we can enhance students' self-efficacy/confidence in Mathematics and Statistics education using SGs. There were three focus group interviews (FGIs) with 20 participants, 11 males, and 9 females. Of these, 16 did the pre- and post-test surveys. The results (both quantitative and qualitative) presented in this book focus on the usefulness of the gameplay for teaching and learning Mathematics and Statistics education (Chapter 9). The quantitative results using pairwise t-test showed that participants' confidence increased immensely. I also indicated that I would discuss the qualitative nascent findings in Chapter 9, see Chapter 9 for further details. I then moved on to "Numbas."

The second game that I discussed is "Numbas," an educational quiz game. It is a Mathematics shooting game, and this game is utilized for comparison reasons (Chapter 5). Furthermore, I mentioned that my experience as a HE senior lecturer suggests that this type of learning (question-answer formats, a recall of facts, or quiz learning) is widely used as a part of HE curriculum teaching. It is sometimes employed in lecture/lab/seminar work to complement learning. I also mentioned that I will return to examine this further in Chapter 7 when I discuss my GBL approach to teaching Statistics with "SimCity." I furthermore introduced the curriculum, the Statistics lecture, and the two games, and how "SimCity" was creatively embedded in teaching Statistics.

The next chapters, **Part III (Chapters 6–10)**, explore the empirical investigations and its findings in this empirical book research journey.

Bibliography

Boaler, J. (2009) *The Elephant in the Classroom*. London: Souvenir Press.

Brown, J.S., Collins, A. and Duguid, P. (1989) Situated Cognition and the Culture of Learning. *Educational Researcher*, 18(1), 32–42.

Bryman, A., (2016) '*Social Research Methods*', 5th ed. United States: Oxford University Press.

Caillois, R. (1961) *Man, Play, and Games*. New York: Free Press.

Cardinot, A. and Fairfield, J.A. (2019) Game-Based Learning to Engage Students with Physics and Astronomy Using a Board Game. *International Journal of Game-Based Learning (IJGBL)*, 9(1), pp. 42–57.

Chen, M.P., Wong, Y.T. and Wang, L.C. (2014) Effects of Type of Exploratory Strategy and Prior Knowledge on Middle School Students' Learning of Chemical Formulas from a 3D Roleplaying Game. *Educational Technology Research and Development*, 62, pp. 163–185. DOI: 10.1007/s11423-013-9324-3

Clark, D.B., Tanner-Smith, E.E. and Killingsworth, S.S. (2015) Digital Games, Design, and Learning a Systematic Review and Meta-Analysis. *Review of Educational Research*, 86(1), pp. 79–122.

Claxton, G. (2002) *Building Learning Power*, 1st Edition. Bristol: TLO Limited.

Cook, J. (2005) Negotiating the Curriculum: Programming for Learning. In *Negotiating the Curriculum*, pp. 15–31. Routledge.

Cooper, P.A. (1993) Paradigm Shifts in Designed Instruction: From Behaviorism to Cognitivism to Constructivism. *Educational Technology*, 33(5), pp. 12–19.

Cordova, D.I. and Lepper, M.R. (1996) Intrinsic Motivation and the Process of Learning: Beneficial Effects of Contextualization, Personalization, and Choice. *Journal of Educational Psychology*, 88, pp. 715–730.

Coulthard, G.J. (2009) A Review of the Educational Use and Learning Effectiveness of Simulations and Games [online]. PhD Thesis, Purdue University. Available at: http://www.coulthard.com/library/paper-simulation.html (Accessed 15/12/2015).

Craft, A. (1999) Creative Development in the Early Years: Some Implications of Policy for Practice. *The Curriculum Journal*, 10(1), pp. 135–150.

Craft, A. (2002) *Creativity and Early Years Education*. London: Continuum.

Crawford, C. (1984) *The Art of Computer Game Design*. Berkeley: California, McGraw-Hill/Osborne Media.

Csikszentmihalyi, M. (1997) *Creativity, Flow and the Psychology of Discovery and Invention*. New York: Harper.

Dweck, C.S. (2006) *Mindset: The New Psychology of Success*. United States: Ballantine Books Trade Paperback Edition.

Dwyer, J. and Slyman, S. (2016) *An Introduction to Quantitative Methods for Business Research*. London: Algana Publishing. 200 pages; ISBN−13: 9781907934094.

Early, R.E. (1992) The Alchemy of Mathematical Experience: A Psychoanalysis of Student Writings. *For the Learning of Statistics*, 12(1), pp. 15–20.
Ejsing-Duun, S. and Karoff, H.S. (2016) Creativity and Playfulness: Producing Games as a Pedagogical Strategy Case studies. In: Munkvold, R. (ed.). *Games Based learning for Researchers, Teachers and Students*, Vol. 1. Reading: Academic Conferences and Publishing International Ltd. ISBN:978-1-911278-12-890000.
Fairweather, E. and Cramond, B. (2010) Infusing Creative and Critical Thinking into the Curriculum Together. In: Beghetto, A. and Kaufman, J.C. (ed.). *Nurturing Creativity in the Classroom*. United States: Cambridge University Press.
Falstein, N. and Barwood, H. (2001) The 400 Project. Available at: http://theinspiracy.com/400_project.htm. (Accessed 20/04/2016).
Faria, A.J. (2001) The Changing Nature of Business Simulation/Gaming Research. *Simulation and Gaming*, 32(1), pp. 97–110. DOI: 10.1177/104687810103200108
Faria, A.J., Hutchinson, D. Wellington, W.J. and Gold, S. (2009) Developments in Business Gaming: A Review of the Past 40 Years. *Simulation and Gaming*, 40(4), pp. 464–487. DOI: 10.1177/1046878108327585
Faria, A.J. and Wellington, W.J. (2004) 'A Survey of Simulation Game Users, Former-Users, and Never-Users. *Simulation and Gaming*, 35(2), pp. 178–207. DOI: 10.1177/1046878104263543
Federoff, M.A. (2002) Heuristics and Usability Guidelines for the Creation and Evaluation of Fun in Video Games. Master Thesis, Indiana University, Bloomington.
Fienberg, S.E. (1992) A Brief History of Statistics in Three and One-Half Chapters: A Review Essay. Available at: https://www.jstor.org/stable/2246307?casa_token=Vek3Rmd5Mk0AAAAA%3AJ0zRZC5AxskwvTNY7GlrgsRV6oLrE20DY3FdeaPad0JvNSNrZDav8uZpX-YqY1hXTDtaz6zM87S4vwkQdflDLSzmrLCNlEsaBqWnykepc0jlYXsQXA
Forest, E. (2014) ADDIE Model. Available at: http://educationaltechnology.net/the-addie-model-instructional-design/ (Accessed 20/04/2016).
Fosnot, T.C. (1989) *Enquiring Teachers, Enquiring Learners: A Constructivist Approach for Teaching*. New York: Teachers College Press.
Fosnot, T.C. and Perry, R.S. (2005) Constructivism: A Psychological Theory of Learning. Available at: http://beyondbitsandatoms.org/readings/fosnot_constructivism_1996.pdf (Accessed 20/07/2019).
Garris, R., Ahlers, R. and Driskell, E.J. (2002) Games, Motivation, and Learning: A Research and Practice Model. *Simulation and Gaming*, 33(4), pp. 441–467.
Gee, J.P. (2003a) *What Video Games Have to Teach Us About Learning and Literacy*. New York: Palgrave Macmillan.
Gee, J.P. (2003b) Good Video Games and Good Learning. Available at: https://academiccolab.org/resources/documents/Good_Learning.pdf (Accessed 03/01/2021).
Gee, J.P. (2004) *Situated Language and Learning: A Critique of Traditional Schooling*. London: Routledge.
Gee, J.P. (2005a) Why are Video Games Good for Learning? Retrieved January 1, 2018, from http://academiccolab.org/resources/documents/MacArthur.pdf (Accessed 20/04/2016).
Gee, J.P. (2005b) *Why Video Games Are Good for Your Soul: Pleasure and Learning*. Melbourne: Common Ground.
Gee, J.P. (2007) *Good Video Games and Good Learning: Collected Essays on Video Games, Learning, and Literacy*. New York: Peter Lang.
Gee, J. P. (2008). *Getting over the Slump: Innovation Strategies to Promote Children's Learning*. New York: The Joan Ganz Cooney Center at Sesame Workshop.
Gee, J.P., Hayes, E., Torres, R.J., Games, I.A., Squire, K. and Salen, K. (2008) Playing to Learn Game Design Skills in A Game Context. In: *Proceeding ICLS'08 Proceedings of the 8th International Conference on International Conference for the Learning Sciences*, Vol. 3, pp. 368–374.

Gibbons, J. and Gray, M. (2004) Critical Thinking as Integral to Social Work Practice. *Journal of Teaching in Social Work*, 24, pp. 19–37.

Glaser, B.G. and Strauss, A.L. (1967) *The Discovery of Grounded Theory*. Chicago: Aldine.

Gomes, A. and Mendes, A.J. (2007) Learning to Program-Difficulties and Solutions. In *International Conference on Engineering Education–ICEE*, Vol. 7.

Granic, I., Lobel, A. and Engels, R.C. (2014) The Benefits of Playing Video Games. *American Psychological Association*, 69(1), pp. 66–78. DOI: 10.1037/a0034857

Hackett, G. and Betz, N.E. (1989) An Exploration of the Mathematics Self-Efficacy/Mathematics Performance Correspondence. *Journal for Research in Mathematics Education*, 20(3), pp. 261–273.

Hamari, J., Shernoff, D.J., Rowe, E., Coller, B., Asbell-Clarke, J. and Edwards, T. (2016) Challenging Games Help Students Learn: An Empirical Study on Engagement, Flow and Immersion in Game-Based Learning. *Computers in Human Behavior*, 54, pp. 170–179.

Hamari, J. and Tuunanen, J. (2014) Player Types: A Meta-Synthesis. *Transactions of the Digital Games Research Association*, 1(2), pp. 29–53.

Harel, I. and Papert, S. (1990) *Software Design as a Learning Environment*. Massachusetts Institute of Technology Cambridge, MA: MIT Media Laboratory.

Harel, I. and Papert, S. (1991) *Constructionism*. Norwood, NY: Ablex Publishing Corporation.

Hargreaves, A. (2003) *Teaching in the Knowledge Society: Education in the Age of Insecurity*. New York: Teachers' College Press.

Illeris, K. (2007) *How We Learn: Learning and Non-Learning in School and Beyond*. London: Routledge.

Jackson, N. and Shaw, M. (2007) Developing Perspectives on Creativity in Higher Education. In: N. Jackson, M. Oliver, M. Shaw, and J. Wisdom (ed.). *Developing Creativity in Higher Education: An imaginative curriculum*. London: Routledge.

Jameson, M.M. and Fusco, B.R. (2014) Math Anxiety, Math Self-Concept, and Math Self-Efficacy in Adult Learners Compared to Traditional Undergraduate Students. *Adult Education Quarterly*, 64(4), pp. 306–322.

Jarvis, M. (2005) *'Motivation' in Psychology of Learning and Teaching*. Cheltenham: Nelson Thornes.

Jeffrey, B. and Craft, A. (2004) Teaching Creatively and Teaching for Creativity: Distinctions and Relationships. *Educational Studies*, 30(1), pp. 77–87.

Jenson, J. and Droumeva, M. (2015) Making Games with Game Maker: A Computational Thinking Curriculum Case Study. In *ECGBL 2015–9th European Conference on Games Based Learning: ECGBL2015*. Academic Conferences and Publishing Limited, p. 260.

Jones, M.G. (1998) Creating Engagement in Computer-Based Learning Environments. Available at: http://itech1.coe.uga.edu/itforum/paper30/paper30.html (Accessed 12/04/2016).

Kirriemuir, J. and Mcfarlane, A. (2004) Literature Review in Games and Learning. Available at: https://telearn.archives-ouvertes.fr/hal-00190453/document (Accessed 12/04/2020).

Knight, P.T. (2001) A Briefing on Key Concepts – Formative and Summative, Criterion and Norm Referenced Assessment. *Learning and Teaching Support Network Generic Centre*, Vol. 30. http://www.heacademy.ac.uk

Kolb, A.Y. and Kolb. D.A. (2008) The Learning Way: Meta-Cognitive Aspects of Experiential Learning. *Simulation and Gaming*, 40(3), pp. 297–327.

Kolb, D.A. (1984) *Experiential Learning: Experience as the Source of Learning and Development*, Vol. 1. Englewood Cliffs, NJ: Prentice-Hall.

Kuss, D.J. and Griffiths, M.D. (2012) Internet Gaming Addiction: A Systematic Review of Empirical Research. *International Journal of Mental Health and Addiction*, 10(2), pp. 278–296.

Lave, J. and Wenger E. (1991) *Situated Learning: Legitimate Peripheral Participation*. Cambridge University Press. Available at: https://www.google.co.uk/books/edition/Situated_Learning/CAVIOrW3vYAC?hl=en&gbpv=1&dq=Lave,+J.+and+Wenger+E.+(1991)+Situated+Learning:++Legitimate+Peripheral+Participation.+Cambridge+University+Press&pg=PA11&printsec=frontcover

Lazzaro, N. (2004) Why We Play Games: Four Keys to More Emotion without Story. Available at: http://www.xeodesign.com/xeodesign_whyweplaygames.pdf (Accessed 28/12/2015).

Lee, J.S., Keil, M. and Kasi, V. (2012) The Effects of an Initial Budget and Schedule Goal on Software Project Escalation. *Journal of Management Information System*, 29(1), pp. 53–77. DOI: 10.2753/MIS0742-1222290102

Lenhart, A., Kahne, J., Middaugh, E. Macgill, A.R., Evans, C. and Vitak, J. (2008) Teens, Video Games, and Civics: Teens' Gaming Experiences Are Diverse and Include Significant Social Interaction and Civic Engagement. *Pew Internet & American Life Project*. Available at: https://www.google.co.uk/search?q=Lenhart+et+al.%2C+2008andoq=Lenhart+et+ al.%2C+2008andaqs=chrome..69i57j0.1244j0j8andsourceid=chrome..die=UTF-8 (Accessed 20/8/2017).

Levine, D.M., Krehbiel, T.C. and Berenson, M.L. (2003) *Business Statistics: A First Course*, 3rd Edition. New Jersey: Pearson Education Ltd.

Lewis, G. (2013) Emotion and Disaffection with School Mathematics. *Research in Mathematics Education*, 15(1), pp. 70–86.

Lyons, I.M. and Beilock, S.L. (2012) When Math Hurts: Math Anxiety Predicts Pain Network Activation in Anticipation of Doing Math. *PLoS One*, 7(10), p. e48076.

Marciano J.N., Miranda, L.C. and Miranda, E.E.C. (2014) Evaluating Multiple Aspects of Educational Computer Games. Available at: https://www.hindawi.com/journals/ijcgt/2014/139205/ (Accessed 27/08/2017).

Marklund, B.B. and Taylor, A.S.A. (2016) Creating Games in the Classroom: From Native Gamers to Reflective Designers. In: Munkvold, R. (ed.). *Games Based learning for Researchers, Teachers and Students*, Vol. 1. Reading: Academic Conferences and Publishing International Ltd. ISBN: 978-1-911278-12-890000.

Maslow, A.H. (1970) *Motivation and Personality*, 2nd Edition. New York; London: Harper and Row.

Mayer, E.R. (2014) *Computer Games for Learning: An Evidence Based Approach*. London: MIT.

Mayer, I., Warmelink, H. and Bekebrede, G. (2013) Learning in a Game-Based Virtual Environment: A Comparative Evaluation in Higher Education. *European Journal of Engineering Education*, 38(1), pp. 85–106.

Mayo, M. (2009) Video Games: A Route to Large-Scale STEM education? *Science*, 323, pp. 79–82.

McConnell, D. (2000) *Implementing Computer Supported Cooperative Learning*, 2nd Edition. London: Kogan Page.

Mccrindle, M. (2018) What Comes after Generation Z? Introducing Generation Alpha. Availableat:https://mccrindle.com.au/insights/blogarchive/what-comes-after-generation-z-introducing-generation-alpha/ (Accessed 15/10/2018).

Milane, I. and Rowe, G. (2002) Difficulties in Learning and Teaching Programming-Views of Students and Tutors. *Education and Information Technologies*, 7(1), pp. 55–66.

Monroy-Hernandez, A. and Resnick, M. (2008) Empowering kids to create and share programmable media. *Interactions*, 15(2), pp. 50–53. Retrieved August 30, 2015.

Moreno-Ger, P., Burgos, D. Martínez-Ortiz, I. Sierra, J.L. and Fernåndez-Manjón, B. (2008) Educational Game Design for Online Education. *Computers in Human Behavior*, 24(6), pp. 2530–2540. DOI: 10.1016/j.chb.2008.03.012

Moreno-Ger, P., Torrente, J. Hsieh, Y.G. and Lester, W.T. (2012) Usability Testing for Serious Games: Making Informed Design Decisions with User Data. *Advances in Human-Computer Interaction*, 2012. DOI: 10.1155/2012/369637

Multon, K.D., Brown, S.D. and Lent, R.W. (1991) Relation of Self-Efficacy Beliefs to Academic Outcomes: A Meta-Analytic Investigation. *Journal of Counselling Psychology*, 18, pp. 30–38.

Naik, N. (2017) The Use of GBL to Teach Mathematics in Higher Education. *Innovations in Education and Teaching International*, 54(3), pp. 238–246. DOI: 10.1080/14703297.2015.1108857

Nicholls, D., De Charon, L. and Hutkin, R. (2010) Development of Critical Thinking and Creativity: Practical Guidelines for the Postsecondary Classroom. *ATEA Journal*, 38(1), p. 12.

Pachler, N. (2007) *Mobile Learning: Towards a Research Agenda*. London: WLE Centre, IoE. In: Kress, G. and Pachler, N (ed.) (2007) Thinking about the 'm' in m-learning.

Pajares, F. and Kranzler, J. (1995) Self-Efficacy Beliefs and General Mental Ability in Mathematical Problem-Solving. *Contemporary Educational Psychology*, 20, pp. 426–443.

Palincsar, A.S. (1998) Social Constructivist Perspectives on Teaching and Learning. *Annual Review of Psychology*, 49, pp. 345–375.

Papert, S. (1993) *Mindstorms: Children, Computers and Powerful Ideas*. New York: Basic Books.

Paras, B. and Bizzocchi, J. (2005) Game, Motivation, and Effective Learning: An Integrated Model for Educational Game. In *Proceedings of DiGRA Conference on Changing Views–Worlds of Play*. Available at: https://www.researchgate.net/publication/221217604_Game_Motivation_and_Effective_Learning_An_Integrated_Model_for_Educational_Game_Design (Accessed 27/09/2016).

Park, H.J. and Kim, S.B. (2013) Guidelines of Serious Game Accessibility for the Disabled. In *Proceedings of the International Conference on Information Science and Applications (ICISA '13)*, pp. 1–3, IEEE.

Pinelle, D., Wong, N., Stach, T. and Gutwin, C. (2009) Usability Heuristics for Networked Multiplayer Games. *Proceedings of the ACM 2009 International Conference on Supporting Group Work*, pp. 169–178.

Pollard, A., Anderson, J., Maddock, M. Swaffield, S., Warin, J. and Warwick, P. (2008) *Reflective Teaching*, 3rd Edition. London: Continuum International Publishing Group.

Preconception. (2020) In: Dictionary.com. [Online] Available at: https://www.dictionary.com/browse/preconception (Accessed 01/08/2020).

Resnick, M. (2017) *Lifelong Kindergarten: Cultivating Creativity Through Projects, Passion, Peers, and Play*. London. England: MIT Press.

Rogoff, B. (1998) Cognition as a Collaborative Process. In W. Damon, D. Kuhn and R.S. Seigler (ed.), *Handbook of Child Psychology*, 5th Edition. New York: Wiley.

Roth, W.M. (2000) Authentic School Science: Intellectual Traditions In R. McCormick and C. Paechter (ed.), *Learning & Knowledge*. London, UK: Paul Chapman Publishing, pp. 6–20.

Senge, P. (1990) *The Fifth Discipline: The Art and Practice of Learning Organization*. United Kingdom: Random House Business Books.

Senge, P.M. (2006) *The Fifth Discipline: The Art and Practice of Learning Organisation*. London: Random House Business Books.

Shell, D.F., Murphy, C.C. and Bruning, R.H. (1989) Self-efficacy and Outcome Expectancy Mechanisms in Reading and Writing Achievement. *Journal of Educational Psychology*, 81(1), p. 91.

Shunk, D.H. (2000) *Learning Theories: An Educational Perspective*, 3rd Edition. Upper Saddle River, NJ: Prentice-Hall.

Siegel, R.G., Galassi, J.P. and Ware, W.B. (1985) A Comparison of Two Models for Predicting Mathematics Performance: Social Learning Versus Math Aptitude-Anxiety. *Journal of Counseling Psychology*, 32, pp. 531–538.

Sindre, G., Natvig, L. and Jahre, M. (2009) Experimental Validation of the Learning Effect for A Pedagogical Game on Computer Fundamentals. *IEEE Transactions on Education*, 52 (1), pp. 10–18.

Skemp, R.R. (1986) *The Psychology of Learning Mathematics Great*. Britain: Penguin Books Ltd.

Skemp, R.R. (1989) *Mathematics in the Primary School*. Bristol: Routledge.

Slyman, S. (2014) I Hate Maths! An Investigation into Students' Attitudes Towards Mathematics. MA Thesis, University of Roehampton, London, United Kingdom.

Slyman, S. (2018) *Games Based Learning in Universities, a Life Learning Experience*. New York: Business Expert Press Expert Insights Publishing. ISBN:978-1-94819-801-1.

Slyman, S. (2022) Acumenous Game-Based Learning in Simulation Games and Applied Statistics. PhD Thesis, Goldsmiths, University of London. UK

Slyman, S., Gillies, M. and Jessel, J. (2019a) Acumenous or Inquisitional? Towards a New Theoretical Lens within Games Learning. In: *13th European Conference on Games Based Learning*, University of South Denmark, Odense, Denmark. ISBN:978-1-912764-38-9 or E-book: ISSN:2049-100X.

Slyman, S., Gillies, M. and Jessel, J. (2019b) Game-Based Learning to Engage students with Applied Statistics using a Simulation Roleplay Game. In: *9th Irish Conference on Games Based Learning*, Cork, Ireland. Available at: http://igbl-conference.com/2016/programme/generate_book_of_abstracts.php (Accessed 25/09/2019).

Sobke, H., Chan, E. Buttlar, R. Grobe-Wortmann, J. and Londong, J. (2014) Cat King's Metamorphosis. The Reuse of an Educational Game in a Further Technical Domain. In: Gobel, S., and Wiemeyer, J. (ed.), *Games for Training Education, Health and Sports, 4th Conference on Serious Games, GameDays 2014*, Darmstadt, Germany, April 1–5, 2014, Springer.

Song, S. and Lee, J. (2007) Key Factors of Heuristic Evaluation for Game Design: Towards massively multi-player online Roleplaying game. *International Journal of Human-Computer Studies*, 65, pp. 709–723.

Strauss, A. and Corbin, J. (1990) *Basics of Qualitative Research: Grounded Theory Procedures and Techniques*, Newbury Park, CA: Sage.

Sweetser, P., Johnson, D. and Wyeth, P. (2012) Revisiting the Game Flow Model with Detailed Heuristics. *Journal of Creative Technologies*, 2012(3), pp. 1–16.

Sweetser, P. and Wyeth, P. (2005) GameFlow: A Model for Evaluating Player Enjoyment in Games. *Computers in Entertainment*, 3(3), p. 3. DOI: 10.1145/1077246.1077253

Tamura, H. (2007) Foundational Value of Mathematics Education for Management Curriculum. *International Statistical Review*, 75(3), pp. 397–405. DOI: 10.1111/j.1751-5823.2007.00032.x

Tao, Y.H., Cheng, C.J. and Sun, S.Y. (2012) Alignment of Teacher and Students Perceptions on the Continued Use of Business Simulation Games. *Educational Technology and Society*, 15(3), pp. 177–189.

Tondello, G.F., Kappen, D.L. Mekler, E.D. Ganaba, M. and Nacke, L.E. (2016) Heuristic Evaluation for Gameful Design. *CHI PLAY'16*. ACM. 978-1-4503-4458-6/16/10. DOI: 10.1145/2968120.2987729

Vlachopoulos, D. and Makri, A. (2017) The Effect of Games and Simulations on Higher Education: A Systematic Literature Review. *International Journal of Educational Technology in Higher Education*, 14(1), pp. 22. DOI: 10.1186/s41239-017-0062-1

Vos, L. (2014) Marketing Simulation Games: A Review of Issues in Teaching and Learning. *The Marketing Review*, 14(1), pp. 67–96. DOI: 10.1362/146934714X13948909473220

Vos, L. and Brennan, R., 2010. Marketing Simulation Games: Student and Lecturer Perspectives. *Marketing Intelligence & Planning*, 28(7), pp. 882–897.

Vygotsky, L.S. (1980) *Mind in Society. The Development of Higher Psychological Processes*. Edited by Cole, M., John-Steiner, V., Scribner, S. and Souberman, E. London: Harvard University Press.

Whitton, N. (2011) Games-Based Experiences for Learning. Futurelab Innovation in Education. Available at: https://www.nfer.ac.uk/media/1775/futl11.pdf (Accessed 02/04/2016).

Wolfe, J. (1997) The Effectiveness of Business Games in Strategic Management Coursework. *Simulation and Gaming*, 28(4), pp. 360–376. DOI: 10.1177/1046878197284003

Young, M. (2008) *"Powerful Knowledge" and "Knowledge of the Powerful": Some Reflections*. London: Presentation at the Institute of Education.

Zimmerman, B.J. (2000) Self-Efficacy: An Essential Motive to Learn. *Contemporary Educational Psychology*, 25(1), pp. 82–91. DOI: 10.1006/ceps.1999.1016

Part III

Acumenous Game-Based Learning in Simulation Games and Mathematics and Statistics

Having discussed game design and development, evaluation heuristics, Human-Computer Interaction (HCI), and usability guidelines in Chapter 5. Besides, embarking on an exploration of teaching Statistics in higher education in Chapter 6, employing a creative approach through gameplay investigation, utilizing games such as 'SimCity' as a part of the core curriculum, Part III (the Final Part) will present research findings and new theoretical contributions, backed up by research literature, in Chapters 6–10.

DOI: 10.4324/9781032723426-8

6 Exploring the Essence of Simulation Games

Game Players' Experiences and Design Attributes for Game-Based Learning

To gain a comprehensive understanding of game players' experiences, this chapter embraces a qualitative research methodology. Drawing upon Charmaz's Constructivist Grounded Theory (CGT) (2006, 2014), the empirical investigation unfolds, capturing the essence of players' interactions and insights within the realm of SGs. The analytical approach of CGT, reinforced by a rich body of research literature, serves as the lens through which we explore and interpret the profound impact of SGs on players' experiences.

Furthermore, this chapter takes a pioneering step forward by proposing a game evaluation heuristics model. Addressing a gap in existing theoretical frameworks, this model harmoniously blends elements from both entertainment and educational perspectives. By striking a delicate balance, it offers a comprehensive and nuanced framework for analyzing and evaluating the educational simulation games (ESGs) that lie at the heart of transformative learning experiences.

As we conclude this chapter, we stand on the precipice of a new era in game-based learning. With a profound understanding of game players' experiences and the intricate design attributes that shape these experiences, we are poised to unlock the transformative potential of SGs in education. This chapter sets the stage for future explorations, where the convergence of entertainment and education creates a powerful synergy, empowering learners and revolutionizing the landscape of game-based learning.

Within this section, focusing specifically on Simulation Games (SGs), I will extensively delve into the intricate aspects of gameplay, game characteristics, design attributes, mechanics, and development. Our exploration will particularly highlight various game types, with a special emphasis on MMORPGs (Massive Multiple Online Roleplay Games). MMORPGs, a captivating subgenre of action games, seamlessly blend elements of strategy, fantasy, adventure, and simulation. These games serve as a notable illustration of the synergy between entertainment and educational components, forming a compelling and engaging platform for learning. Adopting the analytical framework of

DOI: 10.4324/9781032723426-9

Charmaz's CGT (Constructivist Grounded Theory) (2006, 2014), supported by a comprehensive body of research literature (refer to Chapter 4), we will scrutinize and interpret the profound impact of SGs on players' experiences. This examination encompasses game analysis, design, and development, as well as game implementation and evaluation, as follows.

Game Analysis

A strong emphasis is placed on achieving effective and enjoyable gameplay, which requires clear objectives, a vision, a showdown system, tutorials, and comprehensive game content, encompassing various levels. These elements contribute to player engagement, embodiment, identity formation, agency, fun, and empowerment. This perspective aligns with the research discussed in previous Chapters (2, 3, and 4), citing the works of Malone (1980), Jones (1998), Garris et al. (2002), Gee (2003a, 2003b, 2004, 2005a, 2005b, 2007, 2008, 2011), Prensky (2002, 2005), Squire (2005), Mayo (2009), Park and Kim (2013), Mayer (2014), Schell (2015, 2020), McGonigal (2015), Jagger et al. (2016), Hamari et al. (2016), Vlachopoulos and Makri (2017), and Cardinot and Fairfield (2019). However, ineffective gameplay can lead to disengagement. The complexity of disappointing gameplay can result in players failing to understand the game, leading to a dislike for the game rather than an enjoyment of the gameplay experience. This implies that players' experiences with SGs may be negative, as games can evoke not only positive emotions but also negative ones, such as preconceptions. Connecting this to game usability, the game may not be self-intuitive, as indicated by Nielsen (1995, 2000). Unmet needs or expectations, as discussed by Mullins (2013, 2019), can lead to dissatisfaction, especially when desirable goals or outcomes are lacking. If players struggle to comprehend the game's rules and tutorials, without adequate reward or risk incentives, dissatisfaction and unmet expectations may follow. This dissatisfaction can serve as a catalyst, influencing preconceptions and behavior, resulting in negative gaming experiences and a lack of confidence to return to the game.

Within the realm of game analysis, the impact of the game story on players is noteworthy. Exceptional storylines, rooted in unique skills and character developments, can positively affect players' gaming experiences. This perspective is supported by research from Garris et al. (2002), Gee (2003a, 2003b, 2004, 2005a, 2005b, 2007, 2008, 2011), McGonigal (2015), Resnick (2017), and Arnab and Clarke (2017). Emphasizing the significance of the game story as a narrative powerhouse that captivates players, encouraging them to return, aligns with the insights provided by Schell (2015, 2020), Sweetser et al. (2012), and the concept of being worth playing (Desurive and Wiberg, 2010). However, some players critique game story, design, mechanics, and storylines as lacking originality, feeling like clones or pretenders, and being repetitive with insufficient actual content. Such criticisms lead to disengagement, and players may not find the motivation or confidence to revisit playing SGs due to the unexciting and disappointing gaming experiences that trigger negative emotions, including preconceptions. This underscores the importance for game educators and designers to focus on refining game elements such as story and storylines, mechanics, and design to create engaging, captivating, and empowering gaming experiences.

Game Design and Development

Game mechanics, graphics, animations, and sounds emerge as having positive impacts on game players' experience in playing the SGs, since they are characterized as fantastic and awesome suggesting that game mechanics, aesthetics, and technology are working in harmony as indicated by Silverman et al. (2003). The synergy of these components is crucial for enhancing player engagement and confidence, as emphasized by Schell (2015, 2020), McGonigal (2015), and Resnick (2017), constituting fundamental characteristics in game design and development (Falstein and Barwood, 2001; Garris et al., 2002; Schell, 2015, 2020; refer to Chapter 4).

In contrast, some players report negative and repetitive experiences due to a lack of meaningful content, poor design, uninteresting training, disappointing stories, and a deficiency of quests and progression. Such experiences arise because players seek originality and innovation in games, rather than mere replicas of existing ones. Creative, inspiring, and challenging gameplay, featuring hidden problems and conflicts, is what players crave.

Game Implementation and Evaluation

Additional noteworthy themes include game implementation and evaluation where user interface and mechanics were not accurately functional, controls and immersion were of low standards. These shortcomings suggest a deviation from fundamental game design principles. Critical problems in game usability, such as inadequate feedback, non-user-friendly interfaces, booting issues, a stagnant interface, and the absence of error handling messages that guide users to undo their actions, resulted in fatal technical errors, causing the game to cease loading and functioning. Consequently, these issues had adverse/negative impacts on game players' experiences and overall their confidence in returning to play leading to preconceptions. This underscores the significance of Human-Computer Interaction (HCI) principles, emphasizing the necessity of adhering to usability-responsive movement and controllers in game design across gameplay, story, design, mechanics, usability principles, and implementation and evaluation. In other words, and as argued, games can have positive/negative impacts on game players' experience and give them the confidence to return/not return to play the SGs.

To sum up, the research reveals various design issues in numerous games, even those featuring advanced Simulation Games (SGs). These issues primarily stem from the absence of Human-Computer Interaction (HCI) principles and usability heuristics, ultimately influencing the overall gaming experience for players. Despite the sophisticated development of SGs across multiple genres, certain players have expressed negative experiences, as the findings suggest. This negativity is attributed to the deficiency in game usability heuristics, HCI principles, and unresolved game design issues. The ensuing section will delve into the ESGAE Model, which has been formulated based on the compilation and guidance provided in Chapter 4.

ESGAE: The Game Evaluation Heuristics Model

The proposed game evaluation heuristics model, the ESGAE (Table 6.1), is designed to address a gap in existing frameworks as previous game evaluation heuristics models were often centered on either entertainment or educational perspectives, lacking a balanced integration of both realms, as argued earlier (Chapter 4).

The research focus, of this section, is dedicated to designing a game evaluation heuristics model specifically tailored for analyzing and evaluating educational simulation games (ESGs). This model will be able to detect low self-efficacy and customize learning (to improve it). The game evaluation heuristics are based on the current research investigation, as previously articulated, and it is in line with previous research literature on game design and development perspectives, HCI principles and usability testing guidelines. Consequently, this results in the ESGAE (Educational Simulation Games Analysis and Evaluation) Model (Table 6.1) consisting of five phases: Game Analysis, Design, Development, Implementation, and Evaluation (ESGAE), since these mechanisms are major ingredients, and the most important features, in game design be it for entertainment or educational games and can have significant impacts on game players' experiences when playing the SGs.

The game analysis phase includes analyzing players' gameplay, gaming characteristics, their current levels, styles, expectations, and needs (taking into consideration the cognitive, emotional, and social factors), the outcomes, and any existing constraints they might encounter. The Design phase encompasses for instance designing how the game is displayed with its visual, technical, and graphical design, which will influence the outcomes. The next phase, Development, may include integrating technologies, testing, debugging, and reviewing it before a final release. Implementation involves training facilitators/game players according to the outcomes or curriculum design we would like to implement, specifying its method of delivery and testing procedures. Finally, Evaluation consists of formative and summative assessments/evaluation of the whole design. This includes testing the design and providing opportunities for feedback. Consequently, the proposed model symbolizes a better game evaluation heuristic categorized according to these phases that were embodied in the model as a game analysis, game design and development, and game implementation and evaluation.

Table 6.1 ESGAE model (Slyman, 2022:115)

Phases and categories	*ESGAE: Evaluation heuristics, its description, and source*	
I. Educational simulation game analysis **a) Gameplay** **b) Game story**	1. (I)	**Clear objectives or outcomes of the game.**
	2. (I)	**Players do not require any preliminary skills to commence. Novice/experts can understand the rules.**
	3. (I)	**Game styles/genres are distinctive: strategy/adventure, etc.**
	4. (I)	**Expectations and needs (behavioral outcomes: cognitive, emotional (attitudes), and psychomotor (skills)) are achieved/met leading to high/low self-efficacy (i.e., no fear) and therefore fulfilment/engagement (i.e., new learning outcomes/transferable skills).**
	5. (I)	**Detect high/low self-efficacy in Mathematical and Statistical learning, picking up signals and customizing learning** (see Chapter 10).
	6. (I)	**Levels are obtainable/easy/hard (i.e., balanced).**
	7. (I)	**The element of curiosity and manipulation are integrated in the game.**
	8. (I)	**Players are engaged in the game. The game is valuable, and players care about it.**
	9. (I)	**Game players are asked to solve challenging hidden problems/conflicts as a part of the gameplay. The game generates more problems to keep players engaged and coming.**
	10. (I)	**The driving instincts of the game and how the story is built are exciting/ positive/boring. The suspense; event; story; characters; details; and dramatic changes are balanced/appealing.**
	11. (I)	**The game is fun and enjoyable. Some parts are/need to be fun. The element of play is presented in the game and it is engaging and immersive.**
	12. (I)	**The game surprises the players when they play. The game story has surprises. The rules of the game give players ways to surprise each other and themselves.**
	13. (I)	**Games can transform players for better/worse. The game is worth playing/ the time/the download.**
	14. (I)	**Risks; choice and chance; punishment; competition; reward is built up. A way to turn losing situation to a winning state.**

(*Continued*)

Table 6.1 (Continued)

Phases and categories	*ESGAE: Evaluation heuristics, its description, and source*
	15. (I) **Constant feedback, clear progress display (scores and levels), and response time.**
	16. (I) **Players can socialize/express themselves. Some/all are playing with friends.**
	17. (I) **Simplicity/complexity is/should be balanced. They are puzzles and scenarios in the game.**
	18. (I) **Tutorials/documentations are clear and comprehensible.**
	19. (I) **There are puzzles/quizzes/scenarios/videos in the game.**
	20. (I) **Online payments—upgrading system.**
	21. (I) **Clear outcomes of the game.**
II. Educational simulation game design and developments	1. (II) **Animations, sound effects, rewards, and feedback increase players' curiosity.**
a) Game mechanics	2. (II) **Rewards and feedback increase players' self-efficacy, motivation, and engagement.**
b) Aesthetic	3. (II) **Game instructional, visual, and technical designs: venue, objects, attributes, dimensions (3D, 2D), novelty, uniqueness; imagination, strategy, constraints/limitations, effects are distinctive.**
c) Technology	4. (II) **After detecting high/low self-efficacy in Mathematical and Statistical learning, provide customized learning.**
	5. (II) **Storyline is easy to understand, clear with engaging outcomes/vision.**
	6. (II) **Game designs with its different elements are in harmony. The game time is short/slow/fast.**
	7. (II) **Physical input and output resources are appealing from aesthetics, mode of change, size, font, and colors.**
	8. (II) **User interface, controllers; feedback/rewards and spending money, currency/points all information are obvious and communicated to the user without interfering with the interface.**
	9. (II) **Technology and programs are harmoniously integrated in the game.**
	10. (II) **Online payments—Upgrading game levels.**

III. Educational simulation game implementation and evaluation **a) Game usability** **b) HCI principles**	1. (III) **Training facilities, outcomes, curriculum, and methods of delivery are clear.** 2. (III) **Curriculum and methods of delivery are clear.** 3. (III) **Progress, feedback, scores, and levels are explicable, challenging and balanced according to levels.** 4. (III) **Detect high/low self-efficacy in Mathematical and Statistical learning and provide customized learning.** 5. (III) **Intuitive accessible user-friendly interface; consistency throughout the game interface, menus, design, sounds, and animation.** 6. (III) **Controllers do not confuse the players.** 7. (III) **Storyline is suitable for the level of complexity.** 8. (III) **Storyline is logical with engaging outcomes/vision.** 9. (III) **Storyline is novel and well-designed.** 10. (III) **Intuitive accessible user-friendly interface; consistency throughout the game interface, menus, design, sounds, and animation.** 11. (III) **Useful error handling message, promoting users to undo their action.** 12. (III) **Tutorials/documentations are easy to read and understand.** 13. (III) **Friendly online customer support.**

Key

I II, III: ESGAE Phases
a., b., c: Sub-categories

The game analysis phase (I) consists of Gameplay. Gameplay has 21 heuristics, from (1. (I) to 21. (I)). It deals with how the player interacts with the game starting with the game objectives, players' preliminary skills to commence the game, levels, styles, and genres of the game, engagement, surprises, and curiosity, as well as expectations and needs and whether the game leads to high/low self-efficacy. The game story is remarkably crucial, including how the scene is built up, and whether the game is fun and transforms its players for the better/worse. I trust this is as the main drive that motivates players to continue with the story or end it. Gameplay also analyzes whether the game has risks, clear tutorials and documentation, constant feedback, simple or complex play/interaction, and if it has an online upgrading system, a pay-in system, and clear outcomes of the game.

The game design and development phases (II) have ten heuristics from 1. (II) to 10. (II). These consist of game mechanics, aesthetics, and technology. This may involve animations, sound effects, game educational contents, visuals, and technical design like menus, attributes to objects and input-output resources. In addition, game design and development phases have game story and storylines, and how the latter is developed and whether the storyline is integrated harmoniously. This phase also tests whether the game interface and controllers are interactive and easy-to-use, as well as this it checks the online-payment system (for upgrading game levels).

The game implementation and evaluation phases (III) have 13 heuristics from 1. (III) to 13. (III). It deals with game usability and how accessible and usable it is for players to interact with the game. It investigates whether training facilities, outcomes, curriculum, and methods of delivery are clear, adding to this progress, feedback, scores, and levels, and checking that they are explicable, challenging, and balanced according to levels. It also examines whether customized learning is provided after low self-efficacy (low confidence) is detected. In addition, it examines if the game storyline is stimulating, novel, logical, and well-designed with engaging outcomes. Furthermore, it assesses whether the game has an accessible user-friendly interface that is consistent throughout the game interface with its menus design, controllers, sounds, and animations that do not confuse players, and can handle error messages easily, with an accessible tutorial and documentations, and a friendly online customer support service. As mentioned previously, the ESGAE stands for Educational Simulation Games Analysis and Evaluation Model. The ESGAE Model is as follows.

As previously mentioned, the objective of the ESGAE is to formulate a comprehensive model for analyzing and evaluating ESGs. For this particular empirical investigation, I selected "Numbas" and "SimCity" to gain a thorough insight into game design and development systems. This approach aims to craft an exceptional gaming experience that aligns with players' expectations and requirements, fostering their progression and satisfaction while enhancing their self-efficacy and facilitating the acquisition of new learning outcomes and transferable skills. The analysis and evaluation of "Numbas" and "SimCity" (Chapter 5) are elaborated in the subsequent section, while further information regarding game design can be found in Chapter 4.

Application of the ESGAE

Numbas Game

As previously stated (Chapters 4 and 5), "Numbas" is a Mathematics shooting game where players are required to select correct answers using a tab of numbers as in calculators. It is a simple type of game. This game is set in space. The players take up the role of a defender whose mission is to keep the planet safe from space threats. The player's weapons are Mathematics and speed. Their mission is to destroy asteroids. It is available online/offline on PCs and mobile phones. It is not called quiz as such, but the aim is to test Mathematical learning, a factual recall. If we consider this game from the ESGAE perspective, it is evaluated and analyzed as shown in Table 6.2.

SimCity Game

Following Chapters 4 and 5, "SimCity" is a 3D entertainment SG developed by Will Wright. "SimCity" is an online/offline game and has many types available like "SimCity BuildIt 2013." "SimCity," the original game, however, is a business expansion, although it is not specified as such, and this is its focus. For example, within "SimCity," there is a trading depot where players can buy and sell items, earn extra cash or "Simoleons," or buy and sell items from neighboring cities. Urban planning, public services, landmarks, strategy, and city infrastructure are part of "SimCity" life. This infrastructure includes facilities such as water supply, electricity, coal, and wind power plants. There are also building supply stores, selling planks and nails, hardware stores producing hammer, measuring items and shovels. In addition, the game has 11 items for the City Storage. When it comes to achievements, the game has different types for each level, for example, "Building the City," "Growing City," "Busy Manufacturing," "Happy City," "Fire safety," "Taxes," "City Achievements," "City Storage," "Sim Offers," and "Trade Depot" and "Global trade." Once the player accomplishes a level, s/he can claim it. The game has also a tax building office and it is called "City Hall Building." This tax office has small factories producing items such as plastics, metals, and woods. Overall, what Sims require and value most is money and happiness.

In the next section, the game evaluation heuristics model with a detailed analysis for "SimCity" will be introduced (Table 6.3).

Table 6.2 Application of the ESGAE: Numbas game (Slyman, 2022:119)

	ESGAE: Evaluation heuristics and its description
NUMBAS as a game I. **Educational simulation game analysis** **(Gameplay** **and** **game story)**	1. (I) **Clear objectives or outcomes of the game**. "Numbas," as a game, has clear simple objectives. These are Mathematical operations (plus, minus, multiply/divide) on the form of asteroids carrying out questions that have to be answered using an on-screen calculator where the user is required to type in the right answers (refer to Chapter 5). 2. (I) **Players do not require any preliminary skills to commence. Novice/experts can understand the rules**. No, players require previous knowledge or experience with this type of learning, that is, Mathematical learning of how to calculate mental arithmetic. I guess this is an educational game and this is what differentiates it from other games. Previous knowledge is the condition of this game. The player has to have a basic understanding of how to calculate. 3. (I) **Game styles/genres are distinctive: strategy/adventure, etc**. The game's strategy is to shoot down asteroids using Mathematical learning and speed as a weapon. 4. (I) **Expectations and needs (behavioral outcomes: cognitive, emotional (attitudes), and psychomotor (skills)) are achieved/met leading to high/low self-efficacy and therefore fulfilment/engagement (i.e., new learning outcomes/transferable skills)**. A sense of enjoyment, happiness, and fulfilment when achievements are made but limited since the game has easy rules to follow. The motivation element is to kill the asteroids and keep the planet earth safe. The game raises self-efficacy when an achievement and progress is made and therefore can improve players' negative emotions toward Mathematics. However, it can also elicit preconceptions (i.e., negative ones as previously indicated) if players lack Mathematical learning. In educational games, there are learning outcomes to be achieved from the play. I suppose in education/learning context, it is obvious or clearly stated at the beginning, but in games it is hidden. What can be transferred to learning? Game-based learning (GBL) can be transferred to real life. With this game, "Numbas," you can use Mathematical learning to kill asteroids and keep planet safe. How does GBL take place? GBL takes place in this context using previous knowledge/ experience, memorization/set of instructions as in quiz-type format. The aim thus is to test Mathematical learning, facts, and speed.

5. (I) **Detect high/low self-efficacy in Mathematical and Statistical learning, picking up signals and customizing learning** (Chapter 10).
   No, this is missing and not available.
6. (I) **Levels are obtainable/easy/hard.**
   Yes, it starts from basic and stays basic. There are three levels “beginner,” “master,” and “IQ>120.”
7. (I) **The element of curiosity and manipulation are integrated in the game.**
   No, this is absent.
8. (I) **Players are engaged in the game. The game is valuable, and players care about it.**
   No, this is absent. Social aspects are missing, since it is based on a factual recall, and it is a solo game.
9. (I) **Game players are asked to solve challenging hidden problems/conflicts as a part of the gameplay and story. The game generates more problems to keep players engaged and coming.**
   No, the game is simple. As soon as the player sees calculations/numbers, they answer them by entering the correct answers through a calculator/keyboard. There is no story element generated or hidden problems that increase engagement/the urgency to keep playing the game.
10. (I) **The driving instincts of the game and how the story is built are exciting/positive/boring. The suspense; event; story; characters; details; and dramatic changes are balanced/appealing.**
   The game has one theme/story which is to kill asteroids using Mathematical knowledge as a weapon.
11. (I) **The game is fun and enjoyable. Some parts are/need to be fun. The element of play is presented in the game and it is engaging and immersive.**
   The game is fun for those who like calculating numbers with speed and shooting enemies (asteroids) to some extent but not immersive and can be repetitive.
12. (I) **The game surprises the players when they play. The game story has surprises. The rules of the game give players ways to surprise each other and themselves.**
   No, there are no surprises within this game.
13. (I) **Games can transform players for better/worse. The game is worth playing/the time.**
   It can if users have previous knowledge of Mathematical learning/know their calculation.
14. (I) **Risks; choice and chance; punishment; competition; reward is built up. A way to turn losing state to a winning state.**
   No, there are not any, but reward is built up through the element of scores and levels.

(Continued)

Table 6.2 (Continued)

	ESGAE: Evaluation heuristics and its description	
	15. (I)	**Constant feedback, clear progress display (scores and levels), and response time.** Yes, it has clear progress and constant feedback, scores, and levels. Like "beginner," "master," and "IQ>120," they are still basic as it has four Mathematics operations (plus, minus, multiply, and divide).
	16. (I)	**Players are able to socialize/express themselves. Some/all are playing with friends.** No, this is a single player game. Social element is missing.
	17. (I)	**Simplicity/complexity is/should be balanced.** The game is simple but the combination of simplicity and complexity as heuristics and its balance are missing, and there is no such thinking.
	18. (I)	**They are puzzles/quizzes/scenarios/videos in the game.** No, this is a simple game that has questions and players have to figure out the answers.
	19. (I)	**Tutorials** and **documentations are clear and comprehensible.** Yes, it is very clear.
	20. (I)	**Online Equipment Payments—Upgrading system.** No, this is not available.
	21. (I)	**Clear outcomes of the game.** Yes, there is on the condition that the player has some knowledge of Mathematical and Statistical learning.
II. **Educational simulation game design and developments (game mechanics, aesthetic and technology)**	1. (II)	**Animations and sound effects work in harmony.** Yes, it does but limited.
	2. (II)	**Rewards and feedback increase players' self-efficacy, motivation, and engagement.** Yes, it can as it shows the player how many asteroids, they have killed but it is limited.
	3. (II)	**Game instructional, visual, and technical design: venue, objects, attributes, dimensions (3D, 2D), novelty, uniqueness;** **imagination, strategy, constraints/limitations, effects are distinctive.** The game instructional, visual, and technical design is unique, simple but limited as the player cannot do much.

	4. (II)	**After detecting high/low self-efficacy in Mathematical and Statistical learning, provide customized learning**. This is missing and unavailable.
	5. (II)	**Storyline is easy to understand, clear with engaging outcomes/vision**. Story has one purpose which is to kill asteroids using Mathematical knowledge and speed as a weapon.
	6. (II)	**Game design with its different elements is in harmony. The game time is short/slow/fast**. Yes, it does but it is a basic game.
	7. (II)	**Physical input and output resources are appealing from aesthetics, mode of change, size, font, and colors**. Yes, it does.
	8. (II)	**User interface, controllers; feedback/rewards and spending money, currency/points all information are obvious and communicated to the user without interfering with the interface**. The currency/points are the users' speed and knowledge of how to calculate arithmetic operations like addition, subtraction, multiplication, and division.
	9. (II)	**Technology and programs are harmoniously integrated**. Yes, they are but limited and simple.
	10. (II)	**Online payments—Upgrading game levels**. No, this is not available.
III. **Educational simulation game implementation and evaluation (game usability and HCI principles)**	1. (III)	**Training facilities and outcomes are distinctive**. No, it is rather simple.
	2. (III)	**Curriculum and methods of delivery are clear**. Yes, using Mathematical knowledge and speed as a weapon, players can shoot asteroids and save the planet earth.
	3. (III)	**Progress, feedback, scores, and levels are explicable, challenging, and balanced according to levels**. Progress, feedback, scores, and levels are present but not challenging. There are no hard/advanced challenges because the game is simple and so this is missing.
	4. (III)	**After detecting high/low self-efficacy in Mathematical and Statistical learning, provide customized learning**. No, this is missing.

(*Continued*)

Table 6.2 (Continued)

ESGAE: Evaluation heuristics and its description
5. (III) **Points/rewards are built up**. Yes, points and rewards are built up. It is displayed clearly on the interface/screen.
6. (III) **Storyline is suitable for the level of complexity**. No, the story is rather simple. Keep the planet earth safe by killing/shooting down the asteroids.
7. (III) **Storyline is logical with engaging outcomes/vision**. In this game there is one vision and one story that is to keep planet earth safe from asteroids.
8. (III) **Storyline is novel and well-designed**. No, it is simple but limited.
9. (III) **Intuitive accessible user-friendly interface**; **consistency throughout the game interface, menus, design, sounds, and animations**. Yes, it is very much user-friendly game.
10. (III) **Controllers do not confuse the players**. No presence of controllers. Player can use PC keyboard or mobiles touch screen.
11. (III) **Useful error handling message**, **promoting users to undo their action**. No undo buttons.
12. (III) **Tutorials** and **documentations are easy to read and understand**. Yes, they are but limited.
13. (III) **Friendly Online Customer Support**. Yes, there is but it is simple.

Table 6.3 Application of the ESGAE: SimCity game (Slyman, 2022:124)

	ESGAE: Evaluation heuristics and its description
SIMCITYas a game I. **Educational Simulation Game Analysis** **(Gameplay** **and** **game story)**	1. (I) **Clear objectives or structure of the game**. "SimCity," as a game, has clear and easy objectives and structures. It starts with an empty map (see Chapter 5). The player is faced with a bird's eye view of a map, like someone high in the sky. S/he must think strategically how to build a city and manage the city's economy, growth, and expansion, building roads, landmarks, and parks. It is about building and constructing the city by listening to residents' needs and building the city infrastructure and what a typical city requires. 2. (I) **Players do not require any preliminary skills to commence. Novices and experts can understand the rules**. It is very much intuitive. But they must manage resources within limited constraints and wait for these resources, like seeds or metals, to be ready for collection before distributing them or using them for expansion and growth. 3. (I) **Game styles/genres are distinctive: strategy/adventure, etc**. The game is distinctive in that players construct/fill a map at the beginning of the game. This will have impacts on the game and on the residents. For instance, if a player builds factories near residents, they may then complain about pollution, and their dissatisfaction will affect taxes. The motto of the city is, "the happier the residents, the more taxes they can pay."

(*Continued*)

Table 6.3 (Continued)

ESGAE: Evaluation heuristics and its description
4. (I) **Expectations and needs (behavioral outcomes: cognitive, emotional (attitudes), and psychomotor (skills)) are achieved/met leading to high/low self-efficacy and therefore fulfilment/engagement (i.e., new learning outcomes/transferable skills).** By achieving points/money and making profits: selling and buying resources; expanding and earning more money; meeting residents' needs and providing services like water, electricity, sewage, waste management, fire station, police, and government buildings; and drawing roads. "SimCity" teaches the player management, leadership, and decision-making skills. For example, sometimes there are constraints within the game as players cannot go up level without selling certain amenities, but the player has to think carefully whether they can sell/buy/expand as their actions can cause problems and cost them money. For example, if they do not have sufficient water/electricity supplies, residents will start complaining or abandon a building if there is no fire station. This also means no funding as residents will not pay taxes. Making residents happy by listening and catering for their needs means more taxes more production, expansion, and growth. The game gives the user a sense of mission and ownership, a hero in a scenario. The game can raise high self-efficacy when achievements and progress are made, and therefore can improve players' preconceptions toward Mathematics/Applied Statistics if this was embedded in the game, but this is missing in the current version. How does GBL take place? GBL takes place in this context using previous knowledge/experience, intuition, discovery, observation reasoning, understanding, and decision-making.
5. (I) **Detect high/low self-efficacy in Mathematical and Statistical learning, picking up signals and customizing learning**. No, this is missing and not available.

6. (I)	**Levels are obtainable/easy/hard**. As players go up, the level of complexity increases and some of the locked items, depending on each level, become available to achieve once certain conditions are met like having two golden keys and extra money to purchase a university library building, for instance. There are various and diverse resources just like in real life. Some levels become available as the player proceeds in the game. For instance, to build housing requires items such as metals, woods, plastics, etc.
7. (I)	**The elements of curiosity and manipulation are integrated in the game**. Yes, some of the game functionalities do not operate until players buy certain "Simoleons"/upgrade at each level.
8. (I)	**Players are engaged in the game. The game is valuable, and players care about it**. "SimCity" is a single player game, but it can be played by multiplayers. However, the only player managing/controlling the city is the mayor taking charge of the city infrastructure, building roads, houses, factories, and all cities infrastructure.
9. (I)	**Game players are asked to solve challenging hidden problems/conflicts as a part of the gameplay. The game generates more problems to keep players engaged and coming**. The game increases in difficulty as players go up and it is hard to obtain levels or facilities, but there are no hidden problems/conflicts which makes the game limited in a way as the player requires. For example, to buy "SimCash" using the online store purchasing system to advance in the game or stay and wait for the resources to become available before selling them and expanding.
10. (I)	**The driving instincts of the game and how the story is built are exciting/positive/boring. The suspense; event; story; characters; details; and dramatic changes are balanced/appealing**. There is no suspense, but the whole story of the game is based on building city infrastructure, management, and trading and increasing wealth through taxes and through trading from neighborhood cities.

(Continued)

Table 6.3 (Continued)

	ESGAE: Evaluation heuristics and its description
11. (I)	**The game is fun and enjoyable. Some parts are/need to be fun. The element of play is presented in the game and it is engaging and immersive.** No, not really but it can be addictive as players keep earning points which can be used to buy more items to help with building the city and meeting residents' needs.
12. (I)	**The game surprises the players when they play. The game story has surprises. The rules of the game give players ways to surprise each other and themselves.** No, this is missing although it is apparent in different ways as the player is rewarded when they watch an advertisement/a video. They are faced with three closed boxes wrapped up like a present which player can choose. The box can obtain one of the items the city requires which ranges, for instance, from woods, extra golden points, "SimCash," golden keys, a shovel or simply a metal. The most precious item is the "golden key" or "SimCash," as they allow the user to upgrade or buy more items the city requires. The "surprise" element is used as an incentive to watch the advert and earn some extra points, such as Golden keys, SimCash, or any item. However, most of the time the player is faced with ordinary items—when they choose the box, like seeds or a camera, which can be annoying and not helpful.
13. (I)	**Games can transform players for better/worse. The game is worth playing/the time.** The game can be worth playing depending on the players and whether they like to construct buildings, planning, management, and leadership within the game and care about city and its people. The game gives a sense of responsibility as a leader/mayor in charge of the entire city and whether to trade/not trade or save up money for another service. From a business/computing perspective, it is worth playing. It teaches various skills like problem-solving, decision-making, strategic thinking, and management (i.e., managing resources and trading, buying, and selling, etc.).

14. (I)	**Risks; choice and chance; punishments; competition; reward is built up. A way to turn losing state to a winning state (for instance, earning extra points by watching online videos/advertisement).** Reward is built up as the more extra points players earn the more items, they can buy to build their city. This is what is called "City achievements" and there are various certificates and awards within the city. For instance, the first one is "Tiny Certificate," "Suburb City," "Importer Population Boost," "Clean and Green City," "Wealth City," "Parks and Rec City," "Big City," "Lumberjack"(produce more woods), "Worker (produce metal),"and "Growing city."
15. (I)	**Constant feedback, clear progress display (scores and levels) and response time.** Yes, there is constant feedback and progress is clearly displayed throughout the game.
16. (I)	**Players are able to socialize/express themselves. Some/all are playing with friends.** No, the element of human interaction is missing in this early version. The only way to socialize is through trading with neighborhood cities.
17. (I)	**Simplicity/complexity is/should be balanced.** Yes, the game can be simple, but as players move up the levels some items become unlocked enabling the player to build and trade more, and thus make more profits. However, unless they have "SimCash," it is hard to obtain the resources for expansion and growth. The user is then faced with a choice either wait for a long time until all production is ready for selling and so upgrading to higher levels is a slow process or buy more "SimCash" through the App store using a credit card. The element of strategy is clearly employed throughout the game. However, most of the time the game forces the player to buy resources and limits the completion of the tasks unless certain items are bought, which can be annoying. A limitation within the game is that the system does not realize that the user does not like to pay for items using their credit card, that is, understanding human behavior. Other times, the resources are required and the system shows notification regarding this, for instance "glues," but what the system does not realize is that the player does not have sufficient "SimCash" to buy and does not like to use their credit card to purchase "SimCash."

(*Continued*)

Table 6.3 (Continued)

	ESGAE: Evaluation heuristics and its description
18. (I)	**There are puzzles/quizzes/videos in the game**. No, this element is missing. "SimCity" is a business/computing-focused game although it is not clearly stated so as all transactions involve buying, selling, and trading and expanding, as aforesaid.
19. (I)	**Tutorials and documentations are clear and comprehensible**. Yes, tutorials, tips, and support are available, and the players are directed to the game main website if they require more support.
20. (I)	**Online Equipment Payments—Upgrading system**. Yes, it asks the player to purchase "Simoleons" which is significant for upgrading and buying more resources.
21. (I)	**Clear outcomes of the game**. Yes, for this type of genre, as residents increase, the city with its building grows too and expands. **What is missing/limitation within this game is that** Excel, spreadsheet calculations, database, Applied Statistics lecture, real-life scenarios about companies' shares, complex problems, public services issues like housing problems, business data analysis, business reputation, famous companies, expensive/cheap low deal companies "Lidl," risk assessment, site assessment are all missing. This game had many errors and issues when it initially came out, but these were fixed and many genres have come out since. There is "SimCity" with people, castles, and festivals, etc. This game can be improved by, for instance, integrating business analysis, human element, story, adventure: VR castles, banks a real-life city, enemies, beggars, zoom in options; project-based, business plan; sketch up software to draw building and services. Another limitation is that it does not recognize that the player does not like to purchase items using their credit cards. The system thus does not understand human thinking/behavior/emotions.

II. Educational simulation game design and developments (Mechanics,aesthetic,technology)	1. (II) **Animations and sound effects work in harmony**. Yes, very much so. Yes, it is effective. 2. (II) **Rewards and feedback increase players' self-efficacy, motivation, and engagement**. Yes, it can increase self-efficacy, motivation, and engagement, and be addictive. However, when the player has no "SimCash" to spend or watching the advert cannot provide them with "SimCash" to continue, it can deter players from playing. In addition, players have to wait until the system notifies them to watch the advert (which can take a day or two) so that they can obtain "SimCash" to continue playing the game. 3. (II) **Game instructional, visual, and technical design: venue, objects, attributes, dimensions (3D, 2D), novelty, uniqueness; imagination, strategy, constraints/limitation, effects are distinctive**. Yes, the genre of the game is suited to its style; it is very much about building and constructing a city. 4. (II) **After detecting high/low self-efficacy in Mathematical and Statistical learning, provide customized learning**. No, this is missing and unavailable. 5. (II) **Storyline is easy to understand, clear with engaging outcomes/vision**. Yes, the player is the mayor and so responsible for building and managing facilities, buildings, and departments to cater for citizens' needs. 6. (II) **Game design with its different elements is in harmony. The game time is short/slow/fast**. Yes, the game design is amazing. 7. (II) **Physical input and output resources are appealing from aesthetics, mode of change, size, font, and colors**. Yes, but there are the same and so predictable and can be iterative at a time. However, the aesthetics keep changing depending on religious festivals for instance Halloween or New Year.

(Continued)

Table 6.3 (Continued)

		ESGAE: Evaluation heuristics and its description
	8. (II)	**User interface, controllers; feedback/rewards and spending money, currency/points all information are obvious and communicated to the user without interfering with the interface**. Yes, very much so but notifications are represented in the form of "bubbles" and can interfere with the gameplay. For instance, if the player does not have any "SimCash," the system notifies him/her that one of the items has run out and is required to be produced. When the player attempts to obtain it, there is insufficient "SimCash," but the system does not recognize that the player does not want to use their credit cards and buy "SimCash." Instead the system keeps notifying the user repeatedly, which can be a weakness and annoying. In this case the system is not advanced and does not read/understand human behavior/thinking/emotions.
	9. (II)	**Technology and programs are harmoniously integrated**. Yes, highly integrated.
	10. (II)	**Online payments—Upgrading game levels**. Not easy as the game is online and to access it, internet access is required. In addition, many items are sold online and sometimes it seems that the game is limiting and forcing the player to buy items in order to unlock next level or items.
III. Educational simulation game implementation and evaluationGame usability,and HCI principles	1. (III)	**Training facilities, outcomes, curriculum, and methods of delivery are clear**. Within "SimCity," there are no training facilities available, but the outcomes, curriculum and methods of presenting are clear in the game. There are many cognitive processes/learning involved within the game like problem-solving, decision-making, reasoning, and strategic thinking as in trading, selling, and buying, etc.
	2. (III)	**Curriculum and methods of delivery are clear**. There is a hidden curriculum that is aimed at teaching users the importance of buying and selling, savings cash, trading, developing strategies and techniques as a mayor.
	3. (III)	**Detect high/low self-efficacy in Mathematical and Statistical learning, provide customized learning**. No, this is not available.
	4. (III)	**Progress, feedback, scores, and levels are explicable, challenging, and balanced according to levels**. Yes, there is. Progress, feedback, scores, and levels are clearly displayed on the screen and as the level increases, it becomes hard to move up levels.

5. (III) **Points/rewards are built up**.
A way to turn a trailing state to a winning state (for instance, earning extra points/power-up by watching online videos/advertisement). Within this game business trading is crucial to the running of the city.
6. (III) **Storyline is suitable for the level of complexity**.
Yes, it is but there are no new storylines to be explored. The whole theme of the game is trading, managing, and constructing more building, earning extra points/cash, and making more profits.
7. (III) **Storyline is logical with engaging outcomes/vision**.
Yes, it can be engaging for those who like building, managing, and leading.
8. (III) **Storyline is novel and well-designed**.
Storyline is not novel. It is outdated but many things are added to the new versions, but it is well-designed and business-like environment is the running theme in the game, just like any mayor of a city.
9. (III) **Intuitive accessible user-friendly interface; consistency throughout the game interface, menus, design, sounds, and animations**.
Yes, they are. Consistency is present throughout the game.
10. (III) **Controllers do not confuse the players**.
Yes, very much so.
11. (III) **Useful error handling message, promoting users to undo their action**.
Not really evident from the beginning but there is a "bulldoze" button that can delete roads. In addition, players can move buildings/amenities to different parts of the city if they wish. Buildings can also be moved to another location, but unlike real-life transactions cannot be undone. Also, if the user made a mistake in terms of transactions, they cannot undo it, which can be a limitation as this is only a simulation and the user ought to learn and be entertained.
12. (III) **Tutorials and documentations are easy to read and understand**.
Yes, there are tutorials within "SimCity" to assist the users, and the player is directed to the game's main website with more information and details.
13. (III) **Friendly Online Customer Support**.
Yes, it has and there is a link to the main website if the player requires more information.

Numbas and SimCity

In conclusion, this chapter has taken a pioneering step forward by proposing a game evaluation heuristics model (ESGAE). Addressing a gap in existing theoretical frameworks, this model harmoniously blends elements from both entertainment and educational perspectives. By striking a delicate balance, it offers a comprehensive and nuanced framework for analyzing and evaluating the educational simulation games (ESGs) that lie at the heart of transformative learning experiences.

The aim of the ESGAE is to generate a comprehensive model for analyzing and evaluating ESGs. Furthermore, this model will be able to detect low self-efficacy and customize learning.

Findings show the significance of gameplay, game story and storylines, game design and development (mechanics, aesthetics and technology), and usability and HCI principles in increasing self-efficacy, sustaining engagement and motivation, leading players to feel confident to return to play the game each time. Game characteristics may include clear objectives, self-intuitive game contents, character developments, and the novelty of the story. It may also include creativity, dramatic changes, suspense, events, and hidden conflicts, as well as levels, achievements, reward systems, graphics, animations, and sounds. These are essential characteristics of the gaming experience, and they can have significant impacts on players' experience and confidence.

As I explained above, I chose "Numbas" and "SimCity" in order to obtain a detailed understanding of game design and development systems, and thus create a great game experience. Having analyzed and evaluated SimCity & Numbas games from the ESGAE perspectives, both games have similarities and differences. For example, SimCity has clear objectives and structures and it is very much self-intuitive. Players do not require any preliminary skills to commence. Novices/experts can understand the rules. Whereas Numbas has clear simple objectives but it requires players to have previous knowledge with Mathematical learning, which is the condition of this game. Numbas teaches players calculations and problem-solving skills while SimCity teaches players management, leadership, problem-solving, and decision-making skills. Therefore, GBL in both games are different. GBL in SimCity takes place in this context using intuition, discovery, observation, reasoning, understanding, and decision-making while with Numbas, it requires players to have previous knowledge of Mathematical skills (memorization) and speed: a set of instructions as in quiz type of learning. Both games cannot detect high/low self-efficacy in Mathematical/Statistical learning, picking up signals and customizing learning. This is because both games are not designed to do so in their current version.

When it comes to gameplay, story and storylines, game design and development, game usability and HCI principles (ESGAE Model), SimCity is distinctive and has many hidden problems, events, and tasks. SimCity's story and storylines are about building and managing the city like a mayor. Its game design & development is sophisticated, consistent and user friendly. SimCity can increase players' confidence to return to play the game because of its game levels, achievements, and reward systems. Users' expectations and needs (behavioral outcomes: cognitive, emotional (attitudes), and psychomotor (skills)), for this type of games, can be achieved/met, and lead to high self-efficacy and therefore fulfilment/engagement (i.e., new learning outcomes/transferable skills). Whereas Numbas is distinctive, and it can meet the learning outcomes: testing knowledge, but it has a condition, users must have previous knowledge of Mathematics. Numbas can be therefore limited. Its gameplay, story and storylines are sophisticated. Numbas has also achievements and reward systems but it does not contain any hidden conflicts, events, and challenging tasks. In terms of game design and development, and game usability and HCI principles, Numbas is sophisticated and has a consistent, accessible, and user-friendly interface just like SimCity. However, users' expectations and needs (behavioral outcomes: cognitive, emotional (attitudes), and psychomotor (skills)) of this type of games (Numbas) can be achieved/met on the condition that they know Mathematics. Numbas, in this aspect, is limited. It can therefore decrease users' confidence/self-efficacy because fulfilment/engagement (i.e., new learning outcomes/transferable skills) will not be met.

The next **chapter**, **Chapter 7**, will explore the student experience through a new theoretical lens in mathematics and game-based learning.

Bibliography

Acumen (2017) In: Cambridge University Press. [Online] Available at: http://dictionary.cambridge.org/dictionary/english/acumen (Accessed 30/05/2017).

Adams, A., Lunt, P. and Cairns, P. (2008) A Qualitative Approach to HCI Research. In: Cairns, P. and Cox, A. (ed.). *Research Methods for Human-Computer Interaction*. Cambridge, UK: Cambridge University Press, pp. 138–157. Available at: http://www.cambridge.org/catalogue/catalogue.asp?isbn=9780521870122andss=toc (Accessed 20/11/2016).

ADDIE Model (2016) Available at: http://www.learning-theories.com/addie-model.html (Accessed 20/04/2016).

Anderson, S. (2020) Quantitative Analysis. Available at: https://www.investopedia.com/terms/q/quantitativeanalysis.asp (Accessed 20/02/2021).

Anon (2018) Efficacy. In: Cambridge University Press. [Online] Available at: https://dictionary.cambridge.org/dictionary/english/efficacy (Accessed 30/09/2018).

Applefield, J.M., Huber, R. and Moallem, M. (2000) Constructivism in Theory and Practice: Toward a Better Understanding. *The High School Journal*, 84(2), pp. 35–53.

Arnab, S. and Clarke, S., 2017. Towards a Trans-disciplinary Methodology for a Game-based Intervention Development Process. *British Journal of Educational Technology*, 48(2), pp. 279–312.

Arthur, J., Waring, M. Coe, R. and Hedges, L.V. (2012) *Research Methods and Methodologies in Education*. California: Sage Publications Inc.

Bandura, A. (1986) *Social Foundations of Thought and Action: A Social Cognitive Theory*. Englewood Cliffs, NJ: Prentice-Hall.

Bandura, A. (1994) Self-efficacy. In Ramachaudra, V.S. (ed.). *Encyclopedia of Human Behaviour*, 4, pp. 71–81.

Bandura, A. (1997) *Self-efficacy: The Exercise of Control*. New York: W.H. Freeman and Company.

Barbosa, T., Lopes, S.P. Leao. C. Soares, F. and Carvalho, V. (2019) Serious Game for Teaching Statistics in Higher Education: Storyboard Design. In: *7th EAI International Conference, Braga, Portugal, on 'Interactivity, Game Creation, Design, Learning, and Innovation'* (ed.) Springer, pp. 169–175. DOI: 10.1007/978-3-030-06134-0_18

Bartholomew, H., Darragh, L., Ell, F. and Saunders, J. (2011) "I am Natural, and I Do It for a Love!": Exploring Students' Accounts of Studying Mathematics. *International Journal of Mathematical Education in Science and Technology*, 42(7), pp. 915–924.

Bassey, M. (1999) *Case Study Research in Educational Settings*. Buckingham: Open University Press.

Bellotti, F., Kapralos, B. Lee, K. Moreno-Ger, P. and Berta, R. (2013) Assessment in and of Serious Games: An overview. *Advances in Human-Computer Interaction*, 2013, p. 1.

Belluigi, D.Z. (2009) Exploring the Discourses Around 'Creativity' and 'Critical Thinking' in a South African Creative Arts Curriculum. *Studies in Higher Education*, 34(6), pp. 699–717.

Bennedesen, J. and Caspersen, M.E. (2004) Programming in Context: A Model-First Approach to CS1. In *Proceedings of the 35th SIGCSE Technical Symposium on the Computer Science Education*, pp. 477–481. ACM Press.

Bevan, N. (2001) International Standards for HCl and Usability. *International Journal of Human Computer Studies*, 55(4), pp. 533–552.

Boaler, J. (2009) *The Elephant in the Classroom*. London: Souvenir Press.

British Education Research Association (BERA) (2011) BERA Ethical Guidelines for Educational Research. Available at: http://www.bera.ac.uk/ (Accessed 15/10/2016).

Brown, J.S., Collins, A. and Duguid, P. (1989) Situated Cognition and the Culture of Learning. *Educational Researcher*, 18(1), 32–42.

Bruner, J. (1973) *Going Beyond the Information Given*. New York: Norton.

Bryant, A. (2017) *Grounded Theory and Grounded Theorizing: Pragmatism in Research Practice*. New York, NY: Oxford University Press.

Bryman, A. (2008) *Social Research Methods*, 3rd Edition. Oxford: Oxford University Press (first published: 2001).

Bryman, A. (2012) *Social Research Methods*. 4th Edition. United States: Oxford University Press.

Bryman, A. (2016) *Social Research Methods*. 5th Edition. United States: Oxford University Press.

Caillois, R. (1961) *Man, Play, and Games*. New York: Free Press.

Cardinot, A. and Fairfield, J.A. (2019) Game-Based Learning to Engage Students with Physics and Astronomy Using a Board Game. *International Journal of Game-Based Learning (IJGBL)*, 9(1), pp. 42–57.

Charmaz, K. (2006) *Constructing Grounded Theory*, 1st Edition. London: Sage Publications Ltd.

Charmaz, K. (2014) *Constructing Grounded Theory*, 2nd Edition. London: Sage Publications Ltd.

Chen, M.P., Wong, Y.T. and Wang, L.C. (2014) Effects of Type of Exploratory Strategy and Prior Knowledge on Middle School Students' Learning of Chemical Formulas from a 3D Roleplaying Game. *Educational Technology Research and Development*, 62, pp. 163–185. DOI: 10.1007/s11423-013-9324-3

Clark, D.B., Tanner-Smith, E.E. and Killingsworth, S.S. (2015) Digital Games, Design, and Learning a Systematic Review and Meta-Analysis. *Review of Educational Research*, 86(1), pp. 79–122.

Claxton, G. (2002) *Building Learning Power*, 1st Edition. Bristol: TLO Limited.

Cohen, L., Manion, L. and Morrison, K. (2011) *Research Methods in Education*. London: Routledge.

Conlon, C., Carney, G., Timonen, V. and Scharf, T. (2015) Emergent Reconstruction in Grounded Theory: Learning from Team-Based Interview Research. *Qualitative Research*, 15, pp. 39–56. DOI: 10.1177/1468794113495038

Cooper, P.A. (1993) Paradigm Shifts in Designed Instruction: From Behaviorism to Cognitivism to Constructivism. *Educational Technology*, 33(5), pp. 12–19.

Corbin, J. and Strauss, A. (2015) *Basics of Qualitative Research: Techniques and Procedures for Developing Grounded Theory*, 4th Edition. Thousand Oaks, CA: Sage.

Cordova, D.I. and Lepper, M.R. (1996) Intrinsic Motivation and the Process of Learning: Beneficial Effects of Contextualization, Personalization, and Choice. *Journal of Educational Psychology*, 88, pp. 715–730.

Coulthard, G.J. (2009) A Review of the Educational Use and Learning Effectiveness of Simulations and Games [online]. PhD Thesis, Purdue University. Available at: http://www.coulthard.com/library/paper-simulation.html (Accessed 15/12/2015).

Craft, A. (1999) Creative Development in the Early Years: Some Implications of Policy for Practice. *The Curriculum Journal*, 10(1), pp. 135–150.

Craft, A. (2002) *Creativity and Early Years Education*. London: Continuum.

Crawford, C. (1984) *The Art of Computer Game Design*. Berkeley: California, McGraw-Hill/Osborne Media.

Creswell, I.W. (2009) *Research Design: Qualitative, Quantitative, and Mixed Methods Approaches*, 3rd Edition. London: Sage.

Crockett, L. (2018) Bogost's War on Gamification: Irony and Play on Grandma's Sofa. In: *12th European Conference on Games Based Learning (ECGBL 2018)*. Reading: Academic Conferences and Publishing International Ltd.

Cronan, T.P. and Douglas, D.E. (2012) A Student ERP Simulation Game: A Longitudinal Study. *Journal of Computer Information Systems*, 53(1), pp. 3–13.

Csikszentmihalyi, M. (1990) Psychology Today: Finding Flow. Available at: https://www.researchgate.net/publication/200026151_Finding_Flow_The_Psychology_of_Engagement_With_Everyday_Life/link/55ad6c2f08aed9b7dcdae222/download (Accessed 03/05/2016).

Csikszentmihalyi, M. (1997) *Creativity, Flow and the Psychology of Discovery and Invention*. New York: Harper.

Dann, W.P., Cooper, S. and Pausch, R. (2008) *Learning to Program with Alice, Brief Edition*. USA: Prentice-Hall, Inc.

Data Protection Act (1998). Available at: http://www.legislation.gov.uk/ukpga/1998/29/pdfs/ukpga_19980029_en.pdf (Accessed 10/12/2016).

Data Protection Act (2018). Available at: http://www.legislation.gov.uk/ukpga/2018/12/pdfs/ukpga_20180012_en.pdf (Accessed 10/09/2019).

Davis, F.D. (1989) Perceived Usefulness, Perceived Ease of Use, and User Acceptance of Information Technology. *MIS Quarterly*, 13(3), pp. 319–340.

De Bono, E. (1992) *Teach Your Child How to Think*. England: McQuaig Group Inc.

Dede, C. (2005) Planning for Neomillennial Learning Styles. *Educause Quarterly*, 28(1), pp. 7–12.

Derryberry, A. (2007) Serious Games: Online Games for Learning. White Paper. [Online]. Available at: http://www.adobe.com/resources/elearning/pdfs/serious_games_wp.pdf (Accessed 10/12/2015).

Desurive, H., Caplan, M. and Toth, J. (2004) Using Heuristics to Evaluate the Playability of Games. *CHI Conference*, 32(1), pp. 97–110. DOI: 10.1177/104687810103200108

Desurive. H. and Wiberg, C. (2009) Game Usability Heuristics (PLAY) for Evaluation and Designing Better Games: The Next Iteration. In *Online Communities and Social Computing: Third International Conference, OCSC 2009, Held as Part of HCI International 2009*, San Diego, CA, USA, July 19–24, 2009. Proceedings 3, pp. 557–566. Springer Berlin Heidelberg.

Desurive, H. and Wiberg, C. (2010) User Experience Design for Inexperienced Gamers: GAP – Game Approachability Principles. In: Regina Bernhaupt (ed.). *Evaluating User Experience in Games*. Springer, pp. 131–148. DOI: 10.1007/978-1-84882-963-3_1

Dondlinger, J.M. (2007) Educational Video Game Design: A Review of the Literature. *Journal of Applied Educational Technology*, 4(1), Spring/Summer, pp. 21–31. Available at: http://reserachgate.net/publication/238444705 (Accessed 12/12/2015).

Draper, S. (2013) Social Constructivism. Available at: http://www.psy.gla.ac.uk/~steve/courses/archive/CERE12-13-safari-archive/topic3/webarchive-index.html#social%20con%201 (Accessed 23/12/2019).

Dunne, C. (2011) The Place of The Literature Review in Grounded Theory Research. *International Journal of Social Research Methodology*, 14, pp. 111–124. DOI: 10.1080/13645579.2010.494930

Dweck, C.S. (2006) *Mindset: The New Psychology of Success*. United States: Ballantine Books Trade Paperback Edition.

Dwyer, J. and Slyman, S. (2016) *An Introduction to Quantitative Methods for Business Research*. London: Algana Publishing. 200 pages; ISBN−13: 9781907934094.

Early, R.E. (1992) The Alchemy of Mathematical Experience: A Psychoanalysis of Student Writings. *For the Learning of Statistics*, 12(1), pp. 15–20.

Ejsing-Duun, S. and Karoff, H.S. (2016) Creativity and Playfulness: Producing Games as a Pedagogical Strategy Case studies. In: Munkvold, R. (ed.). *Games Based learning for Researchers, Teachers and Students*, Vol. 1. Reading: Academic Conferences and Publishing International Ltd. ISBN: 978-1-911278-12-890000.

Fairweather, E. and Cramond, B. (2010) Infusing Creative and Critical Thinking into the Curriculum Together. In: Beghetto, A. and Kaufman, J.C. (ed.). *Nurturing Creativity in the Classroom*. United States: Cambridge University Press.

Falstein, N. and Barwood, H. (2001) The 400 Project. Available at: http://theinspiracy.com/400_project.htm. (Accessed 20/04/2016).

Faria, A.J. (2001) The Changing Nature of Business Simulation/Gaming Research. *Simulation and Gaming*, 32(1), pp. 97–110. DOI: 10.1177/104687810103200108

Faria, A.J., Hutchinson, D. Wellington, W.J. and Gold, S. (2009) Developments in Business Gaming: A Review of the Past 40 Years. *Simulation and Gaming*, 40(4), pp. 464–487. DOI: 10.1177/1046878108327585

Faria, A.J. and Wellington, W.J. (2004) 'A Survey of Simulation Game Users, Former-Users, and Never-Users. *Simulation and Gaming*, 35(2), pp. 178–207. DOI: 10.1177/1046878104263543

Federoff, M.A. (2002) Heuristics and Usability Guidelines for the Creation and Evaluation of Fun in Video Games. Master Thesis, Indiana University, Bloomington.

Fienberg, S.E. (1992) A Brief History of Statistics in Three and One-Half Chapters: A Review Essay. Available at: https://www.jstor.org/stable/2246307?casa_token=Vek3Rmd5Mk0AAAAA%3AJ0zRZC5AxskwvTNY7GlrgsRV6oLrE20DY3FdeaPad0JvNSNrZDav8uZpX-YqY1hXTDtaz6zM87S4vwkQdflDLSzmrLCNlEsaBqWnykepc0jlYXsQXA

Forest, E. (2014) ADDIE Model. Available at: http://educationaltechnology.net/the-addie-model-instructional-design/ (Accessed 20/04/2016).

Fosnot, T.C. (1989) *Enquiring Teachers, Enquiring Learners: A Constructivist Approach for Teaching*. New York: Teachers College Press.

Fosnot, T.C. and Perry, R.S. (2005) Constructivism: A Psychological Theory of Learning. Available at: http://beyondbitsandatoms.org/readings/fosnot_constructivism_1996.pdf (Accessed 20/07/2019).

Garris, R., Ahlers, R. and Driskell, E.J. (2002) Games, Motivation, and Learning: A Research and Practice Model. *Simulation and Gaming*, 33(4), pp. 441–467.

Gee, J.P. (2003a) *What Video Games Have to Teach Us About Learning and Literacy*. New York: Palgrave Macmillan.

Gee, J.P. (2003b) Good Video Games and Good Learning. Available at: https://academiccolab.org/resources/documents/Good_Learning.pdf (Accessed 03/01/2021).
Gee, J.P. (2004) *Situated Language and Learning: A Critique of Traditional Schooling*. London: Routledge.
Gee, J.P. (2005a) Why are Video Games Good for Learning? Retrieved January 1, 2018, from http://academiccolab.org/resources/documents/MacArthur.pdf (Accessed 20/04/2016).
Gee, J.P. (2005b) *Why Video Games Are Good for Your Soul: Pleasure and Learning*. Melbourne: Common Ground.
Gee, J.P. (2007) *Good Video Games and Good Learning: Collected Essays on Video Games, Learning, and Literacy*. New York: Peter Lang.
Gee, J.P. (2008) *Getting over the Slump: Innovation Strategies to Promote Children's Learning*. New York: The Joan Ganz Cooney Center at Sesame Workshop.
Gee, J.P., Hayes, E., Torres, R.J., Games, I.A., Squire, K. and Salen, K. (2008) Playing to Learn Game Design Skills in A Game Context. In: *Proceeding ICLS'08 Proceedings of the 8th International Conference on International Conference for the Learning Sciences*, Vol. 3, pp. 368–374.
Gee, J.P. and Hayes, E.R., (2011) *Language and Learning in the Digital Age*. Routledge.
Gibbons, J. and Gray, M. (2004) Critical Thinking as Integral to Social Work Practice. *Journal of Teaching in Social Work*, 24, pp. 19–37.
Glaser, B.G. and Strauss, A.L. (1967) *The Discovery of Grounded Theory*. Chicago: Aldine.
Gomes, A. and Mendes, A.J. (2007) Learning to Program-Difficulties and Solutions. In *International Conference on Engineering Education–ICEE*, Vol. 7.
Granic, I., Lobel, A. and Engels, R.C. (2014) The Benefits of Playing Video Games. *American Psychological Association*, 69(1), pp. 66–78. DOI: 10.1037/a0034857
Hackett, G. and Betz, N.E. (1989) An Exploration of the Mathematics Self-Efficacy/Mathematics Performance Correspondence. *Journal for Research in Mathematics Education*, 20(3), pp. 261–273.
Hamari, J., Shernoff, D.J., Rowe, E., Coller, B., Asbell-Clarke, J. and Edwards, T. (2016) Challenging Games Help Students Learn: An Empirical Study on Engagement, Flow and Immersion in Game-Based Learning. *Computers in Human Behavior*, 54, pp. 170–179.
Hamari, J. and Tuunanen, J. (2014) Player Types: A Meta-Synthesis. *Transactions of the Digital Games Research Association*, 1(2), pp. 29–53.
Harel, I. and Papert, S. (1990) *Software Design as a Learning Environment*. Massachusetts Institute of Technology Cambridge, MA: MIT Media Laboratory.
Harel, I. and Papert, S. (1991) *Constructionism*. Norwood, NY: Ablex Publishing Corporation.
Hargreaves, A. (2003) *Teaching in the Knowledge Society: Education in the Age of Insecurity*. New York: Teachers' College Press.
Holbert. N., Penney, L. and Wilensky, U. (2010) *Bringing Constructionism to Action Game-Play, Constructionism*. Paris. Available at: https://www.google.co.uk/search?q=Bringing+Constructionism+to+Action+Game-Playandoq=Bringing+Constructionism+to+Action+Game-Playandaqs=chrome..69i57.7170j0j8andsourceid=chromeandie=UTF-8 (Accessed 20/03/2017).
Hoover, W.A. (1996) The Practice Implications of Constructivism. *SEDL Letter*, 9(3), pp. 1–2.
Horn, B., Hoover, A.K. Barnes, J. Folajimi, Y. Smith, G. and Harteveld, C. (2016) Opening the Black Box of Play: Strategy Analysis of An Educational Game. In *Proceedings of the 2016 Annual Symposium on Computer-Human Interaction in Play*, pp. 142–153.
Hsu, C.-Y., Tsai, C.-C. and Wang, H.-Y. (2012) Facilitating Third Graders' Acquisition of Scientific Concepts Through Digital Game-Based Learning: The Effects of Self-Explanation Principles. *Asia-Pacific Education Researcher*, 21(1), pp. 71–82.

Hunicke, R., LeBlanc M. and Zubek, R. (2004) MDA: A Formal Approach to Game Design and Game Research. Available at: https://www.google.co.uk/search?q=MDA%3A+A+Formal+Approach+to+Game+Design+and+Game+Researchandoq=MDA%3A+A+Formal+Approach+to+Game+Design+and+Game+Researchandaqs=chrome..69i57j69i58.3868j0j8andsourceid=chromeandie=UTF-8 (Accessed 8/10/2016).

Illeris, K. (2007) *How We Learn: Learning and Non-Learning in School and Beyond.* London: Routledge.

ISO (1998) Available at: http://www.standardsglossary.com/iso1.htm (Accessed 8/10/2016).

Jackson, N. and Shaw, M. (2007) Developing Perspectives on Creativity in Higher Education. In: N. Jackson, M. Oliver, M. Shaw, and J. Wisdom (ed.). *Developing Creativity in Higher Education: An imaginative curriculum.* London: Routledge.

Jagger, S., Siala, H. and Sloan, D. (2016) It's All in The Game: A 3D Learning Model for Business Ethics. *Journal of Business Ethics*, 137(2), pp. 383–403.

Jameson, M.M. and Fusco, B.R. (2014) Math Anxiety, Math Self-Concept, and Math Self-Efficacy in Adult Learners Compared to Traditional Undergraduate Students. *Adult Education Quarterly*, 64(4), pp. 306–322.

Jarvis, M. (2005) *'Motivation' in Psychology of Learning and Teaching.* Cheltenham: Nelson Thornes.

Jeffrey, B. and Craft, A. (2004) Teaching Creatively and Teaching for Creativity: Distinctions and Relationships. *Educational Studies*, 30(1), pp. 77–87.

Jenson, J. and Droumeva, M. (2015) Making Games with Game Maker: A Computational Thinking Curriculum Case Study. In *ECGBL 2015–9th European Conference on Games Based Learning: ECGBL2015*. Academic Conferences and Publishing Limited, p. 260.

Jones, M.G. (1998) Creating Engagement in Computer-Based Learning Environments. Available at: http://itech1.coe.uga.edu/itforum/paper30/paper30.htmI (Accessed 12/04/2016).

Kato, P.M., Cole, S. Bradlyn, A.S. and Pollock, B.H. (2008) Video Games Improves Behavioural Outcomes in Adolescents and Young Adults with Cancer: A Randomized Trial. *Pediatrics*, 122(2), pp. e305–e217.

Kim, B. (2001) Social Constructivism. *Emerging Perspectives on Learning, Teaching, and Technology*, 1(1), p. 16.

Kodugamelab (2015) Available at: https://www.kodugamelab.com/ (Accessed 8/11/2018).

Korhonen, H. and Koivisto, E.M.I. (2007) Playability Heuristics for Mobile Multi-Player Games. *Proceedings of the 2nd International Conference on Digital Interactive Media in Entertainment and Arts*, pp. 28–35.

Kraiger, K., Ford, J.K. and Salas, E. (1993) Application of Cognitive, Skill-based, and Affective Theories of Learning Objectives to New Methods of Training Evaluations, *Journal of Applied Psychology*, 78(2), pp. 311–328.

Lenhart, A., Kahne, J., Middaugh, E. Macgill, A.R., Evans, C. and Vitak, J. (2008) Teens, Video Games, and Civics: Teens' Gaming Experiences Are Diverse and Include Significant Social Interaction and Civic Engagement. *Pew Internet & American Life Project*. Available at: https://www.google.co.uk/search?q=Lenhart+et+al.%2C+2008andoq=Lenhart+et+al.%2C+2008andaqs=chrome..69i57j0.1244j0j8andsourceid=chromeandie=UTF-8 (Accessed 20/8/2017).

Loftus, C., Thomas, L. and Zander, C. (2011) Can Graduating Students Design: Revisited. In *Proceedings of the 42nd ACM Technical Symposium on Computer Science Education*, pp. 105–110.

Lyons, I.M. and Beilock, S.L. (2012) When Math Hurts: Math Anxiety Predicts Pain Network Activation in Anticipation of Doing Math. *PLoS One*, 7(10), p. e48076.

Majgaard, G. (2016) Creating Games in the Classroom: From Native Gamers to Reflective Designers. In: Munkvold, R. (ed.). *Games Based Learning for Researchers, Teachers and Students*, Vol. 1. Reading: Academic Conferences and Publishing International Ltd. ISBN:978-1-911278-12-890000.

Malaka, R. (2014) How Computer Games Can Improve Your Health and Fitness. In: Gobel, S., and Wiemeyer, J., (ed.). *Games for Training Education, Health and Sports, 4th Conference on Serious Games, GameDays 2014*, Darmstadt, Germany, April 1–5, 2014, Springer.

Malone, T.W. (1980) What Makes Things Fun to Learn? A Study of Intrinsically Motivating Computer Games. Available at: http://cci.mit.edu/malone/Malone%20 computer%20games%20bibliography.htm

Malone, T.W. and Lepper, M.R. (1987) 'Making Learning Fun: A Taxonomy of Intrinsic Motivations for Learning. Aptitude, Learning, and Instruction': Volume 3: Cognitive and Affective Process Analyses. R. E. Snow and M. J. Farr. Hillsdale, USA, Erlbaum.

Maloney, E.A., Schaeffer, M.W. and Beilock, S.L. (2013) Mathematics Anxiety and Stereotype Threat: Shared Mechanisms, Negative Consequences and Promising Interventions. *Research in Mathematics Education*, 15(2), pp. 115–128.

Marciano, J.N., Miranda, L.C. and Miranda, E.E.C. (2014) Evaluating Multiple Aspects of Educational Computer Games. Available at: https://www.hindawi.com/journals/ijcgt/2014/139205/ (Accessed 27/08/2017).

Marklund, B.B. and Taylor, A.S.A. (2016) Creating Games in the Classroom: From Native Gamers to Reflective Designers. In: Munkvold, R. (ed.). *Games Based learning for Researchers, Teachers and Students*, Vol. 1. Reading: Academic Conferences and Publishing International Ltd. ISBN: 978-1-911278-12-890000.

Mayer, E.R. (2014) *Computer Games for Learning: An Evidence Based Approach*. London: MIT.

Mayer, I., Warmelink, H. and Bekebrede, G. (2013) Learning in a Game-Based Virtual Environment: A Comparative Evaluation in Higher Education. *European Journal of Engineering Education*, 38(1), pp. 85–106.

Mayo, M. (2009) Video Games: A Route to Large-Scale STEM education? *Science*, 323, pp. 79–82.

McConnell, D. (2000) *Implementing Computer Supported Cooperative Learning*, 2nd Edition. London: Kogan Page.

Mccrindle, M. (2018) What Comes after Generation Z? Introducing Generation Alpha. Available at: https://mccrindle.com.au/insights/blogarchive/what-comes-after-generation-z-introducing-generation-alpha/ (Accessed 15/10/2018).

McFarlane, A., Sparrowhawk, A., and Heald, Y. (2002) *Report on the Educational Use of Games: An Exploration by TEEM on the Contribution Which Games Can Make to the Educational Process*. Cambridge, UK: TEEM. http://www.teem.org.uk/publications/teem_gamesined_full.pdf (Accessed 06/12/2015).

McGonigal, J. (2015) *Superbetter: A Revolutionary Approach to Getting Happier Stronger and Happier*. Braver and More Resilient, New York: Penguin Press.

Milane, I. and Rowe, G. (2002) Difficulties in Learning and Teaching Programming-Views of Students and Tutors. *Education and Information Technologies*, 7(1), pp. 55–66.

Monroy-Hernandez, A. and Resnick, M. (2008) Empowering Kids to Create and Share Programmable Media. *Interactions*, 15(2), pp. 50–53. Retrieved August 30, 2015.

Moreno-Ger, P., Burgos, D. Martínez-Ortiz, I. Sierra, J.L. and Fernández-Manjón, B. (2008) Educational Game Design for Online Education. *Computers in Human Behavior*, 24(6), pp. 2530–2540. DOI: 10.1016/j.chb.2008.03.012

Moreno-Ger, P., Torrente, J. Hsieh, Y.G. and Lester, W.T. (2012) Usability Testing for Serious Games: Making Informed Design Decisions with User Data. *Advances in Human-Computer Interaction*, 2012. DOI: 10.1155/2012/369637

Morgan, D.L. (1988) *Focus Groups as Qualitative Research*. Beverly Hills, CA: Sage.
Morrison, K. (2002) *School Leadership and Complexity Theory*. London: Routledge Falmer.
Muijs, D. (2011) *Doing Quantitative Research in Education with SPSS*. London: Sage Publication Ltd.
Mullins, L. (2013) *Management and Organisational Behaviour*, 10th Edition. Harlow: Pearson Education.
Mullins, L. (2019) *Organisational Behaviour in The Workplace*, 12th Edition. Harlow: Pearson Education.
Multon, K.D., Brown, S.D. and Lent, R.W. (1991) Relation of Self-Efficacy Beliefs to Academic Outcomes: A Meta-Analytic Investigation. *Journal of Counselling Psychology*, 18, pp. 30–38.
Naik, N. (2017) The Use of GBL to Teach Mathematics in Higher Education. *Innovations in Education and Teaching International*, 54(3), pp. 238–246. DOI: 10.1080/14703297.2015.1108857
Nicholls, D., De Charon, L. and Hutkin, R. (2010) Development of Critical Thinking and Creativity: Practical Guidelines for the Postsecondary Classroom. *ATEA Journal*, 38(1), p. 12.
Nielsen, J., 1994, April. Usability Inspection Methods. In *Conference Companion on Human Factors in Computing Systems*, pp. 413–414. Available at: https://dl.acm.org/doi/pdf/10.1145/259963.260531
Nielsen, J. (1995) *Usability Inspection Methods, CHI'95 Mosaic of Creativity*, pp. 7–11.
Nielsen, J. (2000) *Designing Web Usability*. Indianapolis, USA: New Riders Publishing.
Nygård, S., Kolås, L. and Sigurdardottir, H. (2015) Teachers' Experiences Using KODU as a Teaching Tool. In: *European Conference on Games Based Learning*, Academic Conferences International Limited. p. 416.
Otkay, J.S. (2012) *Grounded Theory*. Oxford, England: Oxford University Press.
Paavilainen, J. (2010) Critical Review on Video Game Evaluation Heuristics: Social Games Perspective. *Proceedings of the International Academic Conference on the Future of Game Design and Technology*, pp. 56–65. DOI: 10.1145/1920778.1920787
Pachler, N. (2007) *Mobile Learning: Towards a Research Agenda*. London: WLE Centre, IoE. In: Kress, G. and Pachler, N (ed.) (2007) Thinking about the 'm' in m-learning.
Pajares, F. and Kranzler, J. (1995) Self-Efficacy Beliefs and General Mental Ability in Mathematical Problem-Solving. *Contemporary Educational Psychology*, 20, pp. 426–443.
Palincsar, A.S. (1998) Social Constructivist Perspectives on Teaching and Learning. *Annual Review of Psychology*, 49, pp. 345–375.
Papert, S. (1993) *Mindstorms: Children, Computers and Powerful Ideas*. New York: Basic Books.
Paras, B. and Bizzocchi, J. (2005) Game, Motivation, and Effective Learning: An Integrated Model for Educational Game. In *Proceedings of DiGRA Conference on Changing Views–Worlds of Play*. Available at: https://www.researchgate.net/publication/221217604_Game_Motivation_and_Effective_Learning_An_Integrated_Model_for_Educational_Game_Design (Accessed 27/09/2016).
Park, H.J. and Kim, S.B. (2013) Guidelines of Serious Game Accessibility for the Disabled. In *Proceedings of the International Conference on Information Science and Applications (ICISA '13)*, pp. 1–3, IEEE.
Piaget, J., and Inhelder B. (1973) *Memory and Intelligence* London: Routledge and Kegan Paul.
Pinelle, D., Wong, N., Stach, T. and Gutwin, C. (2009) Usability Heuristics for Networked Multiplayer Games. *Proceedings of the ACM 2009 International Conference on Supporting Group Work*, pp. 169–178.

Pollard, A., Anderson, J., Maddock, M. Swaffield, S., Warin, J. and Warwick, P. (2008) *Reflective Teaching*, 3rd Edition. London: Continuum International Publishing Group.
Preconception. (2020) In: Dictionary.com. [Online] Available at: https://www.dictionary.com/browse/preconception (Accessed 01/08/2020).
Preece, Y., Rogers, Y. and Sharp, H. (2015) *Interaction Design: Beyond Human-Computer Interaction*, 4th Edition. Chichester: John Wiley and Sons Ltd.
Prensky, M. (2002) The Motivation of Gameplay: The Real Twenty-First Century Learning Revolution. *On the Horizon*, 10(1), pp. 5–11.
Prensky, M. (2005) Listen to Natives. *Association for Supervision and Curriculum Development Educational Leadership*, 63(4), pp. 9–13.
Prensky, M. (2006) *'Don't Bother Me Mom, I'm Learning!': How Computer and Video Games Are Preparing Your Kids for the Twenty-First Century*. St Paul, MN: Paragon House.
Pritchard, A. and Woollard, J. (2010) *Psychology for the Classroom: Constructivism and Social Learning*. Oxon: Routledge.
Punch, K.F. (2009) *Introduction to Research Methods in Education*, London: Sage Publications Ltd.
Resnick, M. (2017) *Lifelong Kindergarten: Cultivating Creativity Through Projects, Passion, Peers, and Play*. London, England: MIT Press.
Robins, A., Runtree, J. and Rountree, N. (2003) Learning and Teaching Programming: A Review and a Discussion. *Computer Science Education*, 13(2), pp. 137–172.
Rogoff, B. (1998) Cognition as a Collaborative Process. In W. Damon, D. Kuhn and R.S. Seigler (ed.), *Handbook of Child Psychology*, 5th Edition. New York: Wiley.
Roth, W.M. (2000) Authentic School Science: Intellectual Traditions In R. McCormick and C. Paechter (ed.), *Learning & Knowledge*. London, UK: Paul Chapman Publishing, pp. 6–20.
Russ, T., and Drury-Grogan, M.L. (2013) Assessing the Impacts of a Business Communication Simulation on Students' Self-Perceptions. *Eastern Communication Association*, 61(5), pp. 584–595. DOI: 10.1080/01463373.2013.822404
Salen, K. and Zimmerman, E. (2011) The Game Designer Reader, A Rules of Play Anthology. Roger Caillois: The Definition of Play. *The Classification of Games*. Cambridge, Massachusetts. London, England: The MIT Press.
Salisbury, J.H. and Cole. T. (2016) Grounded Theory in Games Research: Making the Case and Exploring the Options. Available at: https://www.academia.edu/27624170/Grounded_Theory_in_Games_Research_Making_the_Case_and_Exploring_the_Options (Accessed 15/10/2015).
Schell, J. (2015) *The Art of Game Design: A Book of Lenses*, 2nd Edition. USA: CRC Press.
Schell, J. (2020) *The Art of Game Design: A Book of Lenses*, 3rd Edition. USA: CRC Press.
Shell, D.F., Murphy, C.C. and Bruning, R.H. (1989) Self-efficacy and Outcome Expectancy Mechanisms in Reading and Writing Achievement. *Journal of Educational Psychology*, 81(1), p. 91.
Shunk, D.H. (2000) *Learning Theories: An Educational Perspective*, 3rd Edition. Upper Saddle River, NJ: Prentice-Hall.
Siegel, R.G., Galassi, J.P. and Ware, W.B. (1985) A Comparison of Two Models for Predicting Mathematics Performance: Social Learning Versus Math Aptitude-Anxiety. *Journal of Counseling Psychology*, 32, pp. 531–538.
Silverman, B.G., Johns, M., Weaver, R. and Mosley, J. (2003) Authoring Edutainment Stories for Online Players (AESOP): Introducing Game play into Interactive Dramas. In: *International Conference on Virtual Storytelling*, Springer.
Silverman, D. (1993) *Interpreting Qualitative Data: Methods for Analysing Talk, Text, and Interaction*. London: Sage Publications Ltd.

Silverman, D. (2014) *Interpreting Qualitative Data*, 5th Edition. London: Sage Publications Ltd.

Sindre, G., Natvig, L. and Jahre, M. (2009) Experimental Validation of the Learning Effect for A Pedagogical Game on Computer Fundamentals. *IEEE Transactions on Education*, 52 (1), pp. 10–18.

Skilbeck, A. (2017) Dewey on Seriousness, Playfulness and the Role of the Teacher. *Education Sciences*, 7(1), p. 16.

Slyman, S. (2014) I Hate Maths! An Investigation into Students' Attitudes Towards Mathematics. MA Thesis, University of Roehampton, London, United Kingdom.

Slyman, S. (2018) *Games Based Learning in Universities, a Life Learning Experience.* New York: Business Expert Press Expert Insights Publishing. ISBN:978-1-94819-801-1.

Slyman, S. (2022) Acumenous Game-Based Learning in Simulation Games and Applied Statistics. PhD Thesis, Goldsmiths, University of London, UK.

Slyman, S., Gillies, M. and Jessel, J. (2019a) Acumenous or Inquisitional? Towards a New Theoretical Lens within Games Learning. In: *13th European Conference on Games Based Learning*, University of South Denmark, Odense, Denmark. ISBN:978-1-912764-38-9 or E-book: ISSN:2049-100X.

Slyman, S., Gillies, M. and Jessel, J. (2019b) Game-Based Learning to Engage students with Applied Statistics using a Simulation Roleplay Game. In: *9th Irish Conference on Games Based Learning*, Cork, Ireland. Available at: http://igbl-conference.com/2016/programme/generate_book_of_abstracts.php (Accessed 25/09/2019).

Sobke, H., Chan, E. Buttlar, R. Grobe-Wortmann, J. and Londong, J. (2014) Cat King's Metamorphosis. The Reuse of an Educational Game in a Further Technical Domain. In: Gobel, S., and Wiemeyer, J. (ed.), *Games for Training Education, Health and Sports, 4th Conference on Serious Games, GameDays 2014*, Darmstadt, Germany, April 1–5, 2014, Springer.

Song, S. and Lee, J. (2007) Key Factors of Heuristic Evaluation for Game Design: Towards massively multi-player online Roleplaying game. *International Journal of Human-Computer Studies*, 65, pp. 709–723.

Sweetser, P., Johnson, D. and Wyeth, P. (2012) Revisiting the Game Flow Model with Detailed Heuristics. *Journal of Creative Technologies*, 2012(3), pp. 1–16.

Sweetser, P. and Wyeth, P. (2005) GameFlow: A Model for Evaluating Player Enjoyment in Games. *Computers in Entertainment*, 3(3), p. 3. DOI: 10.1145/1077246.1077253

Tamura, H. (2007) Foundational Value of Mathematics Education for Management Curriculum. *International Statistical Review*, 75(3), pp. 397–405 DOI: 10.1111/j.1751-5823.2007.00032.x

Tao, Y.H., Cheng, C.J. and Sun, S.Y. (2012) Alignment of Teacher and Students Perceptions on the Continued Use of Business Simulation Games. *Educational Technology and Society*, 15(3), pp. 177–189.

TED (2019) *Souad Slyman: Why Change Attitudes Towards Mathematics [YouTube].* TedxGul, London: Goldsmiths, University of London. Available at: https://www.youtube.com/watch?v=leNDJMNJoNY (Accessed 26/04/2019).

Timonen, V., Foley, G. and Conlon, C. (2018) Challenges when Using Grounded Theory: A pragmatic introduction to Doing GT Research. *International Journal of Qualitative Methods*, 17, pp. 1–10. DOI: 10.1177/1609406918758086

Tondello, G.F., Kappen, D.L. Mekler, E.D. Ganaba, M. and Nacke, L.E. (2016) Heuristic Evaluation for Gameful Design. *CHI PLAY'16*. ACM. 978-1-4503-4458-6/16/10. DOI: 10.1145/2968120.2987729

van der Spek, E.D., van Oostendorp, H. and Meyer, J.-J.C.H. (2013) Introducing Surprising events can stimulate deep learning in a serious game. *British Journal of Educational Technology*, 44(1), 156–169.

Vlachopoulos, D. and Makri, A. (2017) The Effect of Games and Simulations on Higher Education: A Systematic Literature Review. *International Journal of Educational Technology in Higher Education*, 14(1), pp. 22. DOI: 10.1186/s41239-017-0062-1

Vos, L. (2014) Marketing Simulation Games: A Review of Issues in Teaching and Learning. *The Marketing Review*, 14(1), pp. 67–96. DOI: 10.1362/146934714X13948909473220

Walonick, D.S. (2000) *Designing and Using Questionnaire in Survival Statistic*. MN: StatPacInc. Available at Internet WWW page at URL: http://www.statpac.com/surveys/surveys.pdf (Accessed 02/04/2016).

Whitton, N. (2011) Games-Based Experiences for Learning. Futurelab Innovation in Education. Available at: https://www.nfer.ac.uk/media/1775/futl11.pdf (Accessed 02/04/2016).

Wolfe, J. (1997) The Effectiveness of Business Games in Strategic Management Coursework. *Simulation and Gaming*, 28(4), pp. 360–376. DOI: 10.1177/1046878197284003

Women in STEM – 'Facts and Statistics' (Reporting period 2012–2015) IET. Available at: https://communities.theiet.org/files/8042 (Accessed 20/06/2017).

Yin, R.K. (1994) *Case Study Research: Design and Methods*, 2nd Edition. London: Sage.

Young, M. (2008) *"Powerful Knowledge" and "Knowledge of the Powerful": Some Reflections*. London: Presentation at the Institute of Education.

Yusoff, A., Crowder, R. and Gilbert, L. (2010) Validation of Serious Games Attributes Using the Technology Acceptance Model [online]. In: IEEE (ed.), *2010 Second International Conference on Games and Virtual Worlds for Serious Applications*, Braga, Portugal, 25–26 March 2010, pp. 45–51. Available at: http://eprints.soton.ac.uk/270809/ (Accessed 10/05/2015).

Zimmerman, B.J. (2000) Self-Efficacy: An Essential Motive to Learn. *Contemporary Educational Psychology*, 25(1), pp. 82–91. DOI: 10.1006/ceps.1999.1016

7 Exploring the Student Experience through a New Theoretical Lens in Mathematics and Game-Based Learning

This chapter, Chapter 7, takes us on an insightful journey into the realm of students' experiences, focusing on Mathematics and game-based learning (GBL).

Drawing upon the voices of students, this chapter delves into their experiences through the lens of Charmaz's CGT (2006, 2014) and a series of focus-group interviews (FGIs). The research data collection and analysis processes, informed by the extensive body of research literature, provide a robust framework for exploration. By intertwining these methodological aspects, we gain valuable insights into the student perspective, illuminating their thoughts, challenges, and triumphs in the realm of GBL and Mathematics.

The emergent insights and theoretical underpinnings gleaned from the data are carefully examined, with explicit reflections on the relevant literature explored in previous chapters. Building upon the theoretical framework established in Chapters 2–5, this chapter delves into the alignment between the findings and the newly proposed theoretical framework, setting the stage for further exploration in Chapter 9.

With every innovative contribution, this chapter paints a vivid picture of the student experience, providing a foundation for future explorations and progression of GBL in Mathematics education. By capturing the nuances, challenges, and transformative potentials, we inch closer to unlocking the true power of GBL and its role in shaping the future of Mathematics learning.

Within the rich sphere of student experiences, a key theme surfaces: The contrasting dynamics between students' success in learning and playing entertainment games (SGs) and their struggles with Mathematical Educational Games (MEGs). This dichotomy serves as an indicator of preconceptions toward Mathematics, fueled by a lack of self-efficacy and fear. The chapter dissects each emergent theme, shedding light on their intricacies and implications:

- **Impacts of GBL**: Uncovering the transformative effects of GBL on student engagement and learning outcomes.
- **Inquisitional Learning Can Cause Disengagement with MEGs**: Exploring how a purely inquisitional approach to learning can hinder student engagement with MEGs.

DOI: 10.4324/9781032723426-10

- **Acumenous GBL Can Improve MEGs**: Investigating the potential of Acumenous GBL approaches to enhance student comprehension and engagement with MEGs.
- **Mathematics and Programming Are Highly Syntax-Based Subjects**: Recognizing the unique challenges posed by the syntax-driven nature of Mathematics and programming disciplines.
- **Acumenous and Inquisitional GBL**: Introducing a novel framework that combines Acumenous and Inquisitional GBL approaches, offering a comprehensive and nuanced approach to Mathematics education.

I will now delve into each identified theme, drawing upon extracted data from the focus-group interviewers (FGIs). The findings indicate a prevalent existence of negative preconceptions toward Mathematics education, stemming from a lack of self-efficacy and fear.

Impacts of Game-Based Learning (GBL)

Impacts of GBL from cognitive, emotional, and social perspectives were some of the emerging themes from participants' responses about reasons behind engagement with simulation games (SGs) such as perceived usefulness and fun, increased self-efficacy, social interaction, concentration, engagement, challenge, and control. Our interviewees enjoyed playing games since "they are fun, doing things that are fun and you don't think about it as learning you [just] concentrate more and you have a great fun but after a while you realize that you learn something." This implies that our interviewees perceived games as useful and fun; these are the motives that increased their self-efficacy and positively drove their preconception to interact with the SG in a meaningful way, as they seemed to be confident and in control playing the game, that is, "you concentrate more and you have a great fun," they declared. This self-control over the game has engaged students to learn a great deal, as they continued, "[do]ing games is a good way of learning" whilst others noted "I do a lot learning for games." In contrast, our interviewees reported that they cherish the learning outcomes, the challenge and the complexity of the game stating, "I learn a lot [to] achieve the game objective," I like "competition," "challenge," and "real-life" characteristics of the games. These learning elements, goals/storylines are what motivate players to gain from the SG. Then one of the interviewees added, "in [e]very game you learn something, but you don't realize it, I learn a lot [to] achieve the game objective. I think with every new game engine there is something new to learn." Similarly, they noted that games "are great, [and] useful in building skills [like] critical thinking, decision-making, leadership skills, with the FIFA [for example] you start from the bottom, it teaches you a lot of statistical analysis and the best way to make the program running."

Nevertheless, not all the games especially educational games are perceived as enjoyable or effective tools for entertainment and learning. Our interviewees expressed that educational games can be disengaging. Often described as "boring," "not interactive," and "not fun." These game characteristics are crucial considerations in game design as we will explore further. This prompts a call for change, if we aim to boost students' engagement in Mathematical and Statistical learning and address their self-efficacy, Mathematics Educational Games (MEGs) must evolve. Currently these MEGs could potentially drive learners away from Mathematics, thus negatively impacting their learning instead of enhancing it. Several interviewees consistently shared that they "have preconceptions with educational games that when they hear educational, they think it is not going to be fun." This suggests that their experiences with Game-Based Learning (GBL) did not foster a new lifelong learning experience. Instead, it influenced their emotion/feelings, behaviors, and personalities due to preconceived notions that such games would lack fun. These motives and perceptions, characterized by lack of usefulness and fun, influenced their self-efficacy, ultimately leading to negative emotions and experiences in their learning journey.

The reasons behind this preconception were further elucidated by one participant, who noted, "when the objective to the user is to learn a skill, they don't learn because they are there (playing the game) to have fun." Conversely, it was highlighted that "when the user does not realize that they are learning, they learn" as they "are playing the game to have fun." Another interviewee emphasized the significance of exploration within the game, stating, "when you are in, you have to explore the game [...] you learn new tactics [...] you explore and solve problems yourself." This underscores the notion that the primary purpose of playing the game is enjoyment/fun; these are the perceptions (usefulness and fun) that positively influence preconceptions, boosting self-efficacy, and fostering engaging interactions with the game. However, it is worth noting that not all games are perceived as educational, as participants pointed out, "Some education is meaningless, there should be more creative educational games so that we can use it in everyday life."

Inquisitional Learning Can Cause Disengagement with MEGs

Findings indicate that there are many game characteristics that make MEGs disengaging such as lack of: Surprise, competition, rewards, creativity, problem-solving, social interaction, options, challenging environment, and missions. According to our interviewees, "if [a game] does not go anywhere [or] completely repeating same things over and over again [has] [n]o amazements no competition, no self-satisfaction, no mission, no variety, no direction, and no hints." These would make the game less appealing/disengaging. In contrast, they recommended that "games should involve other members, [since] in real life you work with other people." When looking closely at which level these disengaging games deter our members of generation "Z," it is found that across primary, secondary, or even A level, these learners experience outdated educational games, due to the factors suggested above, which led to disengagement/

lack of self-efficacy. In particular, MEGs were found to be dominant because most students found Mathematics, memory learning, hard as a subject. As they explained: "maths [games] used to be quite dominant when we were younger because students in general found more memory [learning] hard to pick up." "Maths is sort of memory stuff." Then they noted that MEGs are hard and "[b]oring" while others stressed that Mathematics "just gets different as you progress. When you are younger, Maths educational games were more interactive. As maths gets harder, games have few multiple options. Not as interactive as it was. Not appealing. No animation. It [just] gets harder. Not as appealing as it is used to be." This implies that MEGs at a younger age were stimulating but as Mathematics becomes advanced, educational games become hard and less appealing. This is also because the type of Mathematics they contain is disengaging. This I propose is what I have referred to as "Inquisitional Learning" (discussed next) or a "quiz-like" type of learning which has "few[er] multiple options" to choose from, it is not "as interactive as it was. No animation [, and it just] gets harder. Not as appealing as it is used to be," continued one of the interviewees. Similarly, others referred to this type of Inquisitional Learning as "repeating same things over and over again. There are no amazements no competition, no self-satisfaction, no mission, no variety, no direction, and no hints."

This furthermore indicates that some students do not feel at ease with Mathematics at an advanced level, but in the case of our interviewees, and in particular MEGs at a higher level, implying that they have a lack of self-efficacy in Mathematical learning. Whereas in a primary school, MEGs were enjoyable and rewarding and so students had high self-efficacy playing them but as students move on to a higher level, MEGs began to become difficult and "[n]ot as interactive as it was. Not appealing. No animation." Hence, preconceptions with MEGs can cause a lack of self-efficacy. In contrast, while these students "were young they played the game since it was fun" and used their own intuition to interact with the games. As they explained: "When you are younger you bind in the game because you learn without knowing, you just play. However, when you are older you know it is for educational purposes because you play other games, like entertainment games." This suggests that participants' interaction with the game, at a younger age, was mainly intuitive and fun, adapting and adopting these GBL. They played and learnt without realizing since games at a primary to upper secondary level were stimulating and fun. However, at a higher level, they struggled to understand MEGs because Mathematics becomes harder and has fewer multiple options, indicating Inquisitional "quiz-"like type of GBL, and because they have grown up now and so are their perceptions of the world. Furthermore, they play other entertainment games, leading their experiences to change. Similarly, they elucidated that their generation would like more autonomy and prefer an Acumenous type of GBL (discussed next) while they indicated that they want control over their learning because: "You just want to do it without knowing. It should be something that we want to [be] part of. We want to be part of it."

This suggests that learners' perceptions of IT (information technology) as useful and fun can have positive impacts on their preconceptions, increasing their self-efficacy, and therefore behavior and interaction with the game. Game design and its characteristics positively impact learners' skills like decision-making and learning outcomes, and therefore a significant part of the learning experience. But how can we make MEGs engaging? Here is another identified theme.

Acumenous GBL Can Improve MEGs

Based on insights from our interviewees, Mathematics Educational Games (MEGs) can effectively captivate and educate when imbued with transferable skills such as a strong focus on "problem-solving, interaction, decision-making and experiment." While others underscore the significance of incorporating "storylines, which you have to follow and continue from one place to another place." "Two outcomes and you can pick one." This will "make user thinks of the outcomes." For example, some interviewees highlighted the appeal of offering multiple outcomes, akin to choices presented in popular video games like GTA and Mafia III where decisions shape the narrative and lead to diverse consequences, noting that, such games help you to "make choice whether to kill someone or join [their] team." It has "different learning and can lead to different outcomes, and you start all over again when you are bored." Moreover, other interviewees referenced a "roller skate game" that involves "flipping objects, making money." They highlighted that it "involves a lot of test run to do financing and create as much wealth as possible in short spin." Others noted that, "Stock markets," for example, involves "profits-making and gold-farming." Those skills were transferable to real-life skills. "Our goal in the simulation games," they emphasized, "were to create wealth in a short time."

Other interviewees suggested that university-style educational games "in a project where they were learning about history and science," stimulate scientific and historical processes, for instance, "how vaccine evolves over time." They then added, "the game was that if you add this percentage of chemical there were different variable[s]/chemicals involved": "What will happen? If this chemical increase/decrease, you see the outcome straightaway. You know you would go for creating the perfect outcome, if you add this percentage of chemicals, you can see the outcome [effects] straightaway and you will create the best outcome," they emphasized.

These skills in simulation games, I suggest, are relevant to what I have referred to as "Acumenous **G**ame-**B**ased **L**earning (**AGBL**)." What they are referring to is an **"Acumenous"** type of GBL, which is the ability to understand, reason, and make decisions, as in the example of understanding how the vaccine evolves over time, what will happen, why and what the impacts of their decision are. This also implies that transferable skills like experimentation and financing are significant real-life skills in GBL. These perceptions of usefulness and fun are what motivate players to achieve in SGs, having positive impacts on their preconceptions, behavior, and how they interact with these games. Consequently, this can be utilized to improve Mathematical learning and as a learning tool to address self-efficacy in MEGs.

Mathematics and Programming are Highly Syntax-Based Subjects

Findings illustrate that the relationship between Mathematics and programming is that both are highly syntax-based subjects. They consist of patterns, structure, logic, and symbols. The challenging issue is that most learners find Mathematics and programming challenging and extremely difficult to handle since they cannot think in abstract ways of how to apply them in real life, and so not every student finds these STEM subjects stimulating. On the other hand, learning to program is a significant skill to obtain in a Computer Science degree. Nevertheless, many students found programming a daunting and complex process, and while some express their feelings of discomfort and a lack of self-efficacy to program, others develop strategies for dealing with this complexity; stating that they found programming "complex and challenging." "Programming is extremely difficult," they added. "Programming is a different language, it has patterns and letters." Programming is "extremely difficult," but "tutorials" are a great tool for learning and seeking help. Tutorials, they suggested, can facilitate the spark of novel ideas of how to create things from scratch; read definitions; and learn how to start from scratch. They "manage to understand, spot their own problems and input functions." Then when asking about best practice in programming they suggested, "watching online tutorials" to add something "innovative, getting ideas from online, researching different visualizations." Using tutorials to help breakdown the complexities of learning and understand what it is required, they reinforced. Programming is difficult because there are "rules to identify, to practice preciseness, and syntax."

In contrast, some learners develop strategies for how to deal with difficulties in programming, building high self-efficacy with this subject. This signifies that students' perception of usefulness and fun have led them to interact with programming in a more constructive way by building skills, competence, and knowledge. These are the perceptions (usefulness and fun) that drove positively their preconceptions and self-efficacy, leading them to interact with programming in fun ways, adding a useful new experience to their learning journey. This implies that some students have attained a strong sense of self-efficacy in their learning, leading to heightened confident. They demonstrate adept problem-solving and decision-making skills akin to those on in AGBL. Meanwhile, other students have developed fresh coping strategies to manage complexities, such as utilizing tutorials and breaking down problems into logical sequences to enhance their learning experience and consequently elevate their self-efficacy. Some learners describe this self-efficacy as a form of "empowerment," as one interviewee articulated, recognizing how computational learning "can be applied to everyday life." Conversely, other interviewees delineated programming as a distinct "language with patterns and symbols." Echoing these sentiments, some summarized their programming journey as "something novel" that grants them "empowerment" as they perceive its practical applications in daily life. They described it as a "completely different experience" yet ultimately "rewarding." Several individuals admitted to having "no prior programming experience," yet persisted, acknowledging the necessity of adhering to rules and the challenge of debugging. They emphasized the importance of practice and iteration, highlighting the frustration of troubleshooting errors in code. Additionally, they noted the difficulty in problem-solving, emphasizing the necessity of breaking down complex tasks into logical steps, a skill many find challenging despite understanding the overarching concepts.

This ability to break down problems into logical steps, I suggest, is similar to what I call **AGBL** since one has to understand, reason, and apply the rules (as indicated above). These are the connections between GBL and Mathematics. This is computational learning (programming) that is embedded in GBL and which, if applied, can improve Mathematical learning within MEGs, challenging any preconceptions and perhaps changing how students perceive MEGs.

To this point, research findings indicate that there are two main types of GBL, which I have associated with, "**Inquisitional**" and "**Acumenous.**" These preliminary novel ideas are my own and were further established in Chapters 8 and 9. I will discuss each concept in detail in the next section. Examining participants' discourse unfolds this type of learning. For instance, one interviewee referred to this stating that MEGs while they were "younger were more interactive." But as Maths "gets harder, games have few multiple options." "Not as interactive as it was." "Not appealing. No animation. It just gets harder. Not as appealing as it is used to be."

It is worth noting that this type of learning is observable within some MEGs that contain **Inquisitional Learning**, such as quiz games like "MyMaths," "Business Data Analysis game" (Online, 2017, 2024), or "Calculus quiz game" (https://www.studiosity.com/student-resources/practice-tests/year-12-maths-calculus-quiz), where players face a disengaging interface during which factual/fictional statements are introduced. Then questions are asked in a quiz format type of GBL. According to the findings from this investigation, this creates problems, as lack of interaction in the user interface leads to disengagement with these types of games (MEGs). Learners become disconnected from learning and thus lose their interest and self-efficacy with it as a result.

The second type of GBL which surfaced from participants' discussions is what I have referred to as **"Acumenous Game-Based Learning (AGBL)"** (from acumen). Acumen is a noun, it means the ability to understand, reason, and make decisions. Acumenous refers to the ability to understand, reason, and decide why learning is occurring such as in a business. AGBL is a cognitive process, a thinking cap that goes through game players' minds while interacting with the games. This process involves understanding, reasoning, and deciding the next action the player is required to undertake. In participants' discourses, this was manifested. For instance, some interviewees emphasized this ability to understand, reason, and decide, that is, understanding how the vaccine evolves over time, what will happen, why and what the impacts of their decision are, noting: "The game was that if you add a chemical … What will happen? If this chemical increases/decreases, you see the outcome/effects straightaway and you will create the best outcome."

Furthermore, this type of GBL is manifested within some entertainment SGs, for instance "SimCity" (see Chapter 5 for more info about the game) "Harvard Business Simulation" (https://hbsp.harvard.edu/simulations/), "Assassins' Creed" games (https://www.playstation.com/en-gb/games/assassins-creed-odyssey/), and some educational games such as Harel's and Papert's (1990) "Instructional Software Design Research Project" in schools and Jagger et al.'s (2016) "3D Business Simulation" game. Examples of AGBL include SimCity (Online, 2017), Harvard Business game (Online, 2016). Assassins' Creed game (Online, 2017).

What these SGs have in common is that they nurture players' creativity, immersing them in missions and decision-making processes. Thus, they offer players the opportunity to engage in a real-life experience, in a safe environment, where they can understand, reason, and decide. As some participants declared: We grow up playing these games and we learn games mechanics already playing them so we know how to create them. We are the generations who grow up with games, identified with games, you know as you play games you learn game mechanics. Subconsciously we bring what we play; so, we know how to play games and the rules we just need to think of another way, you know how it works we played them before.

Empirical Discussion

In this chapter, I will delve into participants' experiences with Mathematics and GBL. I will analyze the emergent insights drawing explicit connections to relevant literature (Chapters 2–5), where applicable. Additionally, I will discuss how my findings align with the new theoretical framework which will be further examined in Chapter 9. Furthermore, findings indicate how interviewees succeed in learning and playing entertainment games (SGs) while they struggle to understand MEGs, indicating preconceptions toward Mathematics due to lack of self-efficacy and fear. I will examine each emergent theme.

The themes are as follows:

- **Impacts of GBL**: Uncovering the transformative effects of GBL on student engagement and learning outcomes.
- **Inquisitional Learning Can Cause Disengagement with MEGs**: Exploring how a purely inquisitional approach to learning can hinder student engagement with MEGs.
- **Acumenous GBL Can Improve MEGs**: Investigating the potential of Acumenous GBL approaches to enhance student comprehension and engagement with MEGs.
- **Mathematics and Programming are Highly Syntax-Based Subjects**: Recognizing the unique challenges posed by the syntax-driven nature of Mathematics and programming disciplines.
- **Acumenous and Inquisitional GBL**: Introducing a novel framework that combines Acumenous and Inquisitional GBL approaches, offering a comprehensive and nuanced approach to Mathematics education.

Impacts of GBL: Uncovering the Transformative Effects of GBL on Student Engagement and Learning Outcomes

Impacts of GBL from cognitive, emotional, and social perspectives were some of the emerging themes from participants' responses about reasons behind engagement with SGs such as perceived usefulness and fun, increased self-efficacy, engagement, empowerment, identity, agent, embodiment, concentration, more learning, social interaction, competition, fantasy, challenge, goals/storylines, real-life characteristics of the games, control and game mechanics, as well as building skills like critical thinking, decision-making, problem-solving, leadership and statistical analysis. This can be linked to previous research such as Malone (1980), Jones (1998), Faria (2001), Garris et al. (2002), Prensky (2002, 2005), Gee (2003a, 2003b, 2004, 2005a, 2005b, 2007, 2008, 2011), Faria and Wellington (2004), Gee et al. (2008), Lenhart et al. (2008), Moreno-Ger et al. (2008), Kolb and Kolb (2008), Mayo (2009), Vos and Brennan (2010), Tao et al. (2012), Li et al. (2013), Granic et al. (2014), Vos (2014), McGonigal (2015), Schell (2015, 2020), Jagger et al. (2016), Hamari et al. (2016), and Vlachopoulos and Makri (2017). In other words, all these research studies (above; see also Chapters 2–4) have emphasized the positive impacts of SGs and its role in engaging users, offering a positive experience. Li et al. (2013), Vos (2014), Jagger et al. (2016), and Vlachopoulos and Makri (2017) go as far as to emphasize the key importance of SG's abilities, as a teaching tool, to model skills such as decision-making and numerical data.

In contrast, findings from this research (Chapter 7) indicate that not all games, especially educational games, are considered to be fun or a good tool for entertainment and learning. Educational games are reported as disengaging, boring, and not fun. This is because educational games are games that are designed with a learning purpose in mind other than just entertainment. This thus leads to the following call for a change: If we would like to boost students' engagement in Mathematical learning and so address their self-efficacy, "MEGs have to change" because currently they are putting learners off Mathematics, and hence are likely to have detrimental impacts on their learning rather than enhancing it. Moreover, findings suggest that participants have preconceptions about educational games. Their perceptions that the games are neither useful nor fun drove their self-efficacy and therefore their preconceptions, leading to negative experiences added to their learning journey. This is because games can transform players for the better or worse. When learners are left to explore and have fun playing games, they learn new skills without even realizing (as this investigation indicated) because their fulfilment is achieved. However, if they are restricted/limited, they do not learn. Therefore, there should be more creative educational games with real-life scenarios.

Inquisitional Learning Can Cause Disengagement with MEGs: Exploring How a Purely Inquisitional Approach to Learning Can Hinder Student Engagement with MEGs

Interviewees found MEGs disengaging because there is a lack of: Surprise, competition, rewards, creativity, problem-solving, social interaction, challenging environment, and missions. It is evident (from our interviewees) that across elementary/primary, secondary, and even A level/high schools, these learners have experienced some of these disengaging games leading to preconceptions toward Mathematical learning.

At a younger age, MEGs were fun and appealing, and students played them using their own intuition, embracing GBL without realizing. This led to high self-efficacy. In primary schools, games were dominant because most students found Mathematics, memory learning, hard as a subject but as Mathematics becomes advanced, they become complex and not stimulating. At a higher level, students struggle to understand MEGs because Mathematics becomes complex and not interactive. Mathematics just becomes different as learners progress, it has few multiple options, no appeal, no animation, and it is just hard. This is what I have referred to as ***"Inquisitional Learning"*** (discussed next) or "quiz-like" type of learning where learners face fewer options to choose from, "no amazements, no competition, no self-satisfaction, no mission, no variety, no direction and no hints." This indicates that some students do not feel at ease with Mathematics at an advanced level and in our case with MEGs at a higher level, implying preconceptions about these MEGs that lead to a lack of self-efficacy. Furthermore, they play other entertainment games (e.g., SGs) and this leads their experiences to change. Interviewees indicated that their generation would like more autonomy and prefer *"Acumenous"* type of GBL, discussed next.

Acumenous GBL Can Improve MEGs: Investigating the Potential of Acumenous GBL Approaches to Enhance Student Comprehension and Engagement with MEGs

Interviewees indicated that MEGs can be engaging if we embody them with transferable skills like encouraging high emphasis on problem-solving, creativity, deep understanding, decision-making, and storytelling. They can have two story outcomes where users can decide and choose the best suitable option for them. In addition, scientific experimentation should be considered and added to these MEGs. For example, users can be given an opportunity to investigate different chemicals/formulae and see the results or the effects of their investigation straightaway. This could help them in creating the perfect outcome. These skills in SGs, I suggest, are what I have referred to as "***Acumenous GBL***." "*Acumenous*" is the ability to understand, reason, and decide, such as understanding how the vaccine evolves over time, what will happen, why and what the impacts of their decision are. These are real-life skills in simulation GBL. These perceptions of usefulness and fun are what motivate players to achieve in games. Game design with its attributes positively impacts learners' skills like decision-making and learning outcomes and therefore a significant part of the learning experience. Consequently, this can be utilized to improve Mathematical learning using SGs and as a learning tool for enhancing self-efficacy in MEGs.

Mathematics and Programming are Syntax-Based Subjects: Recognizing the Unique Challenges Posed by the Syntax-Driven Nature of Mathematics and Programming Disciplines

Findings indicate that the relationship between Mathematics and GBL is that we could use the latter (GBL) to empower Mathematical learning since Mathematics and programming are highly abstract syntax-based subjects. They consist of patterns, structure, logic, and symbols. The issue is that most learners find these subjects challenging and extremely difficult to handle because they cannot think in abstract ways of how to apply them in real life, and therefore not every student finds these STEM subjects stimulating.

Many interviewees found programming complex, daunting and a new different language. While some express their feelings of discomfort and a lack of self-efficacy to programs, others developed fresh coping strategies to manage complexities such as utilizing tutorials, watching YouTube videos, conducting research, and breaking down problems into a logical order to facilitate their learning experience. In this way, they raise their self-efficacy/confidence with this subject. This further signifies that students' perceptions of usefulness and fun have led them to interact with programming in a more constructive way, that is, building skills, competence, and knowledge. These are the perceptions (usefulness and fun) that increased their self-efficacy, leading them to interact with programming in fun ways and thus adding a useful new experience to their learning journey. It is the knowledge that subsequently leads to new feelings and actions, which comes from their experience. This implies that some students have reached high self-efficacy in their learning and become highly confident. This self-efficacy is what some learners have referred to as "empowerment" as they can see how computational learning can be applied to everyday life. This ability to break down problems into logical steps is what I referred as AGBL, discussed next. This is computational learning (programming) that is embedded in GBL and which, if applied, can improve Mathematical learning within MEGs challenging any preconceptions and perhaps how students perceive MEGs.

Acumenous or Inquisitional GBL: Introducing a Novel Framework That Combines Acumenous and Inquisitional GBL Approaches, Offering a Comprehensive and Nuanced Approach to Mathematics Education

As enlightened previously, how players think or see IT as useful and fun can affect their preconceptions and subsequently their self-efficacy. How players behave and interact with the game can lead to transferable skills (new skills/

learning outcomes) is what I suggest are new types of GBL, I refer to these as ***"Acumenous"*** or ***"Inquisitional"*** **GBL**. This will be further discussed in Chapter 9. These preliminary novel ideas are my own and were further established in Chapters 7 and 9. I will discuss each concept in detail starting with Inquisitional.

"Inquisitional Learning" is "quiz-like" type of learning that can occur in conversations such as investigations in court/police station: Question answer formats and immediate feedback. They are used for testing memory, knowledge, and facts. *Inquisitional* has "fewer multiple options," "no creativity," "no mission," or "real-life scenarios."

It is worth noting that *Inquisitional Learning* is observable within some MEGs such as quiz games like "MyMaths," (https://www.mymaths.co.uk), where players face a rigid disengaging interface during which factual/fictional statements are introduced. Then questions are asked in a quiz type of learning. According to my findings, this creates problems. Lack of interaction in the user interface leads to disengagement with these types of games (MEGs), where learners become distracted from learning using these MEGs, thus losing their interest and confidence with it as a result.

The second type of GBL which was investigated within participants' discourse is what I have referred to as ***"Acumenous Game-Based Learning"*** (AGBL). **AGBL** is a cognitive process, a thinking cap that goes through game players' minds while interacting with the games such as understanding how a vaccine evolves over time, what will happen, why and what the impacts of their decision are.

Furthermore, this type of AGBL is manifested within some entertainment SGs such as "SimCity." What these SGs have in common is that they nurture players' creativity by immersing them in missions, decision-making processes, and experimental experiences where they understand, reason, and solve real-world problems in a safe environment. This ability can only be accumulated from AGBL. This, if incorporated in MEGs, can enhance Mathematical learning considerably because generation Z are likely to be sociable digital-native learners. They use technology such as "Instagram" and "Snapchat" with utter flare. They need games and immerse themselves in the games' missions as purpose, goal, and duty. These are perceived as useful and fun. They engage when they problem-solve and interact with these SGs in playful ways, because some identify themselves as the game generation who grew up with games, and as they play them, they learn games mechanics. This therefore leads to the following call for a change: In order to engage students' further in Mathematics and address their self-efficacy, MEGs should be designed with users' needs in mind. Since these games could be failing their purpose for these interviewees, damaging their learning rather than enhancing it. This type of "MEGs are at fault" because they are likely to deter these students from learning, so there is a necessity for MEGs to be changed. However, a change in the game design of these MEGs alone will not increase learning. The way Mathematics is integrated in the curriculum should also be re-examined and not left unnoticed, since some interviewees reported that they have "preconceptions with educational games that it is not going to be fun," signifying lack of self-efficacy toward this type of GBL. However, we do not know how high or low this self-efficacy is. Thus, this is still an area yet to be explored further in Chapters 8 and 9.

The next chapter, **Chapter 8**, will explore the Dynamics of Inquisitional and Acumenous GBL in Students' Experiences with Mathematics and Statistics.

Bibliography

Acumen (2017) In: Cambridge University Press. [Online] Available at: http://dictionary.cambridge.org/dictionary/english/acumen (Accessed 30/05/2017).

Adams, A., Lunt, P. and Cairns, P. (2008) A Qualitative Approach to HCI Research. In: Cairns, P. and Cox, A. (ed.). *Research Methods for Human-Computer Interaction*. Cambridge, UK: Cambridge University Press, pp. 138–157. Available at: http://www.cambridge.org/catalogue/catalogue.asp?isbn=9780521870122andss=toc (Accessed 20/11/2016).

Anon (2018) Efficacy. In: Cambridge University Press. [Online] Available at: https://dictionary.cambridge.org/dictionary/english/efficacy (Accessed 30/09/2018).

Applefield, J.M., Huber, R. and Moallem, M. (2000) Constructivism in Theory and Practice: Toward a Better Understanding. *The High School Journal*, 84(2), pp. 35–53.

Arthur, J., Waring, M. Coe, R. and Hedges, L.V. (2012) *Research Methods and Methodologies in Education*. California: Sage Publications Inc.

Barbosa, T., Lopes, S.P. Leao. C. Soares, F. and Carvalho, V. (2019) Serious Game for Teaching Statistics in Higher Education: Storyboard Design. In: *7th EAI International Conference, Braga, Portugal, on 'Interactivity, Game Creation, Design, Learning, and Innovation'* (ed.) Springer, pp. 169–175. DOI: 10.1007/978-3-030-06134-0_18

Bartholomew, H., Darragh, L., Ell, F. and Saunders, J. (2011) "I am Natural, and I Do It for a Love!": Exploring Students' Accounts of Studying Mathematics. *International Journal of Mathematical Education in Science and Technology*, 42(7), pp. 915–924.

Bassey, M. (1999) *Case Study Research in Educational Settings*. Buckingham: Open University Press.

Bellotti, F., Kapralos, B. Lee, K. Moreno-Ger, P. and Berta, R. (2013) Assessment in and of Serious Games: An overview. *Advances in Human-Computer Interaction*, 2013, p. 1.

Belluigi, D.Z. (2009) Exploring the Discourses Around 'Creativity' and 'Critical Thinking' in a South African Creative Arts Curriculum. *Studies in Higher Education*, 34(6), pp. 699–717.

Bennedesen, J. and Caspersen, M.E. (2004) Programming in Context: A Model-First Approach to CS1. In *Proceedings of the 35th SIGCSE Technical Symposium on the Computer Science Education*, pp. 477–481. ACM Press.

Bevan, N. (2001) International Standards for HCl and Usability. *International Journal of Human Computer Studies*, 55(4), pp. 533–552.

Boaler, J. (2009) *The Elephant in the Classroom*. London: Souvenir Press.

British Education Research Association (BERA) (2011) BERA Ethical Guidelines for Educational Research. Available at: http://www.bera.ac.uk/ (Accessed 15/10/2016).

British Education Research Association (BERA) (2018) BERA Ethical Guidelines for Educational Research. Available at: https://www.bera.ac.uk/wp-content/uploads/2018/06/BERA-Ethical-Guidelines-for-Educational-Research_4thEdn_2018.pdf?noredirect=1 (Accessed 25/03/2019).

Brown, J.S., Collins, A. and Duguid, P. (1989) Situated Cognition and the Culture of Learning. *Educational Researcher*, 18(1), 32–42.

Bruner, J. (1973) *Going Beyond the Information Given*. New York: Norton.

Bryant, A. (2017) *Grounded Theory and Grounded Theorizing: Pragmatism in Research Practice*. New York, NY: Oxford University Press.

Bryman, A. (2008) *Social Research Methods*, 3rd Edition. Oxford: Oxford University Press (first published: 2001).
Bryman, A. (2012) *Social Research Methods*. 4th Edition. United States: Oxford University Press.
Bryman, A. (2016) *Social Research Methods*. 5th Edition. United States: Oxford University Press.
Caillois, R. (1961) *Man, Play, and Games*. New York: Free Press.
Cardinot, A. and Fairfield, J.A. (2019) Game-Based Learning to Engage Students with Physics and Astronomy Using a Board Game. *International Journal of Game-Based Learning (IJGBL)*, 9(1), pp. 42–57.
Charmaz, K. (2006) *Constructing Grounded Theory*, 1st Edition. London: Sage Publications Ltd.
Charmaz, K. (2014) *Constructing Grounded Theory*, 2nd Edition. London: Sage Publications Ltd.
Chen, M.P., Wong, Y.T. and Wang, L.C. (2014) Effects of Type of Exploratory Strategy and Prior Knowledge on Middle School Students' Learning of Chemical Formulas from a 3D Roleplaying Game. *Educational Technology Research and Development*, 62, pp. 163–185. DOI: 10.1007/s11423-013-9324-3
Clark, D.B., Tanner-Smith, E.E. and Killingsworth, S.S. (2015) Digital Games, Design, and Learning a Systematic Review and Meta-Analysis. *Review of Educational Research*, 86(1), pp. 79–122.
Claxton, G. (2002) *Building Learning Power*, 1st Edition. Bristol: TLO Limited.
Conlon, C., Carney, G., Timonen, V. and Scharf, T. (2015) Emergent Reconstruction in Grounded Theory: Learning from Team-Based Interview Research. *Qualitative Research*, 15, pp. 39–56. DOI: 10.1177/1468794113495038
Cook, J. (2005) Negotiating the Curriculum: Programming for Learning. In *Negotiating the Curriculum*, pp. 15–31. Routledge.
Cooper, P.A. (1993) Paradigm Shifts in Designed Instruction: From Behaviorism to Cognitivism to Constructivism. *Educational Technology*, 33(5), pp. 12–19.
Corbin, J. and Strauss, A. (2015) *Basics of Qualitative Research: Techniques and Procedures for Developing Grounded Theory*, 4th Edition. Thousand Oaks, CA: Sage.
Cordova, D.I. and Lepper, M.R. (1996) Intrinsic Motivation and the Process of Learning: Beneficial Effects of Contextualization, Personalization, and Choice. *Journal of Educational Psychology*, 88, pp. 715–730.
Coulthard, G.J. (2009) A Review of the Educational Use and Learning Effectiveness of Simulations and Games [online]. PhD Thesis, Purdue University. Available at: http://www.coulthard.com/library/paper-simulation.html (Accessed 15/12/2015).
Craft, A. (1999) Creative Development in the Early Years: Some Implications of Policy for Practice. *The Curriculum Journal*, 10(1), pp. 135–150.
Craft, A. (2002) *Creativity and Early Years Education*. London: Continuum.
Crawford, C. (1984) *The Art of Computer Game Design*. Berkeley: California, McGraw-Hill/Osborne Media.
Creswell, I.W. (2009) *Research Design: Qualitative, Quantitative, and Mixed Methods Approaches*, 3rd Edition. London: Sage.
Crockett, L. (2018) Bogost's War on Gamification: Irony and Play on Grandma's Sofa. In: *12th European Conference on Games Based Learning (ECGBL 2018)*. Reading: Academic Conferences and Publishing International Ltd.
Cronan, T.P. and Douglas, D.E. (2012) A Student ERP Simulation Game: A Longitudinal Study. *Journal of Computer Information Systems*, 53(1), pp. 3–13.
Csikszentmihalyi, M. (1990) Psychology Today: Finding Flow. Available at: https://www.researchgate.net/publication/200026151_Finding_Flow_The_Psychology_of_Engagement_With_Everyday_Life/link/55ad6c2f08aed9b7dcdae222/download (Accessed 03/05/2016).

Csikszentmihalyi, M. (1997) *Creativity, Flow and the Psychology of Discovery and Invention*. New York: Harper.

Dann, W.P., Cooper, S. and Pausch, R. (2008) *Learning to Program with Alice, Brief Edition*. USA: Prentice-Hall, Inc.

Davis, F.D. (1989) Perceived Usefulness, Perceived Ease of Use, and User Acceptance of Information Technology. *MIS Quarterly*, 13(3), pp. 319–340.

De Bono, E. (1992) *Teach Your Child How to Think*. England: McQuaig Group Inc.

Dede, C. (2005) Planning for Neomillennial Learning Styles. *Educause Quarterly*, 28(1), pp. 7–12.

Dondlinger, J.M. (2007) Educational Video Game Design: A Review of the Literature. *Journal of Applied Educational Technology*, 4(1), Spring/Summer, pp. 21–31. Available at: http://reserachgate.net/publication/238444705 (Accessed 12/12/2015).

Draper, S. (2013) Social Constructivism. Available at: http://www.psy.gla.ac.uk/~steve/courses/archive/CERE12-13-safari-archive/topic3/webarchive-index.html#social%20con%201 (Accessed 23/12/2019).

Dunne, C. (2011) The Place of The Literature Review in Grounded Theory Research. *International Journal of Social Research Methodology*, 14, pp. 111–124. DOI: 10.1080/13645579.2010.494930

Dweck, C.S. (2006) *Mindset: The New Psychology of Success*. United States: Ballantine Books Trade Paperback Edition.

Dwyer, J. and Slyman, S. (2016) *An Introduction to Quantitative Methods for Business Research*. London: Algana Publishing. 200 pages; ISBN−13: 9781907934094.

Early, R.E. (1992) The Alchemy of Mathematical Experience: A Psychoanalysis of Student Writings. *For the Learning of Statistics*, 12(1), pp. 15–20.

Ejsing-Duun, S. and Karoff, H.S. (2016) Creativity and Playfulness: Producing Games as a Pedagogical Strategy Case studies. In: Munkvold, R. (ed.). *Games Based learning for Researchers, Teachers and Students*, Vol. 1. Reading: Academic Conferences and Publishing International Ltd. ISBN:978-1-911278-12-890000.

Fairweather, E. and Cramond, B. (2010) Infusing Creative and Critical Thinking into the Curriculum Together. In: Beghetto, A. and Kaufman, J.C. (ed.). *Nurturing Creativity in the Classroom*. United States: Cambridge University Press.

Falstein, N. and Barwood, H. (2001) The 400 Project. Available at: http://theinspiracy.com/400_project.htm. (Accessed 20/04/2016).

Faria, A.J. (2001) The Changing Nature of Business Simulation/Gaming Research. *Simulation and Gaming*, 32(1), pp. 97–110. DOI: 10.1177/104687810103200108

Faria, A.J., Hutchinson, D. Wellington, W.J. and Gold, S. (2009) Developments in Business Gaming: A Review of the Past 40 Years. *Simulation and Gaming*, 40(4), pp. 464–487. DOI: 10.1177/1046878108327585

Faria, A.J. and Wellington, W.J. (2004) 'A Survey of Simulation Game Users, Former-Users, and Never-Users. *Simulation and Gaming*, 35(2), pp. 178–207. DOI: 10.1177/1046878104263543

Federoff, M.A. (2002) Heuristics and Usability Guidelines for the Creation and Evaluation of Fun in Video Games. Master Thesis, Indiana University, Bloomington.

Fienberg, S.E. (1992) A Brief History of Statistics in Three and One-Half Chapters: A Review Essay. Available at: https://www.jstor.org/stable/2246307?casa_token=Vek3Rmd5Mk0AAAAA%3AJ0zRZC5AxskwvTNY7GlrgsRV6oLrE20DY3FdeaPad0JvNSNrZDav8uZpX-YqY1hXTDtaz6zM87S4vwkQdflDLSzmrLCNlEsaBqWnykepc0jlYXsQXA

Forest, E. (2014) ADDIE Model. Available at: http://educationaltechnology.net/the-addie-model-instructional-design/ (Accessed 20/04/2016).

Fosnot, T.C. (1989) *Enquiring Teachers, Enquiring Learners: A Constructivist Approach for Teaching*. New York: Teachers College Press.

Fosnot, T.C. and Perry, R.S. (2005) Constructivism: A Psychological Theory of Learning. Available at: http://beyondbitsandatoms.org/readings/fosnot_constructivism_1996.pdf (Accessed 20/07/2019).

Garris, R., Ahlers, R. and Driskell, E.J. (2002) Games, Motivation, and Learning: A Research and Practice Model. *Simulation and Gaming*, 33(4), pp. 441–467.
Gee, J.P. (2003a) *What Video Games Have to Teach Us About Learning and Literacy*. New York: Palgrave Macmillan.
Gee, J.P. (2003b) Good Video Games and Good Learning. Available at: https://academiccolab.org/resources/documents/Good_Learning.pdf (Accessed 03/01/2021).
Gee, J.P. (2004) *Situated Language and Learning: A Critique of Traditional Schooling*. London: Routledge.
Gee, J.P. (2005a) Why are Video Games Good for Learning? Retrieved January 1, 2018, from http://academiccolab.org/resources/documents/MacArthur.pdf (Accessed 20/04/2016).
Gee, J.P. (2005b) *Why Video Games Are Good for Your Soul: Pleasure and Learning*. Melbourne: Common Ground.
Gee, J.P. (2007) *Good Video Games and Good Learning: Collected Essays on Video Games, Learning, and Literacy*. New York: Peter Lang.
Gee, J.P. (2008) *Getting over the Slump: Innovation Strategies to Promote Children's Learning*. New York: The Joan Ganz Cooney Center at Sesame Workshop.
Gee, J.P., Hayes, E., Torres, R.J., Games, I.A., Squire, K. and Salen, K. (2008) Playing to Learn Game Design Skills in A Game Context. In: *Proceeding ICLS'08 Proceedings of the 8th International Conference on International Conference for the Learning Sciences*, Vol. 3, pp. 368–374.
Gee, J.P. and Hayes, E.R. (2011) *Language and Learning in the Digital Age*. Routledge.
Gibbons, J. and Gray, M. (2004) Critical Thinking as Integral to Social Work Practice. *Journal of Teaching in Social Work*, 24, pp. 19–37.
Glaser, B.G. and Strauss, A.L. (1967) *The Discovery of Grounded Theory*. Chicago: Aldine.
Gomes, A. and Mendes, A.J. (2007) Learning to Program-Difficulties and Solutions. In *International Conference on Engineering Education–ICEE*, Vol. 7.
Granic, I., Lobel, A. and Engels, R.C. (2014) The Benefits of Playing Video Games. *American Psychological Association*, 69(1), pp. 66–78. DOI: 10.1037/a0034857
Hackett, G. and Betz, N.E. (1989) An Exploration of the Mathematics Self-Efficacy/Mathematics Performance Correspondence. *Journal for Research in Mathematics Education*, 20(3), pp. 261–273.
Hamari, J., Shernoff, D.J., Rowe, E., Coller, B., Asbell-Clarke, J. and Edwards, T. (2016) Challenging Games Help Students Learn: An Empirical Study on Engagement, Flow and Immersion in Game-Based Learning. *Computers in Human Behavior*, 54, pp. 170–179.
Hamari, J. and Tuunanen, J. (2014) Player Types: A Meta-Synthesis. *Transactions of the Digital Games Research Association*, 1(2), pp. 29–53.
Harel, I. and Papert, S. (1990) *Software Design as a Learning Environment*. Massachusetts Institute of Technology Cambridge, MA: MIT Media Laboratory.
Harel, I. and Papert, S. (1991) *Constructionism*. Norwood, NY: Ablex Publishing Corporation.
Hargreaves, A. (2003) *Teaching in the Knowledge Society: Education in the Age of Insecurity*. New York: Teachers' College Press.
Holbert. N., Penney, L. and Wilensky, U. (2010) *Bringing Constructionism to Action Game-Play, Constructionism*. Paris. Available at: https://www.google.co.uk/search?q=Bringing+Constructionism+to+Action+Game-Playandoq=Bringing+Constructionism+to+Action+Game-Playandaqs=chrome..69i57.7170j0j8andsourceid=chromeandie=UTF-8 (Accessed 20/03/2017).
Hoover, W.A. (1996) The Practice Implications of Constructivism. *SEDL Letter*, 9(3), pp. 1–2.
Horn, B., Hoover, A.K. Barnes, J. Folajimi, Y. Smith, G. and Harteveld, C. (2016) Opening the Black Box of Play: Strategy Analysis of An Educational Game. In

Proceedings of the 2016 Annual Symposium on Computer-Human Interaction in Play, pp. 142–153.

Hsu, C.-Y., Tsai, C.-C. and Wang, H.-Y. (2012) Facilitating Third Graders' Acquisition of Scientific Concepts Through Digital Game-Based Learning: The Effects of Self-Explanation Principles. *Asia-Pacific Education Researcher*, 21(1), pp. 71–82.

Hunicke, R., LeBlanc M. and Zubek, R. (2004) MDA: A Formal Approach to Game Design and Game Research. Available at: https://www.google.co.uk/search?q=MDA%3A+A+Formal+Approach+to+Game+Design+and+Game+Researchandoq=MDA%3A+A+Formal+Approach+to+Game+Design+and+Game+Researchandaqs=chrome..69i57j69i58.3868j0j8andsourceid=chromeandie=UTF-8 (Accessed 8/10/2016).

Illeris, K. (2007) *How We Learn: Learning and Non-Learning in School and Beyond*. London: Routledge.

Jackson, N. and Shaw, M. (2007) Developing Perspectives on Creativity in Higher Education. In: N. Jackson, M. Oliver, M. Shaw, and J. Wisdom (ed.). *Developing Creativity in Higher Education: An imaginative curriculum*. London: Routledge.

Jagger, S., Siala, H. and Sloan, D. (2016) It's All in The Game: A 3D Learning Model for Business Ethics. *Journal of Business Ethics*, 137(2), pp. 383–403.

Jameson, M.M. and Fusco, B.R. (2014) Math Anxiety, Math Self-Concept, and Math Self-Efficacy in Adult Learners Compared to Traditional Undergraduate Students. *Adult Education Quarterly*, 64(4), pp. 306–322.

Jarvis, M. (2005) *'Motivation' in Psychology of Learning and Teaching*. Cheltenham: Nelson Thornes.

Jeffrey, B. and Craft, A. (2004) Teaching Creatively and Teaching for Creativity: Distinctions and Relationships. *Educational Studies*, 30(1), pp. 77–87.

Jenson, J. and Droumeva, M. (2015) Making Games with Game Maker: A Computational Thinking Curriculum Case Study. In *ECGBL 2015–9th European Conference on Games Based Learning: ECGBL2015*. Academic Conferences and Publishing Limited, p. 260.

Jones, M.G. (1998) Creating Engagement in Computer-Based Learning Environments. Available at: http://itech1.coe.uga.edu/itforum/paper30/paper30.htmI (Accessed 12/04/2016).

Kafai, Y. (2006) *Playing and Making Games for Learning: Instructionist and Constructionist Perspectives for Game Studies. Games and Culture*, 1(1), pp. 36–40.

Kafai, Y.B. and Resnick, M. eds. (2012) *Constructionism in Practice: Designing, Thinking, and Learning in a Digital World*. Routledge. Available at: https://www.google.co.uk/books/edition/Constructionism_in_Practice/KGkIEQAAQBAJ?hl=en&gbpv=1&dq=Kafai,+Y.B.+and+Resnick,+M.+eds.+(2012)+Constructionism+in+Practice:+Designing,+Thinking,+and+Learning+in+a+Digital+World.+Routledge.&pg=PP1&printsec=frontcover

Kato, P.M., Cole, S. Bradlyn, A.S. and Pollock, B.H. (2008) Video Games Improves Behavioural Outcomes in Adolescents and Young Adults with Cancer: A Randomized Trial. *Pediatrics*, 122(2), pp. e305–e217.

Kessner, M., Wood, J. Dillon, R.F. and West, R.L. (2001) On the Reliability of Usability Testing. In *Proceedings of the Extended Abstracts on Human Factors in Computing Systems (CHI '01)*, p. 97.

Kim, B. (2001) Social Constructivism. *Emerging Perspectives on Learning, Teaching, and Technology*, 1(1), p. 16.

Kirriemuir, J. and Mcfarlane, A. (2004) Literature Review in Games and Learning. Available at: https://telearn.archives-ouvertes.fr/hal-00190453/document (Accessed 12/04/2020).

Knight, P.T. (2001) A Briefing on Key Concepts – Formative and Summative, Criterion and Norm Referenced Assessment. *Learning and Teaching Support Network Generic Centre*, Vol. 30. http://www.heacademy.ac.uk

Kodugamelab (2015) Available at: https://www.kodugamelab.com/ (Accessed 8/11/2018).
Kolb, A.Y. and Kolb. D.A. (2008) The Learning Way: Meta-Cognitive Aspects of Experiential Learning. *Simulation and Gaming*, 40(3), pp. 297–327.
Kolb, D.A. (1984) *Experiential Learning: Experience as the Source of Learning and Development*, Vol. 1. Englewood Cliffs, NJ: Prentice-Hall.
Lenhart, A., Kahne, J., Middaugh, E. Macgill, A.R., Evans, C. and Vitak, J. (2008) Teens, Video Games, and Civics: Teens' Gaming Experiences Are Diverse and Include Significant Social Interaction and Civic Engagement. *Pew Internet & American Life Project*. Available at: https://www.google.co.uk/search?q=Lenhart+et+al.%2C+2008 andoq=Lenhart+et+al.%2C+2008andaqs=chrome..69i57j0.1244j0j8andsourceid= chromeandie=UTF-8 (Accessed 20/8/2017).
Levine, D.M., Krehbiel, T.C. and Berenson, M.L. (2003) *Business Statistics: A First Course*, 3rd Edition. New Jersey: Pearson Education Ltd.
Lewis, G. (2013) Emotion and Disaffection with School Mathematics. *Research in Mathematics Education*, 15(1), pp. 70–86.
Li, F.W.B. and Watson, C. (2011) *Game-Based Concept Visualization for Learning Programming*. Scottsdale, Arizona, USA: ACM. 978-1-4503-0994-3/11/12.
Li, Z.Z., Cheng, Y.B. and Liu, C.C. (2013) A Constructionism Framework for Designing Game-Like Learning Systems: Its Effect on Different Learners. *British Journal of Educational Technology*, 44(2), pp. 208–224.
Lister, R., Adams, E.S. Fitzgerald, S. Fone, W. Hamer, J. Lindholm, M. McCartney, R. Moström, J.E. Sanders, K. Seppälä, O. and Simon, B. (2004) A Multi-National Study of Reading and Tracing Skills in Novice Programmers. In *ACM SIGCSE Bulletin*, 36(4), pp. 119–150. ACM.
Litosseliti, L. (2005) *Using Focus Groups in Research*. London: Continuum.
Loftus, C., Thomas, L. and Zander, C. (2011) Can Graduating Students Design: Revisited. In *Proceedings of the 42nd ACM Technical Symposium on Computer Science Education*, pp. 105–110.
Lyons, I.M. and Beilock, S.L. (2012) When Math Hurts: Math Anxiety Predicts Pain Network Activation in Anticipation of Doing Math. *PLoS One*, 7(10), p. e48076.
Majgaard, G. (2016) Creating Games in the Classroom: From Native Gamers to Reflective Designers. In: Munkvold, R. (ed.). *Games Based Learning for Researchers, Teachers and Students*, Vol. 1. Reading: Academic Conferences and Publishing International Ltd. ISBN:978-1-911278-12-890000.
Malaka, R. (2014) How Computer Games Can Improve Your Health and Fitness. In: Gobel, S., and Wiemeyer, J., (ed.). *Games for Training Education, Health and Sports, 4th Conference on Serious Games, GameDays 2014*, Darmstadt, Germany, April 1–5, 2014, Springer.
Malone, T.W. (1980) What Makes Things Fun to Learn? A Study of Intrinsically Motivating Computer Games. Available at: http://cci.mit.edu/malone/Malone%20 computer%20games%20bibliography.htm
Malone, T.W. and Lepper, M.R. (1987) 'Making Learning Fun: A Taxonomy of Intrinsic Motivations for Learning. Aptitude, Learning, and Instruction': Volume 3: Cognitive and Affective Process Analyses. R. E. Snow and M. J. Farr. Hillsdale, USA, Erlbaum.
Maloney, E.A., Schaeffer, M.W. and Beilock, S.L. (2013) Mathematics Anxiety and Stereotype Threat: Shared Mechanisms, Negative Consequences and Promising Interventions. *Research in Mathematics Education*, 15(2), pp. 115–128.
Marciano, J.N., Miranda, L.C. and Miranda, E.E.C. (2014) Evaluating Multiple Aspects of Educational Computer Games. Available at: https://www.hindawi.com/ journals/ijcgt/2014/139205/ (Accessed 27/08/2017).
Marklund, B.B. and Taylor, A.S.A. (2016) Creating Games in the Classroom: From Native Gamers to Reflective Designers. In: Munkvold, R. (ed.). *Games Based*

learning for Researchers, Teachers and Students, Vol. 1. Reading: Academic Conferences and Publishing International Ltd. ISBN: 978-1-911278-12-890000.

Maslow, A.H. (1970) *Motivation and Personality*, 2nd Edition. New York; London: Harper and Row.

Mayer, E.R. (2014) *Computer Games for Learning: An Evidence Based Approach*. London: MIT.

Mayer, I., Warmelink, H. and Bekebrede, G. (2013) Learning in a Game-Based Virtual Environment: A Comparative Evaluation in Higher Education. *European Journal of Engineering Education*, 38(1), pp. 85–106.

Mayo, M. (2009) Video Games: A Route to Large-Scale STEM education? *Science*, 323, pp. 79–82.

McConnell, D. (2000) *Implementing Computer Supported Cooperative Learning*, 2nd Edition. London: Kogan Page.

Mccrindle, M. (2018) What Comes after Generation Z? Introducing Generation Alpha. Available at: https://mccrindle.com.au/insights/blogarchive/what-comes-after-generation-z-introducing-generation-alpha/ (Accessed 15/10/2018).

McFarlane, A., Sparrowhawk, A., and Heald, Y. (2002) *Report on the Educational Use of Games: An Exploration by TEEM on the Contribution Which Games Can Make to the Educational Process*. Cambridge, UK: TEEM. http://www.teem.org.uk/publications/teem_gamesined_full.pdf (Accessed 06/12/2015).

McGonigal, J. (2015) *Superbetter: A Revolutionary Approach to Getting Happier Stronger and Happier*. Braver and More Resilient, New York: Penguin Press.

Milane, I. and Rowe, G. (2002) Difficulties in Learning and Teaching Programming-Views of Students and Tutors. *Education and Information Technologies*, 7(1), pp. 55–66.

Monroy-Hernandez, A. and Resnick, M. (2008) Empowering Kids to Create and Share Programmable Media. *Interactions*, 15(2), pp. 50–53. Retrieved August 30, 2015.

Moreno-Ger, P., Burgos, D. Martínez-Ortiz, I. Sierra, J.L. and Fernández-Manjón, B. (2008) Educational Game Design for Online Education. *Computers in Human Behavior*, 24(6), pp. 2530–2540. DOI: 10.1016/j.chb.2008.03.012

Moreno-Ger, P., Torrente, J. Hsieh, Y.G. and Lester, W.T. (2012) Usability Testing for Serious Games: Making Informed Design Decisions with User Data. *Advances in Human-Computer Interaction*, 2012. DOI: 10.1155/2012/369637

Naik, N. (2017) The Use of GBL to Teach Mathematics in Higher Education. *Innovations in Education and Teaching International*, 54(3), pp. 238–246. DOI: 10.1080/14703297.2015.1108857

Nicholls, D., De Charon, L. and Hutkin, R. (2010) Development of Critical Thinking and Creativity: Practical Guidelines for the Postsecondary Classroom. *ATEA Journal*, 38(1), p. 12.

Nielsen, J. (1995) Usability Inspection Methods, *CHI'95 Mosaic of Creativity*, pp. 7–11.

Nielsen, J. (2000) *Designing Web Usability*. USA: New Riders Publishing.

Nygård, S., Kolås, L. and Sigurdardottir, H. (2015) Teachers' Experiences Using KODU as a Teaching Tool. In: *European Conference on Games Based Learning*, Academic Conferences International Limited. p. 416.

Otkay, J.S. (2012) *Grounded Theory*. Oxford, England: Oxford University Press.

Paavilainen, J. (2010) Critical Review on Video Game Evaluation Heuristics: Social Games Perspective. *Proceedings of the International Academic Conference on the Future of Game Design and Technology*, pp. 56–65. DOI: 10.1145/1920778.1920787

Pajares, F. and Kranzler, J. (1995) Self-Efficacy Beliefs and General Mental Ability in Mathematical Problem-Solving. *Contemporary Educational Psychology*, 20, pp. 426–443.

Palincsar, A.S. (1998) Social Constructivist Perspectives on Teaching and Learning. *Annual Review of Psychology*, 49, pp. 345–375.

Papert, S. (1993) *Mindstorms: Children, Computers and Powerful Ideas*. New York: Basic Books.

Paras, B. and Bizzocchi, J. (2005) Game, Motivation, and Effective Learning: An Integrated Model for Educational Game. In *Proceedings of DiGRA Conference on Changing Views–Worlds of Play*. Available at: https://www.researchgate.net/publication/221217604_Game_Motivation_and_Effective_Learning_An_Integrated_Model_for_Educational_Game_Design (Accessed 27/09/2016).
Park, H.J. and Kim, S.B. (2013) Guidelines of Serious Game Accessibility for the Disabled. In *Proceedings of the International Conference on Information Science and Applications (ICISA '13)*, pp. 1–3, IEEE.
Pinelle, D., Wong, N., Stach, T. and Gutwin, C. (2009) Usability Heuristics for Networked Multiplayer Games. *Proceedings of the ACM 2009 International Conference on Supporting Group Work*, pp. 169–178.
Preconception. (2020) In: Dictionary.com. [Online] Available at: https://www.dictionary.com/browse/preconception (Accessed 01/08/2020).
Preece, Y., Rogers, Y. and Sharp, H. (2015) *Interaction Design: Beyond Human-Computer Interaction*, 4th Edition. Chichester: John Wiley and Sons Ltd.
Prensky, M. (2002) The Motivation of Gameplay: The Real Twenty-First Century Learning Revolution. *On the Horizon*, 10(1), pp. 5–11.
Prensky, M. (2005) Listen to Natives. *Association for Supervision and Curriculum Development Educational Leadership*, pp. 9–13. Available at: https://www.scirp.org/reference/referencespapers?referenceid=633456
Prensky, M. (2006) *'Don't Bother Me Mom, I'm Learning!': How Computer and Video Games Are Preparing Your Kids for the Twenty-First Century*. St Paul, MN: Paragon House.
Punch, K.F. (2009) *Introduction to Research Methods in Education*, London: Sage Publications Ltd.
Robins, A., Runtree, J. and Rountree, N. (2003) Learning and Teaching Programming: A Review and a Discussion. *Computer Science Education*, 13(2), pp. 137–172.
Rogoff, B. (1998) Cognition as a Collaborative Process. In W. Damon, D. Kuhn and R.S. Seigler (ed.), *Handbook of Child Psychology*, 5th Edition. New York: Wiley.
Roth, W.M. (2000) Authentic School Science: Intellectual Traditions In R. McCormick and C. Paechter (ed.), *Learning & Knowledge*. London, UK: Paul Chapman Publishing, pp. 6–20.
Russ, T., and Drury-Grogan, M.L. (2013) Assessing the Impacts of a Business Communication Simulation on Students' Self-Perceptions. *Eastern Communication Association*, 61(5), pp. 584–595. DOI: 10.1080/01463373.2013.822404
Salen, K. and Zimmerman, E. (2011) The Game Designer Reader, A Rules of Play Anthology. Roger Caillois: The Definition of Play. *The Classification of Games*. Cambridge, Massachusetts. London, England: The MIT Press.
Schell, J. (2015) *The Art of Game Design: A Book of Lenses*, 2nd Edition. USA: CRC Press.
Schell, J. (2020) *The Art of Game Design: A Book of Lenses*, 3rd Edition. USA: CRC Press.
Senge, P. (1990) *The Fifth Discipline: The Art and Practice of Learning Organization*. United Kingdom: Random House Business Books.
Senge, P.M. (2006) *The Fifth Discipline: The Art and Practice of Learning Organisation*. London: Random House Business Books.
Shell, D.F., Murphy, C.C. and Bruning, R.H. (1989) Self-efficacy and Outcome Expectancy Mechanisms in Reading and Writing Achievement. *Journal of Educational Psychology*, 81(1), p. 91.
Shunk, D.H. (2000) *Learning Theories: An Educational Perspective*, 3rd Edition. Upper Saddle River, NJ: Prentice-Hall.
Siegel, R.G., Galassi, J.P. and Ware, W.B. (1985) A Comparison of Two Models for Predicting Mathematics Performance: Social Learning Versus Math Aptitude-Anxiety. *Journal of Counseling Psychology*, 32, pp. 531–538.

Sindre, G., Natvig, L. and Jahre, M. (2009) Experimental Validation of the Learning Effect for A Pedagogical Game on Computer Fundamentals. *IEEE Transactions on Education*, 52 (1), pp. 10–18.

Skilbeck, A. (2017) Dewey on Seriousness, Playfulness and the Role of the Teacher. *Education Sciences*, 7(1), p. 16.

Slyman, S. (2014) I Hate Maths! An Investigation into Students' Attitudes Towards Mathematics. MA Thesis, University of Roehampton, London, United Kingdom.

Slyman, S. (2018) *Games Based Learning in Universities, a Life Learning Experience*. New York: Business Expert Press Expert Insights Publishing. ISBN:978-1-94819-801-1.

Slyman, S. (2022) Acumenous Game-Based Learning in Simulation Games and Applied Statistics. PhD Thesis, Goldsmiths, University of London, UK.

Smith Report (2017) Available online at: https://www.gov.uk/government/publications/smith-review-of-post-16-maths-report-and-government-response (Accessed 20/12/2018).

Sobke, H., Chan, E. Buttlar, R. Grobe-Wortmann, J. and Londong, J. (2014) Cat King's Metamorphosis. The Reuse of an Educational Game in a Further Technical Domain. In: Gobel, S., and Wiemeyer, J. (ed.), *Games for Training Education, Health and Sports, 4th Conference on Serious Games, GameDays 2014*, Darmstadt, Germany, April 1–5, 2014, Springer.

Song, S. and Lee, J. (2007) Key Factors of Heuristic Evaluation for Game Design: Towards massively multi-player online Roleplaying game. *International Journal of Human-Computer Studies*, 65, pp. 709–723.

Sweetser, P., Johnson, D. and Wyeth, P. (2012) Revisiting the Game Flow Model with Detailed Heuristics. *Journal of Creative Technologies*, 2012(3), pp. 1–16.

Sweetser, P. and Wyeth, P. (2005) GameFlow: A Model for Evaluating Player Enjoyment in Games. *Computers in Entertainment*, 3(3), p. 3. DOI: 10.1145/1077246.1077253

Tamura, H. (2007) Foundational Value of Mathematics Education for Management Curriculum. *International Statistical Review*, 75(3), pp. 397–405. DOI: 10.1111/j.1751-5823.2007.00032.x

Tao, Y.H., Cheng, C.J. and Sun, S.Y. (2012) Alignment of Teacher and Students Perceptions on the Continued Use of Business Simulation Games. *Educational Technology and Society*, 15(3), pp. 177–189.

TED (2019) *Souad Slyman: Why Change Attitudes Towards Mathematics [YouTube]*. TedxGul, London: Goldsmiths, University of London. Available at: https://www.youtube.com/watch?v=leNDJMNJoNY (Accessed 26/04/2019).

Timonen, V., Foley. G. and Conlon. C. (2018) Challenges when Using Grounded Theory: A pragmatic introduction to Doing GT Research. *International Journal of Qualitative Methods*, 17, pp. 1–10. DOI: 10.1177/1609406918758086

Tondello, G.F., Kappen, D.L. Mekler, E.D. Ganaba, M. and Nacke, L.E. (2016) Heuristic Evaluation for Gameful Design. *CHI PLAY'16*. ACM. 978-1-4503-4458-6/16/10. DOI: 10.1145/2968120.2987729

van der Spek, E.D., van Oostendorp, H. and Meyer, J.-J.C.H. (2013) Introducing Surprising events can stimulate deep learning in a serious game. *British Journal of Educational Technology*, 44(1), 156–169.

Vlachopoulos, D. and Makri, A. (2017) The Effect of Games and Simulations on Higher Education: A Systematic Literature Review. *International Journal of Educational Technology in Higher Education*, 14(1), pp. 22. DOI: 10.1186/s41239-017-0062-1

Vos, L. (2014) Marketing Simulation Games: A Review of Issues in Teaching and Learning. *The Marketing Review*, 14(1), pp. 67–96. DOI: 10.1362/146934714X13948909473220

Vos, L. and Brennan, R. (2010) Marketing Simulation Games: Student and Lecturer Perspectives. *Marketing Intelligence and Planning*, 28(7), pp. 882–897.

Whitton, N. (2011) Games-Based Experiences for Learning. Futurelab Innovation in Education. Available at: https://www.nfer.ac.uk/media/1775/futl11.pdf (Accessed 02/04/2016).

Wolfe, J. (1997) The Effectiveness of Business Games in Strategic Management Coursework. *Simulation and Gaming*, 28(4), pp. 360–376. DOI: 10.1177/1046878197284003

Women in STEM – 'Facts and Statistics' (Reporting period 2012–2015) IET. Available at: https://communities.theiet.org/files/8042 (Accessed 20/06/2017).

Yin, R.K. (1994) *Case Study Research: Design and Methods*, 2nd Edition. London: Sage.

Young, M. (2008) *"Powerful Knowledge" and "Knowledge of the Powerful": Some Reflections*. London: Presentation at the Institute of Education.

Zimmerman, B.J. (2000) Self-Efficacy: An Essential Motive to Learn. *Contemporary Educational Psychology*, 25(1), pp. 82–91. DOI: 10.1006/ceps.1999.1016

8 Exploring the Dynamics of Inquisitional and Acumenous GBL in Students' Experiences with Mathematics and Statistics

Unlike the predominantly qualitative nature of Chapters 6 and 7, which warranted a detailed analysis approach, Chapter 8's investigation adopts a quantitative research method, employing questionnaires. Quantitative research offers a quick and effective means to capture a representative sample of the population and identify overarching patterns within it (Bryman, 2016, 2022; Dwyer & Slyman, 2016; Cohen et al., 2018; Anderson, 2020). Through this methodological shift, we delve into the realm of students' experiences in learning Mathematics and Statistics using the two GBL approaches.

Chapter 8 acts as a gateway to understanding the multifaceted dynamics of *Inquisitional* and *Acumenous* GBL. It sheds light on the experiences of students, their preferences, and the impact of each approach on their learning journey in Mathematics and Statistics. Through a comprehensive exploration of the emergent themes, this chapter provides a deeper understanding of the strengths, weaknesses, and unique contributions of both *Inquisitional* and *Acumenous* GBL approaches. As we navigate the complexities of GBL, Chapter 8 sets the stage for a broader understanding of how these two approaches shape students' experiences, learning outcomes, and perceptions. By unraveling the layers of *Inquisitional* and *Acumenous* GBL, we inch closer to harnessing their transformative potential and paving the way for more effective and engaging Mathematics and Statistics education.

Following findings from this chapter, which suggest that quiz GBL (*Inquisitional*) and simulation GBL (*Acumenous*) are equivalent due to the lack of the game element, this chapter concludes with a *theoretical cycle* (The Game Cognitive Model, Figure 8.4). This model justifies the cognitive processes that users experience while interacting with the game. This Game Cognitive model is developed from the literature (see Chapters 2 and 3) and is validated by the research findings we will explore.

Within this chapter, the similarities and differences between *Inquisitional* and *Acumenous* GBL take the center stage. The fundamental characteristics of

DOI: 10.4324/9781032723426-11

both approaches are meticulously examined, exploring students' experiences in learning Mathematics and Statistics through quiz-based and simulation-based GBL. By delving into their perceptions and understandings of these two modes of GBL, key insights and emergent themes surface, illuminating the intricacies of *Inquisitional* and *Acumenous* GBL.

I will now delve into each emergent theme, drawing upon discoveries from the surveys, starting with simulation-based/*Acumenous* GBL then moving on to quiz-based/*Inquisitional* GBL.

Exploring the Potential of Acumenous GBL/Simulation based in Mathematics and Statistics

Findings reveal that when it comes to participants' experiences in learning Mathematics and Statistics using simulation GBL (see Figure 8.1), many people found simulation GBL intrinsically engaging and motivating (Malone, 1980; Faria, 2001; Gee, 2003a, 2003b, 2004, 2005a, 2005b, 2007, 2008, 2011; Prensky, 2005; Dondlinger, 2007; Gee et al., 2008; Moreno-Ger et al., 2008; Mayo, 2009; Li et al., 2013; Mayer et al., 2013; Vos, 2014; Malaka, 2014; McGonigal, 2015; Jagger et al., 2016; Hamari et al., 2016; Vlachopoulos and Makri, 2017; Cardinot and Fairfield, 2019; Schell, 2020). The survey findings presented offer a multifaceted perspective on the utilization of simulation GBL as a tool for learning and understanding Statistics. A resounding 89 percent of participants express keen interest in leveraging simulation games for educational purposes, underscoring the widespread recognition of its potential in enhancing learning experiences. This overwhelming enthusiasm signifies a paradigm shift toward embracing innovative teaching methodologies. Innovative teaching methodologies where Mathematics and Statistical learning can create positive experience (Garris et al., 2002; Lazzaro, 2004; Mayer et al., 2013; Jagger et al., 2016; Cardinot and Fairfield, 2019), engaging and empowering learners in problem-based solving and deep understanding (Gee, 2003a, 2003b, 2004, 2005a, 2005b, 2007, 2011).

An intriguing aspect highlighted by the data is the engagement factor inherent in simulation games, particularly in working with numbers. Fifty-eight percent of respondents find this aspect both intriguing and stimulating, indicating the effectiveness of gamified approaches in captivating learners' interest and fostering active participation. Moreover, the majority, 69 percent, recognize simulation games as an effective means of improving problem-solving skills, affirming its value in cultivating critical thinking abilities (Gee, 2003a, 2003b, 2004, 2005a, 2005b, 2007, 2011).

However, divergent perspectives emerge regarding the efficacy of simulation games in specific skill areas, notably Mathematical problem-solving. While 41 percent acknowledge its benefits, there exists a notable proportion, 27 percent, who express skepticism. This disparity underscores the need for further exploration and refinement of simulation game designs to effectively address diverse learning needs and preferences. This will be explored further in Chapter 9.

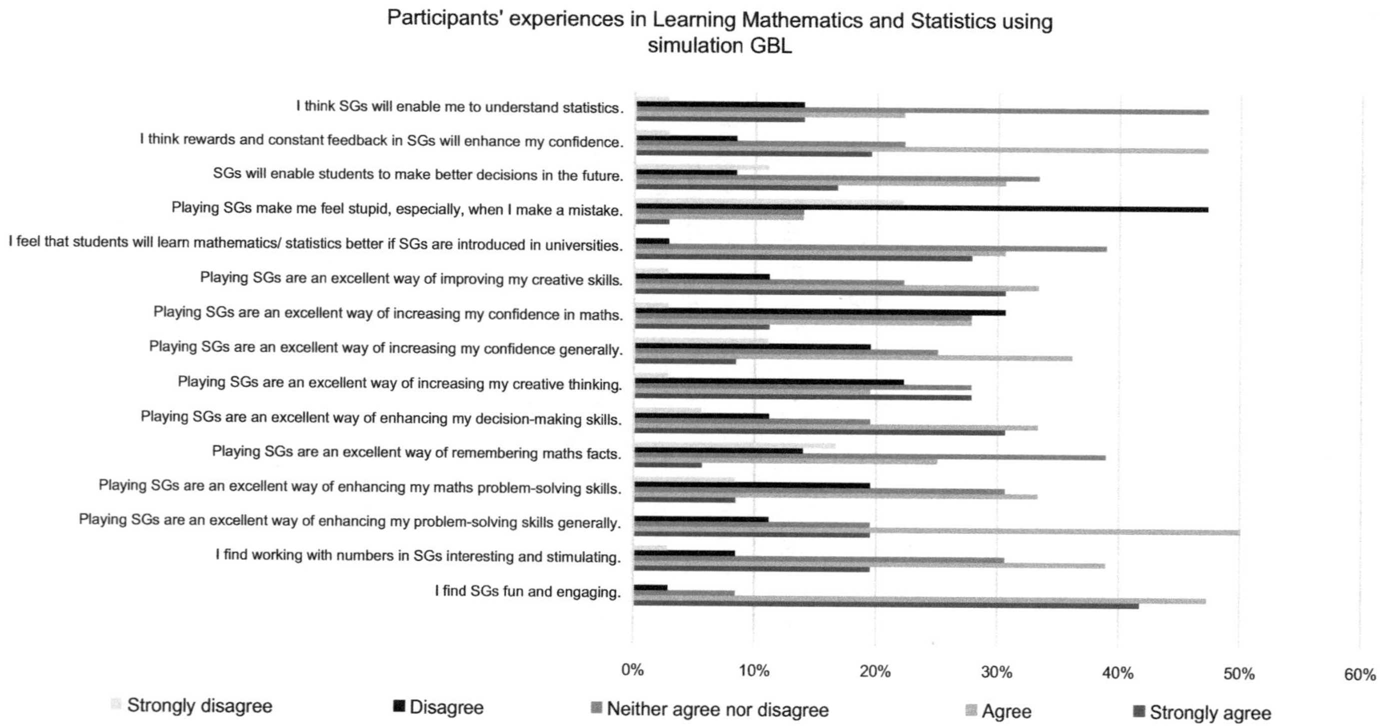

Figure 8.1 Participants' experiences in learning Mathematics and Statistics using simulation GBL (Slyman, 2022:154).

Additionally, participants offer varied insights into the potential of simulation games in fostering creativity and confidence. While 47 percent believe in its efficacy for nurturing creative thinking, only 44 percent perceive it as effective for boosting confidence in general. This highlights the nuanced interplay between gameplay experiences and their impact on learners' psychological well-being and self-perception.

Looking toward educational integration of simulation GBL into HE, a significant proportion, 59 percent, advocate for the introduction of simulation games in universities, envisioning its potential to enhance students' grasp of complex Mathematical and Statistical concepts. However, it's crucial to acknowledge the challenges associated with incorporating simulation games into traditional educational settings, particularly concerning the aspects of play, engagement, and social interaction.

Furthermore, the data unveils the pivotal role of trial and error in simulation games, with 69 percent cherishing the freedom to experiment without fear or judgment. This element of exploration fosters a conducive learning environment, encouraging learners to explore concepts and refine their understanding through iterative processes.

To sum up, while simulation games hold promise as a transformative tool in Mathematics and Statistics education, their successful integration requires careful consideration of pedagogical strategies and learner needs. By leveraging the inherent strengths of simulation games and addressing potential challenges, educators can harness their full potential to revolutionize Mathematical and Statistical education and empower learners to thrive in an increasingly complex data-driven world.

Exploring the Potential of Inquisitional/Quiz-Based GBL in Mathematics and Statistics

When considering participants' Mathematics and Statistics learning experiences through quiz-based learning (GBL) (see Figure 8.2), the survey data provided sheds light on the diverse perceptions surrounding quiz GBL, offering a nuanced understanding of its efficacy in enhancing various cognitive skills, particularly in the realm of mathematics and problem-solving. Delving into the findings, a mosaic of viewpoints emerges, presenting both proponents and sceptics of this innovative educational approach.

Initially, it's evident that a significant proportion of respondents, comprising 47 percent, resonated with the notion that quiz games are enjoyable and engaging. This sentiment underscores the potential of GBL to captivate learners, making the educational process inherently more appealing. Moreover, a majority, 60 percent, found the incorporation of numbers within quiz games to be not only interesting but also stimulating, emphasizing the efficacy of this method in fostering engagement with Mathematical concepts. This finding is interesting and contrary to what Chapter 7 has found out. In other words, quiz GBL is not engaging nor stimulating. Additional investigations have been carried out in Chapter 9, see this chapter for further details.

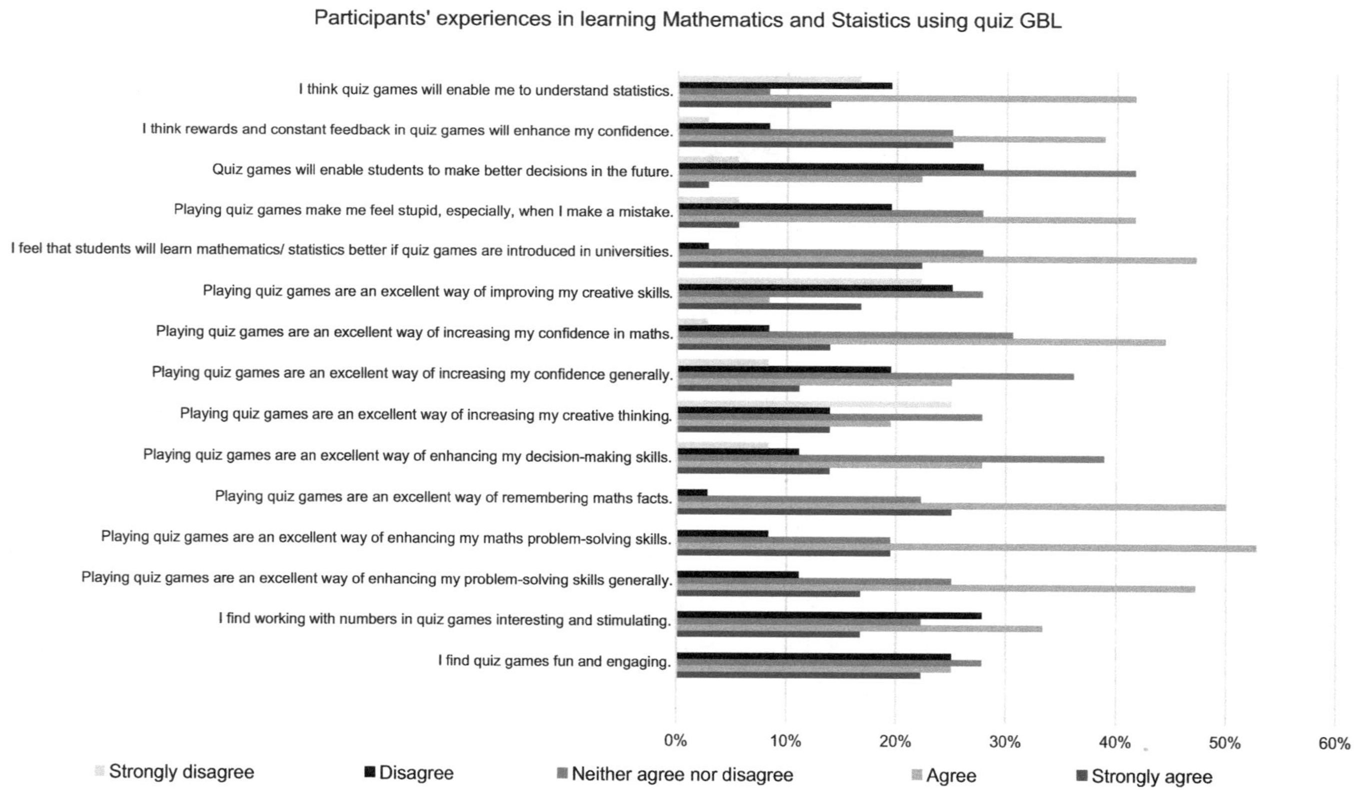

Figure 8.2 Participants' experiences in learning Mathematics and Statistics using quiz GBL (Slyman, 2022:155).

Crucially, the data underscores the perceived benefits of quiz GBL in honing problem-solving skills, with a resounding 64 percent attesting to its efficacy. Furthermore, an overwhelming 72 percent believe in its potential to enhance Mathematical problem-solving skills, reflecting a widespread recognition of its value in this domain unlike Chapter 7. Again, further investigations have been carried out in Chapter 9, see Chapter 9 for further details.

However, it's noteworthy that while the majority recognize the benefits, there exists a subset of respondents who harbor reservations. For instance, a significant portion, 39 percent, expressed skepticism regarding the ability of quiz GBL to enhance creative thinking. This is in line with previous investigations in earlier chapters and would be further explored in Chapter 9. Similarly, only a quarter of respondents view it as a viable means of improving creative skills, indicating a perceived limitation in this aspect of learning.

Furthermore, the survey highlights the potential psychological implications associated with quiz GBL. Notably, nearly half of the participants reported feelings of incompetence, particularly when confronted with mistakes during gameplay. This underscores the importance of addressing emotional well-being alongside cognitive development in educational settings.

Despite these reservations, a notable 75 percent of respondents perceive quiz GBL as an effective tool for retaining Mathematical facts, underscoring its utility in facilitating long-term retention. Additionally, a substantial majority, 69 percent, advocate for its introduction in university curricula, signaling a growing acceptance of GBL as a viable educational strategy. See Figure 8.3.

In summary, while the data present a mixed outlook on quiz GBL, it questions its potential to maintain educational paradigms, particularly in the domain of Mathematics and Statistics education. By leveraging its strengths in engagement and retention, educators can harness the power of quiz GBL to

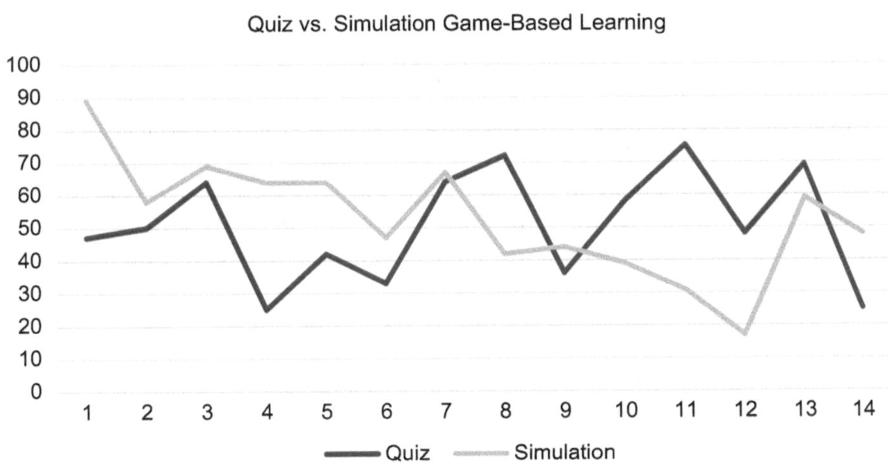

Figure 8.3 A comparison into quiz vs simulation GBL.

cultivate a generation of adept problem-solvers, albeit with due consideration to emotional well-beings and holistic skill development: Innovative teaching pedagogies and more thoughts should be considered when using Quiz GBL, see Chapter 9.

Are There Any Similarities and Differences between Quiz vs. Simulation GBL?

Survey results indicate that there are not many differences and similarities between the two types of GBL unlike research findings from the GBL education literature. For instance, quiz GBL primarily involves interactions centered around answering questions or solving quizzes, often with a competitive or self-paced format. On the other hand, simulation game-based learning typically immerses learners in virtual environments where they engage in experiential learning through simulated scenarios and activities. Therefore, emphasizing the element of interaction, immersion, fun, and play in simulation is superior unlike quiz GBL; however, survey findings indicate that there are the same.

Quiz GBL tends to focus more on testing and reinforcing specific knowledge or skills, often in a structured manner. In contrast, simulation GBL emphasizes contextualized learning experiences, allowing learners to explore and apply concepts in real-world situations.

When it comes to skill development, while both approaches aim to enhance learning outcomes, they often prioritize different sets of skills. Quiz GBL may focus more on memory recall and problem-solving within a confined scope, whereas simulation GBL may emphasize complex decision-making, critical thinking and problem-solving in dynamic environments, and practical application of concepts.

Feedback mechanisms in quiz GBL are often immediate and focused on correctness or performance metrics, such as scores or rankings. In simulation GBL, feedback is integrated into the gameplay experience, providing learners with insights into the consequences of their decisions and actions within the simulated environment.

Another significant finding is the element of *emotional engagement* in both quiz and simulation GBL. For instance, quiz GBL may elicit emotions related to competition, achievement, or frustration, fear and negative preconceptions particularly during assessment moments or even worse when used as a part of learning. Simulation GBL, on the other hand, may evoke a wider range of emotions, including curiosity, excitement, empathy, and perseverance, as learners navigate simulated challenges and experiences in a fun and playful manners where errors are tolerated unlike quiz GBL.

When thinking deeper about the *engagement* aspect in both GBL, both quiz GBL and simulation game-based learning leverage gamification elements to enhance learner engagement and motivation. By incorporating elements such as challenges, rewards, and interactivity, both approaches aim to create immersive learning experiences that capture learners' interest and sustain their attention.

Similarly, in *active learning* both approaches promote active learning by encouraging learners to interact with the learning materials actively. Whether through answering quiz questions or navigating simulated environments, learners are actively involved in the learning process, which can lead to deeper understandings and retention of concepts.

While the nature of *feedback integration* may differ, both quiz GBL and simulation GBL prioritize the integration of feedback into the learning experience. Timely and relevant feedback helps learners assess their progress, identify areas for improvement, and adjust their learning strategies accordingly.

Moreover, further investigations (i.e., Statistical tests/analysis) into participants' perception of quiz vs. simulation GBL reveal that participants did not see any major differences between quiz vs. simulation GBL. In other words, simulation GBL is similar to quiz GBL. Findings also suggest that without the real interaction and gameplay, the researcher and social interaction, it becomes apparent that participants' views show no difference at all between the two types of games (see also Figure 8.3).

As explained in Chapter 7, findings show that interviewees had positive experiences playing simulation games since reasons behind such engagement are perceived usefulness and fun, increased self-efficacy, concentration, more learning, social interaction, competition, fantasy, fun, challenge, goals/storylines, real-life characteristics of the games, control and game mechanics, as well as building skills like critical thinking, decision-making, problem-solving, leadership, and Statistical analysis.

In contrast, in this chapter, Chapter 8, participants did not play any games, since they had just questionnaires to fill in, and therefore they did not experience any engagement with these games. This is also because when interacting with IT, each player/learner comes with some perceptions (external variables) in mind of how they perceive the world and IT. These external variables are perceived usefulness and fun, which affect how preconception is formulated and therefore player's self-efficacy and confidence. This subsequently affects players' behavior in using the game as an interactive system. In this instance, the game was absent, preventing the various perspectives that influence a player's skills, knowledge, learning outcomes, and overall learning experience from unfolding. It's significant to revisit what constitute a game, as well as to discover and explore new theoretical perspectives and thought processes. Discussed next.

As previously outlined in *Chapter 2, a game is a fun playful dynamic learning activity with a fixed set of rules and limited time that can solve real-world problems*. It has specific rules, either win or lose, manipulation, curiosity, secrecy, and mystery. Games consist of game elements (learning objectives or storylines); Game mechanics (the functionalities of the game, the core mechanics and how the game behaves); Goals and objectives (what motivate players to achieve/gain from the game); Game motives: Challenge/competition; Game risks and Game interactions (players' interaction with the game and other players). Games can stimulate curiosity, creativity, challenge, fantasy, and complexity (Malone, 1980). It can also affect cognitive functions and motivation; immerse and engage players. Furthermore, games can improve learning through visualization (or problem-solving), experimentation, and creativity of play (Shell, 2015, 2020). There are many diverse types of games available which range from simulations, first-shooter games, adventure, Mathematics, memory, strategy to **GBL**.

GBL is a type of games that have clear learning outcomes where players can experiment with real-life examples in a risk-free environment. It consists of challenge, fantasy, complexity, and control, and these according to Malone (1980) and Gee (2003a, 2003b) are fundamental tools in game design. **Game characteristics** are: **Fantasy** as a crucial element/characteristic of the game, as it allows users to experience the beauty of imagination, immersing players, in an incredible world of fantasy where unbelievable becomes real; **Rules**/Goals are some of the components that govern how games' rules are determined and played (Garris et al., 2002). These game characteristics provide the structure of the game and contains fixed **rules**, strategies, and unpredictable results; **Sensory** stimuli is the **creative** sensation of the game that eludes the perception of the player into accepting the fantasy of the game world using sound effects, graphics, and other stimulus (Malone & Lepper, 1987; Garris et al., 2002; Shell, 2015); **Challenge**: Challenging games and meaningful goals should be made essential ingredients of games. This could be made according to levels, enabling players to track their performance, review feedback, and progress further up in levels. Mystery/**Curiosity** to unlock some of the game mysteries through discovering the unknown or new knowledge (e.g., adventure games); **Control** is where the learner has control over the game s/he is playing. According to Garris et al. (2002:11), games evoke "a sense of personal control when users are allowed to select strategies," this self-control over the game has increasingly engaged and motivated students to learn a great deal. These game engagements and interactions can be summarized using game theoretical model, discussed next.

Game Cognitive Model

Following Chapters 2 and 3 where we discussed the game literature, this section discusses the processes that goes through users' minds while interacting with the game.

The **Game Cognitive (GC)** Model (Figure 8.4) is a *theoretical cycle* that is developed based on the literature and validated by the research findings, it states that users go through many *cognitive processes* while interacting with the game. Through embracing the game—which involves many cognitive skills such as *planning, making decisions, and choices*—the players *interact, play, and engage* with the game, as a system that has mental challenges. This acts as a driving force in behavior/action where the game player obtains feedback as a winning or losing state. These can be summarized using the **Game Cognitive (GC)** Model, which is an integral part of the main social interaction system: **The Game**.

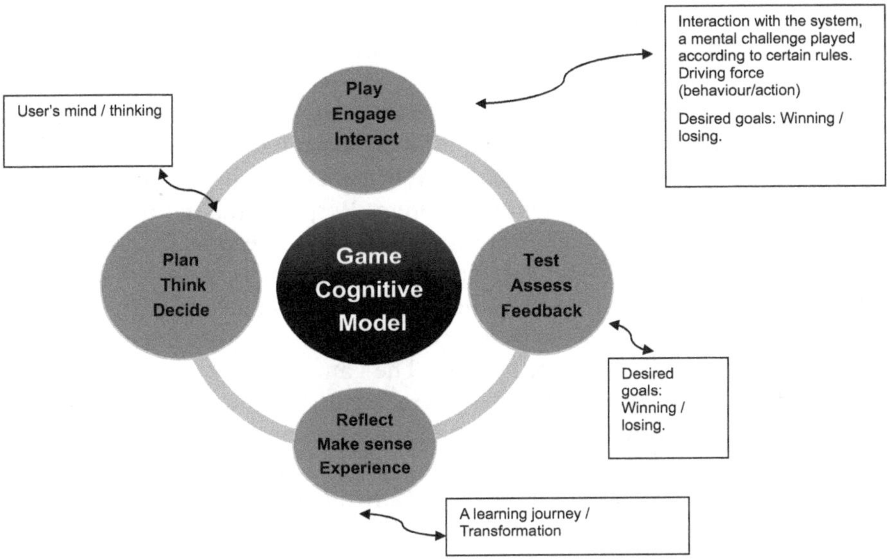

Figure 8.4 Game cognitive model.

The **Game Cognitive (GC)** Model **or** the **game** (itself) as an **interactive system** consists of various processes, that is at the level of the game, there are many cognitive processes that go through *users' minds/thinking*, starting with **"Plan, Think, and Decide."** This is followed by the second process, **"Play, Engage, and Interact"** where the *interaction with the game system, a mental challenge played according to certain rules, a driving force for behavior/action*. The third process is **"Test, Assess, and Feedback,"** this, in contrast, is achieved with the purpose of meeting *desired goals: Winning/losing* where the user obtains **"Feedback."** Moving on to the final phase of this **GC Model**, which is **"Reflect, Make Sense, and Experience,"** *a learning journey/transformation*. The game, therefore, as a motivating tool for learning provides players with life learning skills since cognitively the player goes through many psychological and socio-cognitive phases (*planning, thinking, reasoning, playing to assessment*...etc.), interacting with the game in an iterative cyclic process. The game player consequently experiences transformational/new skills, that is, a learning journey as a game player which can be ***Acumenous/Inquisitional*** depending on the game. This is because one of the motivational tools, as suggested in Chapter 2 employed in designing **gameplay** are levels (according to Schell (2015, 2020)) where player initially starts from level one, working their way up step-by-step to higher levels. This **levelling system** is a process or a strategy where players receive **feedback** and **progress** gradually as they proceed in the game, **leading to full engagement**. Nevertheless, when players cannot progress in the game, that is, due to lack of understanding the curriculum/contents; or unmet expectations/needs; this leads to a driving force or a **change in behavior**, disliking the gaming experience. This subsequently can create problems because if the purpose of the game is to engage, this can cause the opposite effects—disengaging game players, that is ***Inquisitional.***

Having introduced and discussed the game cycle, it appears that without the game, players could not make sense of their learning. Thus, their perceived learning from GBL be it **Inquisitional** or **Acumenous** was similar since there was no social interaction and no games learning. Respondents did not learn any skills/learning outcomes. They were just answering questions based on their experience on a question-answer format type of learning. In other words, respondents did not play and/or interact with any game. This furthermore explains the reasons behind the mean scores. This also implies that without the game, **Acumenous GBL** will not happen. The game was missing and therefore learning did not take place. Since they were just question answers, participants did not learn or conduct any activity with the game, and this did not make any sense to their experiences. This is **"Inquisitional Learning"** (see Chapter 9), which can lead to preconceptions toward Mathematical and Statistical learning.

The next chapter, **Chapter 9**, will unravel the interconnectedness of empirical investigations: Acumenous GBL in Mathematics and Statistics.

Bibliography

Anderson, S. (2020) Quantitative Analysis. Available at: https://www.investopedia.com/terms/q/quantitativeanalysis.asp (Accessed 20/02/2021).

Belluigi, D.Z. (2009) Exploring the Discourses Around 'Creativity' and 'Critical Thinking' in a South African Creative Arts Curriculum. *Studies in Higher Education*, 34(6), pp. 699–717.

Bruner, J. (1973) *Going Beyond the Information Given*. New York: Norton.

Caillois, R. (1961) *Man, Play, and Games*. New York: Free Press.

Cardinot, A. and Fairfield, J.A. (2019) Game-Based Learning to Engage Students with Physics and Astronomy Using a Board Game. *International Journal of Game-Based Learning (IJGBL)*, 9(1), pp. 42–57.

Chen, M.P., Wong, Y.T. and Wang, L.C. (2014) Effects of Type of Exploratory Strategy and Prior Knowledge on Middle School Students' Learning of Chemical Formulas from a 3D Roleplaying Game. *Educational Technology Research and Development*, 62, pp. 163–185. DOI: 10.1007/s11423-013-9324-3

Clark, D.B., Tanner-Smith, E.E. and Killingsworth, S.S. (2015) Digital Games, Design, and Learning a Systematic Review and Meta-Analysis. *Review of Educational Research*, 86(1), pp. 79–122.

Cordova, D.I. and Lepper, M.R. (1996) Intrinsic Motivation and the Process of Learning: Beneficial Effects of Contextualization, Personalization, and Choice. *Journal of Educational Psychology*, 88, pp. 715–730.

Coulthard, G.J. (2009) A Review of the Educational Use and Learning Effectiveness of Simulations and Games [online]. PhD Thesis, Purdue University. Available at: http://www.coulthard.com/library/paper-simulation.html (Accessed 15/12/2015).

Craft, A. (1999) Creative Development in the Early Years: Some Implications of Policy for Practice. *The Curriculum Journal*, 10(1), pp. 135–150.

Craft, A. (2002) *Creativity and Early Years Education*. London: Continuum.

Crawford, C. (1984) *The Art of Computer Game Design*. Berkeley: California, McGraw-Hill/Osborne Media.

Crockett, L. (2018) Bogost's War on Gamification: Irony and Play on Grandma's Sofa. In: *12th European Conference on Games Based Learning (ECGBL 2018)*. Reading: Academic Conferences and Publishing International Ltd.

Cronan, T.P. and Douglas, D.E. (2012) A Student ERP Simulation Game: A Longitudinal Study. *Journal of Computer Information Systems*, 53(1), pp. 3–13.

Dann, W.P., Cooper, S. and Pausch, R. (2008) *Learning to Program with Alice, Brief Edition*. USA: Prentice-Hall, Inc.

Davis, F.D. (1989) Perceived Usefulness, Perceived Ease of Use, and User Acceptance of Information Technology. *MIS Quarterly*, 13(3), pp. 319–340.

De Bono, E. (1992) *Teach Your Child How to Think*. England: McQuaig Group Inc.

Dede, C. (2005) Planning for Neomillennial Learning Styles. *Educause Quarterly*, 28(1), pp. 7–12.

Derryberry, A. (2007) Serious Games: Online Games for Learning. White Paper. [Online]. Available at: http://www.adobe.com/resources/elearning/pdfs/serious_games_wp.pdf (Accessed 10/12/2015).

Desurive, H., Caplan, M. and Toth, J. (2004) Using Heuristics to Evaluate the Playability of Games. *CHI Conference*, 32 (1), pp. 97–110. DOI: 10.1177/104687810103200108

Desurvire, H. and Wiberg, C. (2009) Game Usability Heuristics (PLAY) for Evaluation and Designing Better Games: The Next Iteration. *CHI Conference*, pp. 557–566.

Desurvire, H. and Wiberg, C. (2010) User Experience Design for Inexperienced Gamers: GAP – Game Approachability Principles. In: Regina Bernhaupt (ed.). *Evaluating User Experience in Games*. Springer, pp. 131–148. DOI: 10.1007/978-1-84882-963-3_1

Dondlinger, J.M. (2007) Educational Video Game Design: A Review of the Literature. *Journal of Applied Educational Technology*, 4(1), Spring/Summer, pp. 21–31. Available at: http://reserachgate.net/publication/238444705 (Accessed 12/12/2015).

Draper, S. (2013) Social Constructivism. Available at: http://www.psy.gla.ac.uk/~steve/courses/archive/CERE12-13-safari-archive/topic3/webarchive-index.html#social%20con%201 (Accessed 23/12/2019).

Dunne, C. (2011) The Place of The Literature Review in Grounded Theory Research. *International Journal of Social Research Methodology*, 14, pp. 111–124. DOI: 10.1080/13645579.2010.494930

Dweck, C.S. (2006) *Mindset: The New Psychology of Success*. United States: Ballantine Books Trade Paperback Edition.

Dwyer, J. and Slyman, S. (2016) *An Introduction to Quantitative Methods for Business Research*. London: Algana Publishing. 200 pages; ISBN−13: 9781907934094.

Early, R.E. (1992) The Alchemy of Mathematical Experience: A Psychoanalysis of Student Writings. *For the Learning of Statistics*, 12(1), pp. 15–20.

Ejsing-Duun, S. and Karoff, H.S. (2016) Creativity and Playfulness: Producing Games as a Pedagogical Strategy Case studies. In: Munkvold, R. (ed.). *Games Based learning for Researchers, Teachers and Students*, Vol. 1. Reading: Academic Conferences and Publishing International Ltd. ISBN:978-1-911278-12-890000.

Fairweather, E. and Cramond, B. (2010) Infusing Creative and Critical Thinking into the Curriculum Together. In: Beghetto, A. and Kaufman, J.C. (ed.). *Nurturing Creativity in the Classroom*. United States: Cambridge University Press.

Falstein, N. and Barwood, H. (2001) The 400 Project. Available at: http://theinspiracy.com/400_project.htm. (Accessed 20/04/2016).

Faria, A.J. (2001) The Changing Nature of Business Simulation/Gaming Research. *Simulation and Gaming*, 32(1), pp. 97–110. DOI: 10.1177/104687810103200108

Faria, A.J., Hutchinson, D. Wellington, W.J. and Gold, S. (2009) Developments in Business Gaming: A Review of the Past 40 Years. *Simulation and Gaming*, 40(4), pp. 464–487. DOI: 10.1177/1046878108327585

Faria, A.J. and Wellington, W.J. (2004) 'A Survey of Simulation Game Users, Former-Users, and Never-Users. *Simulation and Gaming*, 35(2), pp. 178–207. DOI: 10.1177/1046878104263543

Federoff, M.A. (2002) Heuristics and Usability Guidelines for the Creation and Evaluation of Fun in Video Games. Master Thesis, Indiana University, Bloomington.

Fienberg, S.E. (1992) A Brief History of Statistics in Three and One-Half Chapters: A Review Essay. Available at: https://www.jstor.org/stable/2246307?casa_token=Vek3Rmd5Mk0AAAAA%3AJ0zRZC5AxskwvTNY7GlrgsRV6oLrE20DY3FdeaPad0JvNSNrZDav8uZpX-YqY1hXTDtaz6zM87S4vwkQdflDLSzmrLCNlEsaBqWnykepc0jlYXsQXA

Forest, E. (2014) ADDIE Model. Available at: http://educationaltechnology.net/the-addie-model-instructional-design/ (Accessed 20/04/2016).

Fosnot, T.C. (1989) *Enquiring Teachers, Enquiring Learners: A Constructivist Approach for Teaching*. New York: Teachers College Press.

Fosnot, T.C. and Perry, R.S. (2005) Constructivism: A Psychological Theory of Learning. Available at: http://beyondbitsandatoms.org/readings/fosnot_constructivism_1996.pdf (Accessed 20/07/2019).

Garris, R., Ahlers, R. and Driskell, E.J. (2002) Games, Motivation, and Learning: A Research and Practice Model. *Simulation and Gaming*, 33(4), pp. 441–467.

Gee, J.P. (2003a) *What Video Games Have to Teach Us About Learning and Literacy*. New York: Palgrave Macmillan.

Gee, J.P. (2003b) Good Video Games and Good Learning. Available at: https://academiccolab.org/resources/documents/Good_Learning.pdf (Accessed 03/01/2021).

Gee, J.P. (2004) *Situated Language and Learning: A Critique of Traditional Schooling*. London: Routledge.

Gee, J.P. (2005a) Why are Video Games Good for Learning? Retrieved January 1, 2018, from http://academiccolab.org/resources/documents/MacArthur.pdf (Accessed 20/04/2016).

Gee, J.P. (2005b) *Why Video Games Are Good for Your Soul: Pleasure and Learning*. Melbourne: Common Ground.

Gee, J.P. (2007) *Good Video Games and Good Learning: Collected Essays on Video Games, Learning, and Literacy*. New York: Peter Lang.

Gee, J.P. (2008) *Getting over the Slump: Innovation Strategies to Promote Children's Learning*. New York: The Joan Ganz Cooney Center at Sesame Workshop.

Gee, J.P., Hayes, E., Torres, R.J., Games, I.A., Squire, K. and Salen, K. (2008) Playing to Learn Game Design Skills in A Game Context. In: *Proceeding ICLS'08 Proceedings of the 8th International Conference on International Conference for the Learning Sciences*, Vol. 3, pp. 368–374.

Gee, J.P. and Hayes, E.R., 2011. *Language and Learning in the Digital Age*. Routledge.

Gibbons, J. and Gray, M. (2004) Critical Thinking as Integral to Social Work Practice. *Journal of Teaching in Social Work*, 24, pp. 19–37.

Glaser, B.G. and Strauss, A.L. (1967) *The Discovery of Grounded Theory*. Chicago: Aldine.

Gomes, A. and Mendes, A.J. (2007) Learning to Program-Difficulties and Solutions. In *International Conference on Engineering Education–ICEE*, Vol. 7.

Granic, I., Lobel, A. and Engels, R.C. (2014) The Benefits of Playing Video Games. *American Psychological Association*, 69(1), pp. 66–78. DOI: 10.1037/a0034857

Hackett, G. and Betz, N.E. (1989) An Exploration of the Mathematics Self-Efficacy/Mathematics Performance Correspondence. *Journal for Research in Mathematics Education*, 20(3), pp. 261–273.

Hamari, J., Shernoff, D.J., Rowe, E., Coller, B., Asbell-Clarke, J. and Edwards, T. (2016) Challenging Games Help Students Learn: An Empirical Study on Engagement, Flow and Immersion in Game-Based Learning. *Computers in Human Behavior*, 54, pp. 170–179.

Hamari, J. and Tuunanen, J. (2014) Player Types: A Meta-Synthesis. *Transactions of the Digital Games Research Association*, 1(2), pp. 29–53.

Harel, I. and Papert, S. (1990) *Software Design as a Learning Environment*. Massachusetts Institute of Technology Cambridge, MA: MIT Media Laboratory.
Harel, I. and Papert, S. (1991) *Constructionism*. Norwood, NY: Ablex Publishing Corporation.
Hargreaves, A. (2003) *Teaching in the Knowledge Society: Education in the Age of Insecurity*. New York: Teachers' College Press.
Holbert. N., Penney, L. and Wilensky, U. (2010) *Bringing Constructionism to Action Game-Play, Constructionism*. Paris. Available at: https://www.google.co.uk/search?q=Bringing+Constructionism+to+Action+Game-Playandoq=Bringing+Constructionism+to+Action+Game-Playandaqs=chrome..69i57.7170j0j8andsourceid=chromeandie=UTF-8 (Accessed 20/03/2017).
Hoover, W.A. (1996) The Practice Implications of Constructivism. *SEDL Letter*, 9(3), pp. 1–2.
Horn, B., Hoover, A.K. Barnes, J. Folajimi, Y. Smith, G. and Harteveld, C. (2016) Opening the Black Box of Play: Strategy Analysis of An Educational Game. In *Proceedings of the 2016 Annual Symposium on Computer-Human Interaction in Play*, pp. 142–153.
Hsu, C.-Y., Tsai, C.-C. and Wang, H.-Y. (2012) Facilitating Third Graders' Acquisition of Scientific Concepts Through Digital Game-Based Learning: The Effects of Self-Explanation Principles. *Asia-Pacific Education Researcher*, 21(1), pp. 71–82.
Hunicke, R., LeBlanc M. and Zubek, R. (2004) MDA: A Formal Approach to Game Design and Game Research. Available at: https://www.google.co.uk/search?q=MDA%3A+A+Formal+Approach+to+Game+Design+and+Game+Researchandoq=MDA%3A+A+Formal+Approach+to+Game+Design+and+Game+Researchandaqs=chrome..69i57j69i58.3868j0j8andsourceid=chromeandie=UTF-8 (Accessed 8/10/2016).
Illeris, K. (2007) *How We Learn: Learning and Non-Learning in School and Beyond*. London: Routledge.
ISO (1998) Available at: http://www.standardsglossary.com/iso1.htm (Accessed 8/10/2016).
Jackson, N. and Shaw, M. (2007) Developing Perspectives on Creativity in Higher Education. In: N. Jackson, M. Oliver, M. Shaw, and J. Wisdom (ed.). *Developing Creativity in Higher Education: An Imaginative Curriculum*. London: Routledge.
Jagger, S., Siala, H. and Sloan, D. (2016) It's All in The Game: A 3D Learning Model for Business Ethics. *Journal of Business Ethics*, 137(2), pp. 383–403.
Jameson, M.M. and Fusco, B.R. (2014) Math Anxiety, Math Self-Concept, and Math Self-Efficacy in Adult Learners Compared to Traditional Undergraduate Students. *Adult Education Quarterly*, 64(4), pp. 306–322.
Jarvis, M. (2005) *'Motivation' in Psychology of Learning and Teaching*. Cheltenham: Nelson Thornes.
Jeffrey, B. and Craft, A. (2004) Teaching Creatively and Teaching for Creativity: Distinctions and Relationships. *Educational Studies*, 30(1), pp. 77–87.
Jenson, J. and Droumeva, M. (2015) Making Games with Game Maker: A Computational Thinking Curriculum Case Study. In *ECGBL 2015–9th European Conference on Games Based Learning: ECGBL2015*. Academic Conferences and Publishing Limited, p. 260.
Jones, M.G. (1998) Creating Engagement in Computer-Based Learning Environments. Available at: http://itech1.coe.uga.edu/itforum/paper30/paper30.htmI (Accessed 12/04/2016).
Kafai, Y. (2006) *Playing and Making Games for Learning: Instructionist and Constructionist Perspectives for Game Studies. Games and Culture*, 1(1), pp. 36–40.
Kafai, Y.B. and Resnick, M. eds. (2012) *Constructionism in Practice: Designing, Thinking, and Learning in a Digital World*. Routledge. Available at: https://www.google.co.uk/books/edition/Constructionism_in_Practice/KGkIEQAAQBAJ?hl=en

&gbpv=1&dq=Kafai,+Y.B.+and+Resnick,+M.+eds.+(2012)+Constructionism+in+Practice:+Designing,+Thinking,+and+Learning+in+a+Digital+World.+Routledge.&pg=PP1&printsec=frontcov

Kato, P.M., Cole, S. Bradlyn, A.S. and Pollock, B.H. (2008) Video Games Improves Behavioural Outcomes in Adolescents and Young Adults with Cancer: A Randomized Trial. *Pediatrics*, 122(2), pp. e305–e217.

Kim, B. (2001) Social Constructivism. *Emerging Perspectives on Learning, Teaching, and Technology*, 1(1), p. 16.

Kirriemuir, J. and Mcfarlane, A. (2004) Literature Review in Games and Learning. Available at: https://telearn.archives-ouvertes.fr/hal-00190453/document (Accessed 12/04/2020).

Knight, P.T. (2001) A Briefing on Key Concepts – Formative and Summative, Criterion and Norm Referenced Assessment. *Learning and Teaching Support Network Generic Centre*, Vol. 30. http://www.heacademy.ac.uk

Kodugamelab (2015) Available at: https://www.kodugamelab.com/ (Accessed 8/11/2018).

Kolb, A.Y. and Kolb, D.A. (2008) The Learning Way: Meta-Cognitive Aspects of Experiential Learning. *Simulation and Gaming*, 40(3), pp. 297–327.

Kolb, D.A. (1984) *Experiential Learning: Experience as the Source of Learning and Development*, Vol. 1. Englewood Cliffs, NJ: Prentice-Hall.

Li, F.W.B. and Watson, C. (2011) *Game-Based Concept Visualization for Learning Programming*. Scottsdale, Arizona, USA: ACM. 978-1-4503-0994-3/11/12.

Li, Z.Z., Cheng, Y.B. and Liu, C.C. (2013) A Constructionism Framework for Designing Game-Like Learning Systems: Its Effect on Different Learners. *British Journal of Educational Technology*, 44(2), pp. 208–224.

Lister, R., Adams, E.S. Fitzgerald, S. Fone, W. Hamer, J. Lindholm, M. McCartney, R. Moström, J.E. Sanders, K. Seppälä, O. and Simon, B. (2004) A Multi-National Study of Reading and Tracing Skills in Novice Programmers. In *ACM SIGCSE Bulletin*, 36(4), pp. 119–150. ACM.

Loftus, C., Thomas, L. and Zander, C. (2011) Can Graduating Students Design: Revisited. In *Proceedings of the 42nd ACM Technical Symposium on Computer Science Education*, pp. 105–110.

Lyons, I.M. and Beilock, S.L. (2012) When Math Hurts: Math Anxiety Predicts Pain Network Activation in Anticipation of Doing Math. *PLoS One*, 7(10), p. e48076.

Majgaard, G. (2016) Creating Games in the Classroom: From Native Gamers to Reflective Designers. In: Munkvold, R. (ed.). *Games Based Learning for Researchers, Teachers and Students*, Vol. 1. Reading: Academic Conferences and Publishing International Ltd. ISBN:978-1-911278-12-890000.

Malaka, R. (2014) How Computer Games Can Improve Your Health and Fitness. In: Gobel, S., and Wiemeyer, J., (ed.). *Games for Training Education, Health and Sports, 4th Conference on Serious Games, GameDays 2014*, Darmstadt, Germany, April 1–5, 2014, Springer.

Malone, T.W. (1980) What Makes Things Fun to Learn? A Study of Intrinsically Motivating Computer Games. Available at: http://cci.mit.edu/malone/Malone%20computer%20games%20bibliography.htm

Malone, T.W. and Lepper, M.R. (1987) 'Making Learning Fun: A Taxonomy of Intrinsic Motivations for Learning. Aptitude, Learning, and Instruction': Volume 3: Cognitive and Affective Process Analyses. R. E. Snow and M. J. Farr. Hillsdale, USA, Erlbaum.

Maloney, E.A., Schaeffer, M.W. and Beilock, S.L. (2013) Mathematics Anxiety and Stereotype Threat: Shared Mechanisms, Negative Consequences and Promising Interventions. *Research in Mathematics Education*, 15(2), pp. 115–128.

Marciano, J.N., Miranda, L.C. and Miranda, E.E.C. (2014) Evaluating Multiple Aspects of Educational Computer Games. Available at: https://www.hindawi.com/journals/ijcgt/2014/139205/ (Accessed 27/08/2017).

Marklund, B.B. and Taylor, A.S.A. (2016) Creating Games in the Classroom: From Native Gamers to Reflective Designers. In: Munkvold, R. (ed.). *Games Based learning for Researchers, Teachers and Students*, Vol. 1. Reading: Academic Conferences and Publishing International Ltd. ISBN: 978-1-911278-12-890000.

Maslow, A.H. (1970) *Motivation and Personality*, 2nd Edition. New York; London: Harper and Row.

Mayer, E.R. (2014) *Computer Games for Learning: An Evidence Based Approach*, London: MIT.

Mayer, I., Warmelink, H. and Bekebrede, G. (2013) Learning in a Game-Based Virtual Environment: A Comparative Evaluation in Higher Education. *European Journal of Engineering Education*, 38(1), pp. 85–106.

Mayo, M. (2009) Video Games: A Route to Large-Scale STEM education? *Science*, 323, pp. 79–82.

Mccrindle, M. (2018) What Comes after Generation Z? Introducing Generation Alpha. Available at: https://mccrindle.com.au/insights/blogarchive/what-comes-after-generation-z-introducing-generation-alpha/ (Accessed 15/10/2018).

McFarlane, A., Sparrowhawk, A., and Heald, Y. (2002) *Report on the Educational Use of Games: An Exploration by TEEM on the Contribution Which Games Can Make to the Educational Process*. Cambridge, UK: TEEM. http://www.teem.org.uk/publications/teem_gamesined_full.pdf (Accessed 06/12/2015).

McGonigal, J. (2015) *Superbetter: A Revolutionary Approach to Getting Happier Stronger and Happier*. Braver and More Resilient, New York: Penguin Press.

Milane, I. and Rowe, G. (2002) Difficulties in Learning and Teaching Programming-Views of Students and Tutors. *Education and Information Technologies*, 7(1), pp. 55–66.

Moreno-Ger, P., Burgos, D. Martínez-Ortiz, I. Sierra, J.L. and Fernández-Manjón, B. (2008) Educational Game Design for Online Education. *Computers in Human Behavior*, 24(6), pp. 2530–2540. DOI: 10.1016/j.chb.2008.03.012

Moreno-Ger, P., Torrente, J. Hsieh, Y.G. and Lester, W.T. (2012) Usability Testing for Serious Games: Making Informed Design Decisions with User Data. *Advances in Human-Computer Interaction*, 2012. DOI: 10.1155/2012/369637

Mullins, L. (2013) *Management and Organisational Behaviour*, 10th Edition. Harlow: Pearson Education.

Mullins, L. (2019) *Organisational Behaviour in The Workplace*, 12th Edition. Harlow: Pearson Education.

Multon, K.D., Brown, S.D. and Lent, R.W. (1991) Relation of Self-Efficacy Beliefs to Academic Outcomes: A Meta-Analytic Investigation. *Journal of Counselling Psychology*, 18, pp. 30–38.

Naik, N. (2017) The Use of GBL to Teach Mathematics in Higher Education. *Innovations in Education and Teaching International*, 54(3), pp. 238–246. DOI: 10.1080/14703297.2015.1108857

Newby, P. (2010) *Research Methods for Education*, Harlow: Pearson Education Ltd.

Nicholls, D., De Charon, L. and Hutkin, R. (2010) Development of Critical Thinking and Creativity: Practical Guidelines for the Postsecondary Classroom. *ATEA Journal*, 38(1), p. 12.

Palincsar, A.S. (1998) Social Constructivist Perspectives on Teaching and Learning. *Annual Review of Psychology*, 49, pp. 345–375.

Pinelle, D., Wong, N., Stach, T. and Gutwin, C. (2009) Usability Heuristics for Networked Multiplayer Games. *Proceedings of the ACM 2009 International Conference on Supporting Group Work*, pp. 169–178.

Prensky, M. (2002) The Motivation of Gameplay: The Real Twenty-First Century Learning Revolution. *On the Horizon*, 10(1), pp. 5–11.

Prensky, M. (2005) Listen to Natives. *Association for Supervision and Curriculum Development Educational Leadership*/issue December 2005, pp. 9–13.

Prensky, M. (2006) *'Don't Bother Me Mom, I'm Learning!': How Computer and Video Games Are Preparing Your Kids for the Twenty-First Century*. St Paul, MN: Paragon House.

Pritchard, A. and Woollard, J. (2010) *Psychology for the Classroom: Constructivism and Social Learning*. Oxon: Routledge.

Robins, A., Runtree, J. and Rountree, N. (2003) Learning and Teaching Programming: A Review and a Discussion. *Computer Science Education*, 13(2), pp. 137–172.

Rogoff, B. (1998) Cognition as a Collaborative Process. In W. Damon, D. Kuhn and R.S. Seigler (ed.), *Handbook of Child Psychology*, 5th Edition. New York: Wiley.

Roth, W.M. (2000) Authentic School Science: Intellectual Traditions In R. McCormick and C. Paechter (ed.), *Learning & Knowledge*. London, UK: Paul Chapman Publishing, pp. 6–20.

Russ, T., and Drury-Grogan, M.L. (2013) Assessing the Impacts of a Business Communication Simulation on Students' Self-Perceptions. *Eastern Communication Association*, 61(5), pp. 584–595. DOI: 10.1080/01463373.2013.822404

Shunk, D.H. (2000) *Learning Theories: An Educational Perspective*, 3rd Edition. Upper Saddle River, NJ: Prentice-Hall.

Siegel, R.G., Galassi, J.P. and Ware, W.B. (1985) A Comparison of Two Models for Predicting Mathematics Performance: Social Learning Versus Math Aptitude-Anxiety. *Journal of Counseling Psychology*, 32, pp. 531–538.

Silverman, B.G., Johns, M., Weaver, R. and Mosley, J. (2003) Authoring Edutainment Stories for Online Players (AESOP): Introducing Game play into Interactive Dramas. In: *International Conference on Virtual Storytelling*, Springer.

Sindre, G., Natvig, L. and Jahre, M. (2009) Experimental Validation of the Learning Effect for A Pedagogical Game on Computer Fundamentals. *IEEE Transactions on Education*, 52 (1), pp. 10–18.

Slyman, S. (2014) I Hate Maths! An Investigation into Students' Attitudes Towards Mathematics. MA Thesis, University of Roehampton, London, United Kingdom.

Slyman, S. (2018) *Games Based Learning in Universities, a Life Learning Experience*. New York: Business Expert Press Expert Insights Publishing. ISBN:978-1-94819-801-1.

Slyman, S. (2022) Acumenous Game-Based Learning in Simulation Games and Applied Statistics. PhD Thesis, Goldsmiths, University of London, UK.

Sweetser, P. and Wyeth, P. (2005) GameFlow: A Model for Evaluating Player Enjoyment in Games. *Computers in Entertainment*, 3(3), p. 3. DOI: 10.1145/1077246.1077253

Tao, Y.H., Cheng, C.J. and Sun, S.Y. (2012) Alignment of Teacher and Students Perceptions on the Continued Use of Business Simulation Games. *Educational Technology and Society*, 15(3), pp. 177–189.

Vlachopoulos, D. and Makri, A. (2017) The Effect of Games and Simulations on Higher Education: A Systematic Literature Review. *International Journal of Educational Technology in Higher Education*, 14(1), pp. 22. DOI: 10.1186/s41239-017-0062-1

Vos, L. (2014) Marketing Simulation Games: A Review of Issues in Teaching and Learning, 14(1), p. 6796. DOI: 10.1362/146934714X13948909473220

Vos, L. and Brennan, R. (2010) Marketing Simulation Games: Student and Lecturer Perspectives. *Marketing Intelligence and Planning*, 28(7), pp. 882–897.

Yusoff, A., Crowder, R. and Gilbert, L. (2010) Validation of Serious Games Attributes Using the Technology Acceptance Model [online]. In: IEEE (ed.), *2010 Second International Conference on Games and Virtual Worlds for Serious Applications*, Braga, Portugal, 25–26 March 2010, pp. 45–51. Available at: http://eprints.soton.ac.uk/270809/ (Accessed 10/05/2015).

9 Unravelling the Interconnectedness of Empirical Investigations

Acumenous GBL in Mathematics and Statistics Education

To this point, and as previously articulated in Chapters 1, 6, 7, and 8, this book consisted of four empirical investigations. These investigations were conducted because many questions emerged and led to each investigation, necessitating further explorations. This book's chapters are linked to each other, building, complementing, and informing all the empirical work in this book (see Chapter 1 for further details).

In Chapter 6, I embarked on a captivating journey into the realm of simulation games (SGs), delving deep into the multifaceted aspects of gameplay, game characteristics, design attributes, mechanics, and development. These immersive and interactive environments provided players with the opportunity to apply their knowledge and skills in solving complex problems, fostering a rich and transformative learning experience.

Within the realm of SGs, a diverse array of game types awaits exploration. We navigated through this vast landscape, shedding light on MMORPGs (Massive Multiple Online Roleplay Games), a captivating subgenre of action games that blend strategy, fantasy, adventure, and simulation. These games served as a prime example of how entertainment and educational elements intertwine, creating a compelling and engaging platform for learning.

Furthermore, this chapter took a pioneering step forward by proposing a game evaluation heuristics model. Addressing a gap in existing theoretical frameworks, this model harmoniously blends elements from both entertainment and educational perspectives. By striking a delicate balance, it offers a comprehensive and nuanced framework for analyzing and evaluating the educational simulation games (ESGs) that lie at the heart of transformative learning experiences.

Chapter 7 took us on an insightful journey into the realm of student experiences, focusing on Mathematics and game-based learning (GBL).

Drawing upon the voices of students, this chapter delved into their experiences through the lens of Charmaz's CGT (2006, 2014) and a series of focus-group interviews (FGIs). The research data collection and analysis processes,

DOI: 10.4324/9781032723426-12

informed by the extensive body of research literature, provided a robust framework for exploration. By intertwining these methodological aspects, we gained valuable insights into the student perspective, illuminating their thoughts, challenges, and triumphs in the realm of GBL and Mathematics.

The emergent insights and theoretical underpinnings gleaned from the data were carefully examined, with explicit reflections on the relevant literature explored in previous chapters. Building upon the theoretical framework established in Chapters 2–5, this chapter delves into the alignment between the findings and the newly proposed theoretical framework, setting the stage for further exploration in Chapter 9. With every innovative contribution, this chapter (Chapter 7) paints a vivid picture of the student experience, providing a foundation for future explorations and progression of GBL in Mathematics education. By capturing the nuances, challenges, and transformative potentials, we inch closer to unlocking the true power of GBL and its role in shaping the future of Mathematics learning.

Chapter 8 delved into the intricate realm of *Inquisitional* and *Acumenous* Game-Based Learning (GBL), building upon the findings of Chapter 7, which shed light on students' experiences with Mathematics and GBL. Through careful analysis, two distinct types of GB learning emerged: *Inquisitional* (quiz-based) and *Acumenous* (simulation-based). The superiority of *Acumenous* GBL over *Inquisitional* GBL became apparent, paving the way for the exploration of the dynamics of *Inquisitional* and *Acumenous* GBL in students' experiences with Mathematics and Statistics (in this chapter) to further enrich our understanding.

Unlike the predominantly qualitative nature of Chapters 6 and 7, which warranted a detailed analysis approach, Chapter 8's investigation adopted a quantitative research method, employing questionnaires. Quantitative research offers a quick and effective means to capture a representative sample of the population and identify overarching patterns within it. Through this methodological shift, we delved into the realm of students' experiences in learning Mathematics and Statistics using the two GBL approaches. Chapter 8 acts as a gateway to understanding the multifaceted dynamics of *Inquisitional* and *Acumenous* GBL. It sheds light on students' experiences, their preferences, and the impact of each approach on their learning journey in Mathematics and Statistics. Through a comprehensive exploration of the emergent themes, this chapter provides a deeper understanding of the strengths, weaknesses, and unique contributions of both *Inquisitional* and *Acumenous* GBL approaches. My findings demonstrated that there were no statistical differences between quiz vs. simulation GBL and how people viewed these games. The game and social interaction were absent because there was no gameplay (Chapter 8).

This chapter, Chapter 9, commences by outlining the structured plan encompassing lecture slides, a gameplay scenario, and Excel activities. Subsequently, the empirical research findings are presented, offering a deep dive into students' feelings, perceptions, and confidence levels. To capture these insights, Zimmerman's (2000) self-efficacy questionnaire (adapted version) and qualitative focus group interviews were employed as research instruments. The chapter aims to explore the efficacy of *Inquisitional* and *Acumenous* GBL approaches, providing a comprehensive analysis of their impact.

Chapter 9 serves as a testament to the interconnectedness of the empirical investigations presented in this book. It illuminates the multifaceted nature of students' experiences, shedding light on the complexities that arise when merging Statistics education and GBL. Through a thorough examination of the empirical findings, this chapter contributes to a deeper understanding of the dynamics at play and lays the groundwork for the subsequent chapters, which further explore the theoretical and practical implications of these insights.

Within the rich sphere of student experiences, the findings reveal a noteworthy disparity. While interviewees demonstrate proficiency in learning and engaging with simulation games (SGs), they encounter challenges in understanding and grasping concepts in Mathematics and Statistics education. This discrepancy highlights preconceived notions and a lack of self-efficacy, and thus fear, in these subjects. Expanding on these, Chapter 9 proceeds to discuss the emerging themes that have surfaced within this investigation.

The emerging themes discussed in this chapter encapsulate critical aspects of students' experiences in learning Statistics through simulation-based GBL. These themes provide a deeper understanding of the challenges, motivations, and intricacies surrounding students' engagement and performance. By examining these key themes, the chapter offers valuable insights into the interplay between self-efficacy, preconceptions, and the learning outcomes of students participating in these investigations. These are discussed next.

In the following section, I will discuss the overall emergent insights that drove and initiated this empirical investigation which is, "What are participants' experiences in learning Statistics using SGs?" then (where applicable, as I previously noted) "What are participants' experiences with Mathematics and GBL?" (Chapter 7). As I afore-indicated (Chapter 1), the emergent themes in Chapter 7 and this chapter complement each other, enlightening our mind and providing further insights into our unique research book journey. Participants' interview responses were then linked back to the relevant Literature Review (Chapters 2–5) and grouped into themes. I will discuss each nascent theme.

The identified themes are:

- **The Pedagogical Benefits of SGs: Mathematical and Statistical Learning, Creative Thinking, Decision-Making, and Problem-Solving Skills.**
- **Factors behind Disengagement with Mathematics and Statistics.**
- **How Mathematics and Statistics Can Be Engaging.**
- **How We Can Improve Mathematics and Statistics using SGs.**
- **Acumenous and Inquisitional Theory.**

The Pedagogical Benefits of SGs: Mathematical and Statistical Learning, Creative Thinking, Decision-Making, and Problem-Solving Skills

Findings have highlighted the beneficial role of SGs as a learning tool for improving various transferable skills like Mathematical and Statistical learning. This is because when it comes to effective ways to learn skills using SGs, participants noted that whenever they play simulation games, they found themselves analyzing their performance and variables. For instance, one participant noted, "in FIFA, I keep thinking how do they create [these] variables? That is itself educational." Another participant added: "You [keep] doing [it or playing the game] and [then] after a while you realize that you learn something." Other interviewees suggested that in order to combat fear and lack of confidence in Mathematics and Statistics, we should extend "the points, the formula people struggle with like calculus and bring it to the game, [a] good way of learning and carry [ing] on through different mathematical problems." We could also add different "element of problems solving at each stage," explained another interviewee. While others suggested that when "we are just building [in SGs] just building maybe we will get a sense of how to build like [,] you will learn other stuff [skills] that we don't recognize now until maybe we analyze it" [the game: SG]. Others shared their gameplay experience stating that when "they play games like racing car games," it "help [them] with reflexes." They emphasized the pedagogical benefits of SGs are that "there are certain skills you can learn from SGs." Simulation games can help with "the memory in adapted way with Statistics and other things like that." Through SGs you could learn "problem-solving" you "get that amount of money and materials, when to build it [SG] and when it is not," explained one interviewee. Other participants suggested that we can stimulate creative thinking in Statistics using SGs "by employing/adding already interactivity", make it "creative," have "many possibilities that you can pick. Where the road is. What you decide to do." However, there should also be "limits so that it is not so difficult," added one interviewee. "Make it happen, help everyone make more decisions," explained another interviewee. This shows that SGs can stimulate creative and analytical skills (as denoted by our interviewees). SGs can improve learning and teach valuable transferable skills in HE such as increasing and building confidence. As one interviewee enlightened, a game "can help people who may have issues with confidence." "I think [SGs like 'SimCity'] can [increase confidence]," clarified two of our participants. Then, "you can change your decisions" and "delete the whole thing [i.e., the game] and restart" (added two of our participants). SGs can stimulate creative thinking, decision-making, problem-solving, Mathematical and Statistical learning since it has many possibilities, real-life scenarios with real-time feedback. As other participants explained,

> I think it [SGs like SimCity] can [increase confidence]. Because you can always delete the whole thing [the game] and restart. Knowing that you can restart. You can learn not to be afraid [fear] of making decisions.

> Creative thinking works with accessibility. At the same time accessibility and restrictions. So being able to access the game itself if you have and the first one is that but if you have done anything. They can make the game. If they can do anything if you have restrictions, they can work with it or get out of it. They can work with it. If it is blank picture it is not going to work. You need certain structure for them to make it work.
>
> Car game is also creative when you think about its games and simulation games is just like an embedded version.
>
> [T]he way it is being used in this experiment is a lot better [...] since it ['SimCity'] is used for entertain[ment] but the way we use it ['SimCity' as pedagogical tool for teaching Statistics] today is different and it is educational.
>
> I think that what we are doing [is] you are building. I am just going to try it and see without even knowing what I am doing. I work on it work on it until it works. Then I realize I use these concepts in this case and now I can use it in other cases. So, it is trial and error, and with games you are just trying it to do things and see what works and answer questions.
>
> [SGs] give you a chance to trial and make your decisions. It makes you feel like you have a chance. You can always redo. It can help people who may have issues with confidence knowing that you can change your decisions.
>
> I think it is a sense of purpose. [SGs] ha[ve] clear productivity and what we are meant to be doing.
>
> [W]e can be like competition who have more property, or I produce more that would be problem-solving in game communication. We just trade.
>
> You can see [that] you learn step by step.
>
> Competition. When you stand everybody and the ranking. What is the most popular city? Ranking is important.

This suggests that SGs have many pedagogical benefits and can overall increase students' confidence as our interviewee noted: "[k]nowing that you can restart. You can learn not to be afraid [scared] of making decisions." Similarly, another interviewee clarified that SGs like "SimCity" can "address problem-solving because you have strategically what we do on everything and how many things we have in bubbles." In other words, residents either complain or acknowledge the efforts of the mayor. To stimulate creative thinking, problem-solving, or decision-making, "[u]se this type of games like 'SimCity', building things," declared other interviewees. While another participant noted that to improve skills "relate [it] to real-world scenarios and [the] economy." So, the game investigation "is a good thing [example]." Similarly, one participant continued, "the way we use it today is different and it is educational." Participants also indicated that they like the game and the element of trial and error, which address problem-solving, enabling them to make decisions in a safe environment where they can make mistakes without the fear of failure as they stated:

> It gives you a chance to trial and make your decisions. It makes you feel like you have a chance. You can always redo. It can help people who may have issues with confidence [fears] knowing that you can change your decisions.

In addition, the use of Excel spreadsheets made calculations comprehensible as they solve problems associated with Statistics with ease. As one participant highlighted, "[b]ecause of spreadsheet[s]...you can see the problem and use the formula to solve it." Others suggested adding more, "[r]ewards, [make it] easy," [and] make more decisions." In contrast one interviewee added, "you can see it [formula] it is not overload" as "you learn step by step." Similarly, others explained,

> I was doing a project where if you add this percentage of chemical [...] You can see the outcome straightaway and create the best outcome. I mean it gives you a chance to trial and make your decisions. You can always redo.
>
> [You] will get a sense of how to build like [,] you will learn other skills that we don't recognize right now until maybe we analyze it [the SG].
>
> I just want to do all these things. I had like goals. Let's get on with these things done. It was very easy [...] and you just being able to click and keep going.

This suggests that SGs play a beneficial role in enhancing learning since players can see the outcomes of their decisions straightaway as "You can always redo," acknowledged other participants. Whilst two interviewees added, "Larping games are a great fun for learning, since [the] learning outcomes in games [...] are similar principles [to education but you are] doing things that are fun. I do think simulation games can be very beneficial." This indicates that SGs are great tools for learning as participants prefer to learn and live life from other players' perspectives, thus stretching their imagination and creativity, and therefore their confidence (self-efficacy). For example, one participant in particular indicated that with "SGs you play as somebody else and you learn from their perspective." In addition, and according to other interviewees, in every SG you "learn something [skills] but you don't realize it, I learn a lot [to] achieve the game objective. I think with every new game engine [...] there is something new to learn." In SGs when

> we are just building maybe we will get a sense of how to build like [,] you will learn other skills that we don't recognize right now until maybe we analyze it [SGs]. This is because transferable skills such as curiosity and creative thinking are what stimulate participants to play SGs. In agreement to this, others argued that SGs are good tools for learning many skills in fact they "are great, [and] useful in building skills [like] critical thinking, decision-making, leadership skills, with the FIFA [for example, one participant clarified] you start from the bottom, it teaches you a lot of Statistical analysis and the best way to make the program running.

This subsequently reinforces the notion of how SGs, as a tool, can benefit learning and teach many skills such as Statistical analysis, critical thinking, leadership, and decision-making. These skills are what nurture players' self-efficacy, leading them to declare that games are "great [and] useful in building [numerous] skills." As participants recalled, "with the FIFA [for example] it teaches you a lot of Statistical analysis and the best way to make the program running." Others reinforced the theory that in games "every decision you make [,] impact the game." This furthermore illustrates that games have impacts on Mathematical skills (and Statistical learning), programming, and decision-making, since it can teach learners "a lot of Statistical analysis and the best way to make the program running" and "leadership," they continued.

Why Mathematics and Statistics Are Disengaging?

The findings from Chapter 7 indicate that MEGs are disengaging, and we learned from the data this is because Mathematics curriculum is *Inquisitional* type of learning. Our interviewees for this chapter shared their Mathematical and Statistical experience as being complex, uneasy, irrelevant to real life, intense and hard to understand or remember and rather fearful and negative experience. This was manifested as follows:

> [L]ong time ago I did not understand [Mathematics and Statistics]. I haven't got any clue.
>
> I did learn it [Statistics] in a high school very briefly, [but] I didn't really remember any kind of the concept.
>
> Maths is extremely difficult [...] and it is hard for me to understand the concepts.
>
> I hate Maths at school. Maths was less my favourite subject because I think I don't see its relevance to anything outside classroom.
>
> I find Maths too complex.
>
> My experience was quite negative. It is not clear.
>
> Not enjoyable. My Mathematics and Statistics I was not sure about how it is related to real life or use it or apply it.
>
> I am ok at Maths, but I hate Statistics the most. Because it is manual and intense you have to do all steps, calculations and memorize formulas.

This indicates that Mathematics curriculum can be unenjoyable and not interactive. This is mainly because the type of Mathematics and Statistics this cohort of students' experience is Inquisitional (discussed next) where learners face fewer multiple options to choose from and the questions are not related to real-life problems. This consequently leads to low self-efficacy and fears in Mathematics and Statistics. **Inquisitional** type of learning is hard to understand or remember. It is complex, manual, and intense where learners have to carry out all the steps. It is just formulas. *Inquisitional* is theoretical, and it is based on memorization. As one interviewee noted, "I did not understand. I haven't got any clue." While another one indicated that this type of learning cannot be remembered stating, "I didn't really remember any kind of the concept." Similarly, others explained that

> Before I did not have any idea of what Statistics is or did not know how it is applied.
>
> [I]t is manual and intense you have to do all steps, calculations and memorize formulas.
>
> If it is just theoretical in your head, it does not feel as real like a game you cannot.
>
> It is just like when you see all these variables Xs, Ys and the sums (sigma) you see it straight away. I get pretty scared. I don't really understand what they all do mean by the formula.
>
> I struggled with Maths it gives me the absolute fear. Terrified. I don't know I am horrified just feel the numbers like. As soon as I see it the right formula. I just shivered up that I am not going to.
>
> It [Mathematics] was always my weakest subject. It still is. I don't know why I have mental block when it comes to Maths. Even today I forget completely everything quickly what you have explained already.
>
> But it becomes like bricks it is [Maths] just hard to find the motivation to do it again.
>
> I had very strict very traditional Maths teacher. If you did not understand the way they explained it. That's was your problem. They don't want to know, and they won't explain it in different way, and they won't take time to break it down. It was your faults [the students' faults] if you did not understand.
>
> With Maths you have to work really hard/If you don't get it you have to work really hard.

From these quotes, it is clear that participants have identified specific difficulties they experienced whilst studying Mathematics and Statistics, and they are representatives of this cohort. From computing (HCI) perspectives, if a system does not cater to users' needs, it is a failing system. The Mathematics and Statistics curriculum that this cohort of students experienced could therefore be failing its purpose, because the type of Mathematics and Statistics it consists of is Inquisitional. This thus can have detrimental impacts on learners' self-efficacy and emotions toward these subjects, as the data from this research indicate. This furthermore suggests that this type of learning is theoretical, consisting of formulas with no real-life examples and it does not feel real as they clarified:

> If you only taught the formula and no concept behind it, no proper example then it is difficult for a lot of people to really understand the point of it.
>
> If it is just theoretical in your head, it does not feel real.

With other learners this type of learning "becomes like bricks" as they struggle to "just find the motivation to do it." However, with some students their struggles to understand Mathematics and Statistics led to fear or even blockage as they explained,

> I struggled with Maths it gives me the absolute fear. Terrified. I don't know I am horrified just feel the numbers like. As soon as I see it the right formula. I just shrunk up that I am not going to.
>
> It [Maths] was always my weakest subject. It still is, I don't know why I have mental block when it comes to Maths. Even today I forget completely everything quickly what you have explained already.

As previously articulated, this is because these participants have grown up now and so are their perceptions of the world. Furthermore, they play other entertaining games (simulation) and their views/experiences have also changed, leading to high self-efficacy playing these games (SGs). Similarly, participants elucidated that this generation would like more autonomy and prefer **Acumenous** type of GBL. They also want a sense of ownership, a sense of purpose and mission since they added,

> Sense of Ownership, [a] sense of purpose, [and] characters.
>
> Sense of Ownership.
>
> Ownership on what you do, which is pretty much in each game. I can't feel same as [other interviewees]. Because of Ownership.

The reason for this as one participant explained is that "If it feels like yours. You can build what you are visualising." Similarly, another participant added, "being able to create something how you wanted it and vision it." This type of GBL is what I have referred as "**Acumenous Game-Based Learning**"—I will return to examine this further next.

According to my interviewees,

> [Y]ou just want to do it without knowing. It should be something that we want to [be] part of. We want to be part of it.
>
> Something personal to you. Formula is not personal, and the test is not personal to you.
>
> [If it was personal] you [will] apply it immediately even if it is perhaps subconsciously. You are using it and when you tell us we are using correlation. We have connection in our brains that we make. We apply the knowledge subconsciously.
>
> [B]eing able to click and keep going.
>
> It makes it easy to remember.
>
> [With Statistics and Mathematics lecture] It's like this [,] you need to know these letters but we are not going to tell you why [you need them] but in the game there is a lot of back stories of why we're using it.

> [B]ut in the game there is a lot of back stories of why we're using it.
> There should be simulated things I feel like come on. If it is like using soccers and building fake tower you feel like building something of your own.
> Making money, paying taxes.
> Doing tasks, instruction, structure, lists.

From cognitive perspectives, there are strong perceptions that IT is fun and useful; these have positive impacts on users' preconceptions, which subsequently affect learners' behavior and interaction with the SG. Games design with its characteristics positively impacts learners' skills such as decision-making, creative thinking, competence, knowledge, and learning outcomes, and therefore a significant part of the learning experience, and ultimately their worldview. Thus, what is the relationship between self-efficacy, Mathematics and Statistics, and the games? There is an increasing necessity to examine this further next.

How Can Mathematics and Statistics Be Engaging?

Mathematics and Statistics can be engaging if we embody them with skills such as:

> [A] sense of purpose… clear productivity.
> [M]ore focused.
> [A] sense of how to build like you will learn other [skills] that we don't recognize until we analyze it.
> Simulation games.
> Competition. Ranking is important.
> Competition [….] Trade.

These transferable skills in SGs, I suggest, are related to what I have referred to, in Chapter 7, as "**Acumenous Game-Based Learning**," I will return to discuss this next. According to my participants,

> [Y]ou can see the outcome straightaway, and you will create the best outcome.
> [I]t gives you a chance to trial and make your decisions. [SGs] make you feel like you have a chance. You can always redo. It can help people who may have issues with confidence [fear] knowing that you can change your decisions.

Similar to this, other participants explained

> Roller skate game [I played] to go for marketing because I want[ed] to do marketing and stock broking. [The game involves] flipping items and objects and make money. It [also] involves a lot of as test run to do

> financing and creating money, create as much wealth as possible to make money in short spin. Stock markets [for example], profits- [making and] Gold-farming. Those skills were transferable to real-life skills. [His] goal was to create wealth in a short spin of time.
>
> Because of like spreadsheet, you can see the problem and see/use the formula to solve it.
>
> Because we can be like compet[e], who have more property, or I produce more that would be problem-solving in game communication.
>
> [T]hat is problem-solving you get that amount of money and materials, when to build it and when it is not. And keep neighbours happy because some of them want to leave.
>
> We just trade.
>
> There has been people using StarCraft like this to learn about what inflation [is], economics. I just think it is quite interesting they go to Maths and Statistics, and they play each probability. [There are people who] will learn each probability to give them an edge because there is always Maths behind it.
>
> Once relate that to real world [scenarios] and [the] economy [we have].

This furthermore suggests that transferable skills like experimentation (test run), problem-solving, building, financing, profit-making, and stock markets are significant real-life skills in GBL. According to the findings from Chapter 7, learners' perceptions of usefulness and fun play a key role in motivating and increasing their self-efficacy; it subsequently drove positively their preconceptions to interact with the game system (SG) in a meaningful way, leading to positive impacts on their skills, that is, transferable skills.

How Can We Improve Mathematics and Statistics using SGs?

Findings indicate that interviewees felt excited, relaxed, and more engaged about learning Mathematics and Statistics using a SG. They have shown a great deal of confidence and learning because they were understanding, reasoning, and deciding (Chapter 7). They were constantly thinking creatively and critically, managing, and leading their learning, that is, "*acumenifying*" their learning, using the SG. These were apparent in some of the participants' discourse as below. For instance, my participants noted,

> It went from having no knowledge at all to some knowledge. I find it interesting. I have confidence [self-efficacy] now. It did not bore me. It is interesting. What element did excite you? I asked. I don't know. The use of formula the different variables you could use, and the game is fun.
>
> It was really helpful, and it just made it a lot easy to understand.
>
> That was interesting. Quite interesting actually, I had no clue at all [before he mentioned that he hated Maths as he becomes terrified/horrified]. I have a better knowledge. That was interesting, the visual and interacti[on] it was very useful.

> [I]t is interesting to see I understand correlation more [now] and sometimes things are not correlated at all. It is interesting.
>
> Now I can. I will try. But like now It is possible I will try.
>
> I can see the necessity of [it]. I see now the points I mean how you get your decisions. [This is Acumenous GBL: understand, reason, and decide what it is. Why and how we are using it].

Other participants liked the real-life example, the teaching pedagogy, a sense of mission, purpose, and ownership, building and creating, collaborating, thinking creatively, and critically of all aspects of the gameplay, and the AGBL process as a whole stating,

> I think the best thing when you were explaining the real-life example when temperature increases then the sales of ice cream will increase.
>
> This is a good way of how it can be applied to real-life scenarios.
>
> It is I guess unlike coding I don't get to see it moving.
>
> If it is just on my own. No, I am not going to study. But I mean being able to have that sense of ownership of building the space and you know you make a decision regarding the space you own but having that sense of ownership I am building something as making a game the space you own. It does make it real.
>
> [If it is] something physical that you can visualize it makes tactile/real because you have a purpose and reason and [a] sense of ownership.

In addition, the use of Excel as an IT tool with the game was perceived as useful, practical, and great fun. For example, our interviewees commented that

> [T]he game was really good thing and the Excel.
>
> Using the Excel was really helpful, and it just made it a lot easy to understand all the different part of the information that we need to understand. It makes calculations easy to grasp.
>
> It only when I kind [of] practise and see all X, Y and X^2 in different columns that's when I understand that they all fit together.
>
> I just want to do all these things. I had like goals. Let's get on with these things done. It was very easy and just being able to click and keep going.
>
> You have steps then [...] it makes sense. It [learning] hides behind it. When looking back it makes the learning quite easy umm than going straight on without knowing or you don't know where it comes from.

When it comes to integrating SGs in the HE curriculum as a pedagogical tool for teaching complex subjects such as Statistics in universities especially in the Business and Computing Schools, interviewees suggested that SGs with built-in Excel could be more engaging, stimulating, and supportive. It could put a stop to students' struggles/fear and thus facilitate the advancement of Mathematical and Statistical learning through the AGBL process. For example, participants announced that

> It would be great to have games like this with Excel. It will be more helpful. Especially for visual leaners.
>
> I think perhaps there could be a way to put more of that [Excel built-in the games] in simulation games.
>
> It simplified learning.
>
> Extending the points now, the formula people struggle with like calculus and bring it to the game, [a] good way of learning [GBL] and carry on through different Mathematical problems.
>
> You have steps then [...] it makes sense. It [learning] hides behind it. When looking back it makes the learning quite easy umm than going straight on without knowing or you don't know where it comes from.
>
> That [feeling if we use SG with built-in in HE] would be great.
>
> It would be great.
>
> It would be a much more engaging way of learning it than not playing the games. It gives you more real-time feedback for example doing market shares. Real-time data. Real-time feedback, learning something about market shares because there is no way we can do this in real life.
>
> [D]ifferent element of problems solving at each stage.
>
> One stage two questions then the way the game stimulated, the way it will stimulate when you move on another focus another type and look up for another stage. Another problem-solving. It will be something else until we reach at another stage now, we are talking about supermarket then learn it at each stage. Now we reach trading we can use algebra or the formula.

Some of the reasons for this, as interviewees explained, are:

> There are people who see formula and get scared so yeah that would be [a] perfect time to put a stop to their struggle [fears] and not to run away from Maths. So yeah if we use simulation game like this it would definitely help.
>
> It may give you motivation if you think you can make your town better. You could sit down and learn how to do some kind of equations more easily because you know what you can use them for.

Looking back at some participants discourse earlier, they noted that they struggled with Mathematics since "it gives [them] the absolute fear. Terrified." When asked about reasons behind [their] fear they replied, "just feel the numbers like. As soon as I see it the right formula. I just shrunk up that I am not going to." They however noted that they "don't know why [they] have mental block when it comes to Mathematics." While another interviewee stated that,

I did learn it in a high school very briefly very quickly, so I didn't really remember any kind of the concept.

Similarly, others noted,

> [e]very time [they] see Mathematics [they] just shut down. [They] never tried and [they are] not a big fan.

Having participated in the SG experiment the confidence of some of the participants appears to have significantly increased and amplified, since their learning have "acumenified," and the results from the pre- and post-test surveys support this (see Chapter 5). For instance, one interviewee declared,

> That was interesting. Quite interesting actually, I had no clue at all [before]. I have a better knowledge. That was interesting, the visual and interactive it was very useful.

They added,

> I reckon it will be an easy way to learn Statistics I mean when you start off it is not an issue. But later on, for example, you have risks vs. rewards you see something it costs this much and it takes this long, is it worth it? Shall I buy it or wait? I think it will be quite useful.
>
> Using the Excel was really helpful, and it just made it a lot easy to understand all the different part of the information that we need to understand. It makes calculations easy to grasp.

Perhaps the main reason for integrating SGs in HE, as they declared, could be:

> When you brain is engaged you can absorb the information more quicker.
>
> [W]hen you tell us we are using correlation. We have connection in our brains that we make. We apply the knowledge subconsciously.

Similarly, our interviewees explained

> It would be a much more engaging way of learning it than not playing the games. It gives you more real-time feedback for example doing market shares. Real-time data. Real-time feedback, learning something about market shares because there is no way we can do this in real life.
>
> [W]hen you have a game it like stimulates you and you are not just passively taking information.

When participants were asked how confident they feel about learning Mathematics and Statistics using a SG like "SimCity" with Excel, there was a high degree of confidence, more engagement, and motivation to learn, as they noted:

> [P]robably more confident. I cannot explain it in words because it is just more engaging.
>
> It may give you motivation if you think you can make your town better. You could sit down and learn how to do some kind of equations more easily because you know what you can use them for.

Other participants felt that SGs can stimulate learners to learn more easily and digest information more inspirationally rather than absorbing knowledge passively, adding:

> It [SG] engages you to the point that sometimes when you are just listening to a lecture it can get quite exhausting to continuously absorb that information. But when you have a game it like stimulates you and you are not just passively taking information. Getting information is not just a chore [routine].
>
> It is easy to learn something if you are doing something fun, because it is your city and you own everything and just want to make it better and better.
>
> I was more relaxed and not feeling stressed about learning. You just [keep] doing [it] and then you realize after that you ohh I was learning.
>
> I just want to do all these things. I had like goals. Let's get on with these things done. It was very easy … umm and you just being able to click and keep going.

This further indicates that we can improve students' confidence in Mathematical and Statistical learning using SGs, like "SimCity," and Excel. This is also because some of the participants' knowledge and confidence increased and as they articulated,

> It went from having no knowledge at all to some knowledge. I find it interesting. I have confidence [self-efficacy] now. It did not bore me. It is interesting. What element did excite you? I don't know. The use formula the different variables you could use, and the game is fun.
>
> Now I can. I will try. But like now It is possible I will try.
>
> I have a better knowledge. That was interesting, the visual and interactive it was very useful.
>
> As I am not sure if it is Excel or the game, but I can see the necessity of [it]. I see now the points I mean how you get your decisions.
>
> This is a good way of how it can be applied to real-life scenarios.
>
> Why? I asked. It is I guess unlike coding I don't get to see it moving.
>
> It only when I kind [of] practise and see all X, Y and X^2 in different columns that's when I understand that they all fit together.
>
> That was interesting. Quite interesting actually, I had no clue at all [before this he mentioned that before he hated Mathematics as he becomes horrified].

Similarly, other participants acknowledged that using SGs will give them motivation to learn complex ideas with ease because they are **"acumenifying"** their learning. SGs can simplify learning as they noted:

> It may give you motivation if you think you can make your town better. You could sit down and learn how to do some kind of equations more easily because you know what you can use them for.
>
> So, the game was really good thing and the Excel.
>
> I was more relaxed and not feeling stressed about learning. You just [keep] doing [it] and then you realize after that you ohh I was learning.
>
> I just want to do all these things. I had like goals. Let's get on with these things done. It was very easy [..] and you just being able to click and keep going.
>
> Using the Excel was really helpful, and it just made it a lot easy to understand all the different part of the information that we need to understand. It makes calculations easy to grasp.
>
> I am not sure if it is Excel or the game, but I can see the necessity of [it]. I see now the points I mean how you get your decisions. I mean may be other things.
>
> It simplified learning.

On the other hand, others prefer the real-life example and the teaching pedagogy. For example,

> I think the best thing when you were explaining the real-life example when temperature increases then the sales of ice cream will increase.
>
> [I]t is interesting to see I understand correlation more [now] and sometimes things are not correlated at all. It is interesting.
>
> This is a good way of how it can be applied to real-life scenarios.

This is because SGs can be a more useful, helpful, and interesting method of teaching complex ideas such as those found in Statistics (Chapter 5). As participants re-stated,

> Relevance to everyday life and how it is applied in real life.
>
> Real-life scenarios, and it is back to experience.
>
> Competition and the ranking. What is the most popular city? Ranking is important.
>
> Yeah, the most popular city, ranking helps.
>
> Simulation games like game programming and the game tied with each module. Spreadsheets bring in the Maths /the numerical Maths and the game tied in with each module. Financial league game so that you can see whatever one is doing.
>
> [C]ombining all modules together like what we are doing currently.
>
> [Y]ou have risks vs. rewards.

Rewards when you get something right. Often when you get something wrong in Maths it feels like the end of the world.

Something that is not stressful, continuity but not perform actually other standards like exams or tests. They are recognisable in other forms.

Four modules in one programme. Project that you build upon throughout the term, but it is the same project that you can move [modules]. We have this Game project for programming, they split it in weeks. First you do this then you do that.

Similarly, adding programs like Excel and SGs to Statistics curriculum teaching would facilitate the digestion and absorption of complex information more easily and coherently. This also indicates that what is required is ***Acumenous Game-Based Learning***, which is discussed next. Learners would like to see how their learning is applied and what impacts it has. They would like to investigate and see the outcomes of their decisions in easy, stimulating, fun ways rather than having just formulas with numbers. Some of the participants' discourse are noted as follows:

People will play spreadsheet without even thinking about what it is or showing how it works. Having more problem-solving. It just will make people play more games and don't even think about it as educational.

[L]evels. If it is too hard it will just discourage people off. For example, start very simple then it [level] slowly gets harder and get more complicated possibly. So, it actually builds your confidence and you say to yourself if I can do the first one then I will do the next one, so it helps going progressively, easy to do, step by step.

I do think that it is kind of it can be a pass/fail mentality rather than grading. When you get a question wrong in Maths you say well, I missed it up completely.

You can incorporate like visual design you may want to make it more relaxing like valley different sounds and aesthetics or something like that.

It can give you feedback halfway and explain to you what is wrong and how you can improve.

I prefer right/wrong feedback because you cannot progress. Because if you are going through the game you have to get it right.

It could be useful for city planning you could create it virtually before moving on to real life you should have a chance to do the same thing in real life.

[Y]ou can almost run a test. It can be a test.

You can simulate [the learning environment using SGs like scientists]. [This is because] [s]cientist[s] do it all the time with what cell might do and how medicine might react. You can simulate it using Statistics and past data. There is correlation in that. People are able to calculate things so precisely like how people will behave. There is so much. I guess that's why we have Statistics.

What Educational Features Could We Add to SGs Like SimCity?

Findings indicate that in relation to the "SimCity" game (Chapter 5), additional educational features could be more real-life scenarios, complicated problem-solving and risk-taking activities, and more complex housing issues so that players have a stimulating experience of this. Other features could be businesses coming to invest in the simulated city, marketing shares, shops, and brands. We could also add more spreadsheet files with calculations, the use of past data to make predictions, run more simulations, and sophisticated calculations and models, lectures and assignments. In other words, there should be *AGBL* that involves more reasoning, and understanding of various issues, thinking creatively and critically of varieties of methods of solving and making learning more meaningful. Participants made the following suggestions:

Pop up spreadsheet with information [that] told you. You are building that house and those people are moving in the area. How this will affect the money, production, and the area? That interactive spreadsheet [can] run alongside the game.

Given options like shops or brands/supermarket chains and like in the Excel spreadsheet, they will be information of how successful they have been. Because a lot of people will pick the ones that have been round the longest/has best reputation. Maybe it will not be the most successful in getting profit. You will be able to get those things and you will be able to make your decision. It will make the game a lot more thoughtful.

More decisions going on of what you have put in which helped us to learn more about Mathematics and Statistics. Make decisions. That Excel built-in will make learning easy then you can figure out which one is the best.

> [T]he data [Excel] built-in the game I would use it in order to progress.
>
> [R]isks vs. rewards you see something it costs this much, and it takes this long, is it worth it? Shall I buy it or wait? I think it will be quite useful.
>
> Probably you need lectures and some actual past data to go with it. Maybe in these standards of applying Statistics or past data of how Statistics were applied and more information was gathered. Just by playing the game you need to look at other things like previous application [like Excel] of people using Statistics in real-world scenarios.
>
> [We] could have spreadsheet features [built-in] in the game whereas a mayor, a pop up come up saying these are supermarket with options and information we have part of the game and what we need. This will be beneficial for learning.
>
> [T]he game like an assessment. Or what we did today with the homework.
>
> [M]ore engaging way of learning it than not playing the games. It gives you more real-time feedback for example doing market shares. Real-time data. Real-time feedback, learning something about market shares because there is no way we can do this in real life.

> [The] motivation if you think you can make your town better. You could sit down and learn how to do some kind of equations more easily because you know what you can use them for.

However, we are required to think about the implications of integrating SGs in HE as some learners might not listen to lectures but would choose to play instead, as one participant noted:

> There could be some limitations for learning now it is time to stop playing the game [SG] and do some work every now and then because me and [other interviewee] will be playing the game and not paying any attention to the rest of it.

Parallel to this and when it comes to making recommendations for policy and practice on how we can employ SGs to improve the way Mathematical and Statistical learning is taught in universities (see Chapter 10), many participants felt that SGs can be employed to teach students many transferable skills. As they emphasized:

> Introduc[e] SGs to universities to show people how to think and side feature because if people know about it, they can access it, learn about, make it accessible.
>
> People [will] know about it and has accessibility, but everyone will get different results.
>
> Contribute to different parts of the game. Collaboration, everybody needs to chip in with different parts. For example, each team they can contribute with something to build in the game.
>
> Take what they learn and apply what they learn to simulation games to see if their results match. If it is positive/negative correlation and see what actually happen.
>
> Make it more visual. You could have tabs of where you open everything and Statistics [is displayed/used] in a table, where you can see everything, but you can and cannot so much. I would like to have another tab to see what is happening/not happening or transactions. You need visual aids.
>
> In universities and I know it is weird criticism to make because Maths is not about being pretty. It is just numbers. But it is usually taught in very auditory written way it can be very unstimulating it can be bulk especially for visual learners. But if you had a [SG] that is well-designed. It could perhaps engage people more especially those who are visual learners to process more information a lot better, make it stimulating.

In the next section, I shall discuss my grounded theoretical framework.

Acumenous and Inquisitional Theory

As previously enlightened in Chapter 7, my findings indicate that there are two types of GBL: *Acumenous* or *Inquisitional* (see Qualitative Findings; Chapter 10 for further clarifications). These theoretical concepts of "*Acumenous"* and "*Inquisitional"* were conceived and found in the research data of Chapter 7, then it re-appeared (and was further developed) in this chapter. In this section, I will present in detail the theoretical concepts behind *Acumenous* and *Inquisitional Theory*, providing examples of where it occurs within my research data justifying my theoretical concepts. I will start with *Inquisitional* then move on to *Acumenous*.

Acumenous and Inquisitional Theory is a perceived game-based theory which I have associated with *Acumenous* and *Inquisitional* GBL. What I term *Inquisitional* Learning is part of a perceived game-based theory/theoretical framework. *Inquisitional* or "quiz like" is a quiz type of learning/a recall that can occur in conversations or discourses such as police investigations or courtroom questioning: Questions and answers. Inquisitional consists of instructions or multiple-choice type of answers.

In participants' discourse *Inquisitional* was described as having "fewer multiple options," "[n]ot as interactive." "Not appealing," "boring," "No animation," more text involved, and if it is "too hard," it "discourage[s] people off." *Inquisitional* is "too complex." It has "no imagination/creativity" and is "not interactive." It relies on memorization and formulas and a set of instructions. *Inquisitional* type of learning is hard to understand its complex concepts. *Inquisitional* consists of "formulas with numbers, all variables Xs and Ys." It is "manual and intense" in that learners have to "do all steps, calculations and memorize formulas." *Inquisitional* has letters and numbers, and it is not related to "real life," as students "do not see its relevance outside classroom," "no mission or real-life scenarios or how it is related to real life," and therefore, it is "theoretical"/feels theoretical. It is "too complex," "intense," and difficult where the learner is "passively taking knowledge." It can be "quite an exhausting process" for listeners/learners since it requires a "considerable amount of effort" and it is "not personal." *Inquisitional* can deter learners who do "not know where all information is coming from," and "why they are required to understand it." For instance, our interviewees referred to this and reinforced it, with comments about their feelings and experiences with learning Maths in the past:

> When [I was] younger [Maths] was more interactive. [But as] Maths gets harder, [games have] few multiple options. Not as interactive as it was. Not appealing. No animation. It [just] gets harder. Not as appealing as it is used to be.
>
> I did learn it in a high school very quickly, so I didn't really remember any kind of the concept.
>
> Maths is extremely difficult, and it is hard for me to understand the concepts.
>
> I am not interested in Statistics at all because I don't see its relevance.

> I hate Maths at school. Maths was less my favourite subject because I think I don't see its relevance to anything outside classroom.
>
> My experience was quite negative. It is not clear.
>
> I find Maths too complex. Why is it complex? I suppose I could work out with time, but it is not something I am not interested in doing it.
>
> Not enjoyable. My Maths and Statistics I was not sure about how it is related to real life or use it or apply it.
>
> [I]t is manual and intense you have to do all steps, calculations and memorize formulas. I hate Statistics the most.
>
> Before I did not have any idea of what Statistics is or did not know how it is applied. Why is that the case? I asked. I am not a number person; it does not interest me. I can't stand any form of Maths.
>
> I did GCSEs [secondary studies] Maths and I always fail it. I never took much of them [Maths.] Every time I see Maths I just shut down. I never tried and I am not a big fan. Why is that the case? I don't know. I never took a lot of interest in Maths.

It is worth noting that this type of learning is used in teaching and learning in the Mathematics and Statistics curriculum. Examples of *Inquisitional* Learning are found in MEGs such as "Calculus quiz game"(https://play.google.com/store/apps/details?id=com.puremath.calculusgame&hl=en&gl=US), "MyMaths," (https://www.mymaths.co.uk/), and "Numbas" quiz games.

The second type of GBL which emerged from participants' discourse is what I have named as "**'Acumenous Game-Based Learning" (AGBL)** from the noun acumen, which means understand, reason, and decide. It is the ability to understand, reason, and decide why something is occurring. In other words, there are several stories of why learners are using it. References to this type of learning are evident in participants' discourse, as can be seen in the extracts from the FGIs both above and below. For instance, they emphasized this ability to understand, reason, and decide noting,

> The game was that if you add [a chemical] [...] What will happen? If this chemical increase[s]/decrease[s], you see the outcome straightaway [...] you can see the outcome [effects] straightaway and you will create the best outcome.
>
> [S]omething new, [programming] give[s] me power I can see how it can be applied to everyday life.
>
> [C]ompletely different experience but [programming/computational learning] is rewarding at the end.
>
> I had no experience in programming [and he continued, but]. There are rules to follow if you miss it. It can be hard. It needs a lot of practice [and] iteration. It can be annoying I notice when I am coding, I don't find my mistake. What is hard is how to break problems down because a lot of people know what to do but they don't know how to break the problem into logical steps.

This ability to break down problems into logical steps, I suggest, is similar to what is referred as AGBL since one has to understand, reason, and apply the rules. Parallel to this, other interviewees (as indicated above) valued the following features: The relevance of real-life scenarios with real-time feedback and rewards; the acumen or "acumenifying" process; the creative and critical thinking/way of solving issues differently; the step-by-step problem-solving aspects of it; the reasoning and many explanations and possibilities of why they are using it; and the collaboration and interactivity. This is evident in the following extracts from the FGIs:

> I can see the necessity of [it]. I see now the points I mean how you get your decisions. I mean may be other things.
>
> It only when I kind practise and see all X, Y and X^2 in different columns that's when I understand that they all fit together.
>
> You have steps then what I did. Oh yeah it makes sense. It [learning] hides behind it. When looking back it makes the learning quite easy than going straight on without knowing or you don't know where it comes from.
>
> So, the game was really good thing and the Excel.
>
> But when you have a game it stimulates you and you are not just passively taking information. Getting information is not just a chore [routine].
>
> It is easy to learn something if you are doing something fun because it is your city and you own everything and just want to make it better and better.
>
> I mean it gives you a chance to trial and make your decisions. It makes you feel like you have a chance. You can always redo. It can help people who may have issues with confidence knowing that you can change your decisions.
>
> Relevance to everyday life and how it is applied in real-life experiments.
>
> Real-life scenarios, and it is back to experience.
>
> Rewards when you get something right.
>
> [A]lready interactivity it can what makes it creative. You have many possibilities that you can pick. Where the road is. What you decide to do. But it also has enough limits so that it is not so difficult of what you can do.
>
> Definitely. Yes. That visual aids that sort of participation, also that explain feature[s] see why you are doing it and seeing why you are doing it.
>
> The real-life example If it is in the game [Excel] and built-in the game. This Excel we have a list, Excel I think it will really help. It can be just basic but that it is extra step, but it will help.
>
> It would be a much more engaging way of learning it than not playing the games. It gives you more real-time feedback for example doing market shares. Real-time data. Real-time feedback, learning something about market shares because there is no way we can do this in real life.
>
> Because it is personal to you. Something personal to you. You are applying it immediately even if it is perhaps subconsciously. You are using

> it and when you tell us we are using correlation. We have connection in our brains that we make. We apply the knowledge subconsciously.
>
> It makes it easy to remember. [With normal Mathematics and Statistics lecture] It's like this you need to know these letters but we are not going to tell you why you need them [but in the game] there is a lot of back stories of why we're using it.

These are the sort of things that are coming out from the data: *Acumenous* and *Inquisitional* GBL. These are discussed next.

Empirical Discussion

In view of the above, this section begins by presenting a detailed representation of the findings (as denoted in participants' quotes), concerning participants' experiences in learning Statistics using SGs, and where applicable participants' experiences with Mathematics and GBL. I will focus on what the emergent insights are with explicit reflection to the relevant literature (Chapters 2–5), where applicable. My findings show that interviewees can succeed in learning and playing SGs while they struggle to understand Mathematics and Statistics, thus indicating preconceptions toward this type of learning, that is, fear and lack of self-efficacy in these subjects. My findings also indicate how these issues were resolved using the gameplay investigation, since interviewees (and using AGBL) have increased confidence in learning Mathematics and Statistics. Thus, the emerging themes (discussed next) are as follows:

- **The Pedagogical Benefits of SGs: Mathematical and Statistical Learning, Creative Thinking, Decision-Making, and Problem-Solving**.
- **Inquisitional Learning Can Cause Disengagement with Mathematics and Statistics**.
- **Acumenous GBL Can Improve Mathematics and Statistics Learning**.
- **Acumenous GBL in SGs Can Increase Self-Efficacy in Mathematics and Statistics Learning**.
- **Acumenous and Inquisitional Theory**.

The Pedagogical Benefits of SGs: Mathematical and Statistical Learning, Creative Thinking, Decision-Making, and Problem-Solving Skills

As previously articulated, interviewees reported that SGs, as a learning tool, have many pedagogical benefits such as enhancing Mathematical and Statistical learning, leadership skills, concentration, control, speed, confidence boosting, strategic thinking to offering full engagement, fun, immersion, creative and critical thinking, decision-making, and problem-solving. A SG can be educational. It has many possibilities including real-life scenarios with real-time feedback, performance analysis, learning to live life from other players' perspectives, and various transferable skills that can be learnt from games. With every new game, there is something new to learn. Just by playing the SG itself users are learning, interacting, and in control of their learning. This can be linked to previous and current research such as Malone (1980), Jones (1998), Faria (2001), Garris et al. (2002), Prensky (2002, 2005), Gee (2003a, 2003b, 2004, 2005a, 2005b, 2007, 2008, 2011), Faria and Wellington (2004), Gee et al. (2008), Lenhart et al. (2008), Moreno-Ger et al. (2008), Kolb and Kolb (2008), Mayo (2009), Pelletier (2009), Vos and Brennan (2010), Tao et al. (2012), Li et al. (2013), Granic et al. (2014), Vos (2014), McGonigal (2015), Schell (2015, 2020), Jagger et al. (2016), Hamari et al. (2016), Naik (2017), Vlachopoulos and Makri (2017), and Barbosa et al. (2019). They emphasize the key importance of simulation games' abilities, as a teaching tool, to model skills such as decision-making and numerical data (see Chapters 2, 3, 5, and 7).

Nevertheless, not all games especially educational games were considered to be fun or a great tool for entertainment and learning. In particular, Mathematics was found to be challenging and rather Inquisitional as we found in Chapter 7, a recurring theme that reappears in this chapter and is discussed next.

Inquisitional Learning Can Cause Disengagement with Mathematics and Statistics

Following Chapter 7 findings have indicated that MEGs are disengaging, and we learned from the data that the reason for this is that the Mathematics and Statistics curriculum is *Inquisitional* type of learning. In this chapter, interviewees described this *Inquisitional* learning experience as being "complex," "uneasy," "not relevant to real life," and "hard to understand or remember." It is "manual and intense with learners having to carry out all the steps." It is just "formulas." *Inquisitional* type of learning is "theoretical," and it is based on "memorization and variables like Xs and Ys."

Based on the research findings from Chapter 7, external variables (not useful and fun) can affect how users perceive the world, which in turn affects their preconceptions and therefore their self-efficacy and behavior. Looking back at participants' discourse, the reason for their struggles with Mathematics and Statistics include fear, mental blockage, not remembering any Mathematical and Statistical concepts and/or simply shutting down. This further illustrates that participants have preconceptions toward Mathematics and Statistics because of its *Inquisitional* type of learning as previously articulated. This furthermore indicates that the way Mathematics and Statistics is taught, to our interviewees, is "hard, not interactive" and has "few multiple options," mainly because the type of Mathematics and Statistics students learn is *Inquisitional* or "quiz type" of learning where "learners face fewer options to choose from" and their learning is "not related to real-life problems." This consequently leads to low self-efficacy and fear in Mathematics and Statistics. As previously articulated, interviewees experienced difficulties while studying Mathematics and Statistics. From HCI perspectives, if a system does not cater to users' needs, it is a failing system (Chapters 7 and 8). Mathematics curriculum could be thus failing because the type of Mathematics and Statistics our interviewees experienced is *Inquisitional*. This thus can have detrimental impacts on interviewees' self-efficacy and emotions toward this numerical subject, as this research indicates. Game design with its characteristics can positively impact learners' skills like decision-making, creative thinking, competence, knowledge, and learning outcomes, and therefore a significant part of the learning experience.

Acumenous GBL Can Improve Mathematics and Statistics

The Mathematics and Statistics curriculum can be engaging and effective in HE Business and Computing Schools if we employ SGs to teach it (Chapter 5). This is because experimentation and transferable skills come originally from SGs. What these generations, Z and Alpha, would like is more autonomy and control over their learning, and they prefer *Acumenous* GBL (*AGBL*) (Chapter 7) where the focus is on transferable skills such as problem-solving, decision-making, competition, a chance to trial and investigate, on giving users a sense of purpose, on creating the best outcomes, and on seeing the outcome straightaway. This type of GBL gives learners a "sense of ownership," a sense of "control" and a "mission" to create artefacts the way they want and imagine it. *AGBL* is stimulating and "easy to remember." There are several stories "of why" users or learners "are using it."

Following research findings from Chapter 7, how learners perceive IT as useful and fun can play a key role in increasing their self-efficacy, and positively drive their preconceptions, leading to positive impacts on their skills, that is, transferable skills as in *AGBL* (discussed next). These skills include decision-making, problem-solving, finance, performance indicators, numerical data, analysis, interpreting historical and factual data to illustrate trends, patterns, and future predictions. Consequently, this can be utilized to improve Mathematical and Computational learning using SGs, and therefore as a learning tool for improving self-efficacy in Mathematics and Statistics.

Acumenous GBL in SGs Can Increase Self-efficacy in Mathematics and Statistics

As previously explained, simulation GBL approach to teaching was employed, that is, "SimCity" gameplay scenario, with Excel (Chapter 5), with the aim of testing participants' confidence in Statistical learning before and after the game.

Findings show (in particular, quantitative results) that participants' confidence significantly increased after playing the SG, as pairwise t-test results demonstrated (see Chapter 5 for further details, Table 5.2; Figure 5.3).

Findings indicate that participants liked the many pedagogical benefits of SGs such as a sense of mission, purpose and ownership, building confidence, solving issues, more motivation to learn and absorb learning quicker rather than passively taking in information, as well as a chance to experience and investigate real-life examples with real-time feedback. Findings also suggest that AGBL can play a beneficial role in enhancing Mathematical and Statistical learning since players can see the outcomes of their decision straightaway, and as they have an opportunity to try it again, they are more likely to create the best outcomes. This is because simulation GBL approach to learning Statistics was perceived as useful and great fun (as aforementioned). These (usefulness and fun) drove interviewees' preconceptions leading to high self-efficacy that acts as a force changing positively their behavior. Participants interacted with the gameplay scenario in playful ways. They felt excited, relaxed, and were more engaged. Participants were understanding, reasoning, and deciding. They were constantly thinking creatively and critically, collaborating, managing, and leading their learning, that is, acumenifying their learning (*AGBL*) and subsequently developing new skills added to their learning journey. This *AGBL* process was apparent in some of the participants' responses as demonstrated above.

Acumenous and Inquisitional Theory: My Theoretical Contribution

Following the research findings from Chapter 7, we learnt that how players think or see IT as useful, and fun can affect their preconceptions and so their self-efficacy. This self-efficacy can affect how players behave and interact with the game. This social interaction with the game leads to transferable skills (new skills/learning outcomes), which is a new theoretical lens. I have referred to this as *Acumenous"* or *Inquisitional. Acumenous and Inquisitional* are two types of GBL, that is, **Acumenous Game-Based Learning (AGBL)** or *Inquisitional* Game-Based Learning (IGBL) (see Chapter 7). I have also defined and explained these theoretical concepts from the grounded data. In this section, and based on research findings, I will present in detail the theoretical concepts behind ***Acumenous and Inquisitional Theory***.

Acumenous and Inquisitional Theory is a perceived game-based theory which I have associated with Acumenous or Inquisitional GBL. I will start with *Inquisitional* then move on to *Acumenous* GBL, providing some examples of where it occurs within my research data justifying my theoretical concepts.

Following Chapter 7, *Inquisitional* emerges as a new theoretical lens and I have defined it. For example, I explained that *Inquisitional* or "quiz-like" type of learning can occur in conversations or discourses such as police interviews or courtroom cross-examinations: Questions and answers. *Inquisitional* consists of instructions or multiple-choice type of answers. According to participants, *Inquisitional* has "fewer multiple options," "no creativity," "no mission," or "real-life scenarios."

In this chapter, Chapter 9, *Inquisitional* re-emerges and was further developed with more explanations, as I will discuss next. "*Inquisitional* Game-Based Learning" (IGBL) is part of a perceived game-based theory.

In participants' discourse, *Inquisitional* was described as having "fewer multiple options." "Not appealing," "boring," "no animation," more text involved, and if it is "too hard," it "discourage[s] people off" (Chapter 7; Chapter 9).

Inquisitional is "too complex" (Chapter 9). It has "no imagination/creativity" and is "not interactive" (Chapter 7). It relies on a set of instructions. *Inquisitional* type of learning is hard to understand its concepts (Chapter 9). *Inquisitional* consists of "formulas with numbers, all variables Xs and Ys" (Chapter 9). It is "manual" in that learners have to "do all steps, calculations and memorize formulas" (Chapter 9). *Inquisitional* has "letters and numbers," and it is "not related to real life" (Chapter 9), as students "do not see its relevance outside classroom," "no mission or real-life scenarios" or how it is "related to real life" (Chapter 9), and therefore, it is "theoretical"/feels theoretical (Chapter 9). It is too complex, "intense," and "difficult where the learner is passively taking knowledge" (Chapter 9). It can be "quite an exhausting process for listeners" or learners since it requires a "considerable amount of effort," and it is "not personal" (Chapter 9). *Inquisitional* can deter learners who "do not know where all information is coming from," and "why they are required to understand it" (Chapter 9).

Examples of *Inquisitional* type of learning are manifested within MEGs such as "Calculus quiz game" (Chapter 7). *Inquisitional* could be associated with behaviorism, since learners are not required to think or understand reasons/implications behind such a learning, they can just learn by repetition and iteration, that is, copy the behavior (Chapter 10).

The second type of GBL which emerged from participants' discourse is what I have named as ***Acumenous Game-Based Learning*** **(AGBL)** from the noun acumen. It is the ability to understand, reason, and decide why something is occurring (Chapter 7). References to this type of learning are evident in participants' discourse, as can be seen in the extracts from the FGIs both above and below. According to participants, *AGBL* simplifies learning, it is "rewarding" (Chapter 9) as you can see "[w]hat will happen?" and "you will create the best outcome" (Chapter 7). *AGBL* was defined as a type of learning that give learners "power" since they "can see how it can be applied to everyday life." *AGBL* involves "break[ing] the problem into logical steps" (Chapter 7). This ability to break down problems into logical steps, I suggest, is similar to what is referred as *AGBL* since one has to understand, reason, and apply the rules. Parallel to this, other interviewees (Chapter 9) value the following features: The "relevance of real-life scenarios" with "real-time feedback" and "rewards," the acumen or "acumenifying" process, the creative and critical thinking/way of solving issues differently, the step-by-step problem-solving aspects of it, the "reasoning" and many explanations and "possibilities of why they are using" it, the "collaboration" and "interactivity" (Chapter 9). Participants noted that they can see the "necessity" of this learning and how they obtain their "decisions" (Chapter 9). They could see all the "X[s], Y[s] and X^2[s]" and "understand that they all fit together" (Chapter 9). *AGBL* have "steps," "makes sense," "easy to learn," "fun," and learning "hides behind it" (Chapter 9). Participants indicated that "when looking back it makes the "learning quite easy" than "going straight on without knowing" where it "comes from." Participants noted that ***Acumenous Game-Based Learning*** is "stimulat[ing]" where the learner is "not passively taking information" (Chapter 9). Obtaining "information is not just a chore." *AGBL* "gives you a chance to trial and make your decisions." It makes learners "feel like [they] have a chance." They "can always redo" their decisions and it can "help people who may have issues with confidence knowing that [they] can change [their] decisions" (Chapter 9). In participants' discourse *AGBL* was described as "more engaging way of learning" and "easy to remember" (Chapter 7). "It gives you more real-time feedback" with "real-time data" and a chance to "learn something" [for instance, market shares] because "there is no way [they] can [experience] this in real life" (Chapter 9). *AGBL* is "relevan[t] to everyday life" and learners know "how it is applied in real-life experiments" with "[r]eal-life scenarios" and it is "personal to [learners]" (Chapter 9). *AGBL* is "interactive," "creative" and you "have many possibilities that you can pick," "that sort of participation and "explain[ed] feature[s]" is useful (Chapter 9). This is because users are "applying" the learning "immediately even if it is perhaps subconsciously." *AGBL* gives learners "connection[s]" in their "brains that [they] make," since they "apply the knowledge subconsciously" (Chapter 9).

Participants also liked the "collaboration" and "many possibilities that they can use." Participants felt they can learn complex ideas in fun easy ways rather than "absorbing knowledge passively." They were "acumenifying" their learning. They were constantly thinking creatively and critically, managing, and leading their learning.

These are the sort of things that emerged from the empirical data: ***Acumenous and Inquisitional***. I will now consider to what extent these things are already are accounted for by existing theories. While Gee and others do consider learners and their learning processes (I will return to discuss Gee' theoretical learning framework in Chapter 10), their approaches are somewhat different and on the margin. What is required is something that brings these facets together more centrally just as in ***Acumenous and Inquisitional Theory***.

As previously illustrated, and in Chapter 7 in detail, I explained that how players think or see IT as useful and fun can subsequently affect their preconceptions and so their self-efficacy. This self-efficacy can affect how players feel, behave, and interact with the game. This is because each player/learner comes with some variables that could determine how they see the world and use IT. These variables are perceived usefulness and fun, which can affect how preconceptions are formulated and therefore player's self-efficacy. This latter can subsequently affect players' behavior /intention of using the actual system: The game, and how it is selected. This game system is influenced by the player's choice of game design/style, which interestingly plays a key role in how the user interacts with the game system. These can affect players' skills and knowledge, the learning outcomes, and thus players' overall learning experience. This social interaction with the game leads to transferable skills (new skills/learning outcomes), which is ***Acumenous or Inquisitional GBL***.

Acumenism/AGBL, as a perceived game-based theory, builds on previous theories (i.e., social constructivism). It comes to highlight the issue of preconceptions toward learning (e.g., Mathematics and Statistics education) that can cause disengagement. *Acumenism* emphasizes that how learners' perceived usefulness and fun can affect their preconceptions and thus their self-efficacy and how they interact with the game system, leading to learning outcomes which can be *Acumenous or Inquisitional.*

Theories of social constructivism and their pioneers such as Gee (2003a, 2003b, 2004, 2005a, 2005b, 2007, 2008, 2011) focused on good video games and the 13 learning principles and how best to assist learners to discover knowledge for themselves, for instance, through game interaction, problem-solving activities, deep understanding, and empowerment (Chapters 2–5). These principles (as I will explain in Chapter 10) are in line with *Acumenous* GBL and present the most fundamental elements of Acumenous GBL. This is because (as explained above) ***Acumenous*** is the ability to understand, reason, and decide. This book, therefore, is building on Gee's principles on GBL (discussed next; see also Chapter 10 for further details).

Acumenism**/AGBL* framework agrees with social constructivism adding that through GBL students develop their problem-solving, decision-making, and reasoning skills because they are constantly thinking, reasoning, and deciding in order to proceed in the gameplay. **Thus, the main contribution and differences from those of Gee** is ***Acumenous **and** ***Inquisitional***—a perceived game-based theory (as previously articulated) and how learners' perceived usefulness and fun can affect their preconceptions (and thus their self-efficacy and interaction with the game system), leading to learning outcomes which can be *Acumenous or Inquisitional.*

Theories of behavior and their pioneers such as Skinner (1978) focuses on the concept of utilizing stimulus to reward/punish positive or negative behavior. Learning can take place by using rewards, repetition, and iteration, and this can thus change behavior. In this vein, learners are considered to be passive (i.e., behaviorism ignores learners' thinking processes), and this is one of its criticisms. Learners, I argue, should be challenged through questioning their ideas and breaking down tasks. Unlike behaviorism, ***Acumenism**/AGBL* recognizes its features like using rewards to reinforce good behavior and add to this theory that it is necessary to consider learners' thinking processes, since cognition has impacts on learners' preconceptions and thus on their behavior and interaction with the game system. Learners are not passive. Learners are active and necessitate social interaction as an element for learning and a complement for grasping knowledge.

Acumenism uses GBL and relies on reasoning, understanding, thinking critically and collaborating, investigating real-life scenarios, creating artefacts, projects and assignments in group work, alone, or in discussions. I propose that by engaging in the AGBL process, with the right game and the right game scenario, and under the right conditions, learners could *acumenify* their learning.

Examples of AGBL, as a type of learning, is manifested within some simulation games, such as "SimCity" (see Chapters 4–6) "Assassins' Creed," "Harvard Business Simulation," and some educational games for instance Harel's and Papert's (1990) "Instructional Software Design Research Project" in schools, Mayer et al.'s (2013) Cyberdam-2D GBL serious game, and Jagger et al.'s (2016) "3D Business Simulation" game (Chapter 3).

The next chapter, **Chapter 10**, will present Outline of the Main Findings and my Theoretical contributions: *Acumenous and Inquisitional Theory.*

Bibliography

Acumen (2017) In: Cambridge University Press. [Online] Available at: http://dictionary.cambridge.org/dictionary/english/acumen (Accessed 30/05/2017).

Adams, A., Lunt, P. and Cairns, P. (2008) A Qualitative Approach to HCI Research. In: Cairns, P. and Cox, A. (ed.). *Research Methods for Human-Computer Interaction*. Cambridge, UK: Cambridge University Press, pp. 138–157. Available at: http://www.cambridge.org/catalogue/catalogue.asp?isbn=9780521870122andss=toc (Accessed 20/11/2016).

ADDIE Model (2016) Available at: http://www.learning-theories.com/addie-model.html (Accessed 20/04/2016).

Anderson, S. (2020) Quantitative Analysis. Available at: https://www.investopedia.com/terms/q/quantitativeanalysis.asp (Accessed 20/02/2021).

Anon (2018) Efficacy. In: Cambridge University Press. [Online] Available at: https://dictionary.cambridge.org/dictionary/english/efficacy (Accessed 30/09/2018).

Applefield, J.M., Huber, R. and Moallem, M. (2000) Constructivism in Theory and Practice: Toward a Better Understanding. *The High School Journal*, 84(2), pp. 35–53.

Arthur, J., Waring, M. Coe, R. and Hedges, L.V. (2012) *Research Methods and Methodologies in Education*. California: Sage Publications Inc.

Atkins, L. and Wallace, W. (2012) *Qualitative Research in Education*. Los Angels, London, New Delhi, Singapore, Washington, DC: Sage: Publications Ltd.

Bandura, A. (1986) *Social Foundations of Thought and Action: A Social Cognitive Theory*. Englewood Cliffs, NJ: Prentice-Hall.

Bandura, A. (1994) Self-efficacy. In Ramachaudra, V.S. (ed.). *Encyclopedia of Human Behaviour*, 4, pp. 71–81. New York: Academic Press. (Reprinted in H. Friedman (Ed.) (1998). Encyclopediaof Mental Health. San Diego: Academic Press).

Bandura, A. (1997) *Self-efficacy: The Exercise of Control*. New York: W.H. Freeman and Company.

Bandura, A. and Schunk, D. (1981) Cultivating Competence, Self-Efficacy, and Intrinsic Interest Through Proximal Self-Motivation. *Journal of Personality and Social Psychology*, 41(3), pp. 586–598.

Barbosa, T., Lopes, S.P. Leao. C. Soares, F. and Carvalho, V. (2019) Serious Game for Teaching Statistics in Higher Education: Storyboard Design. In: *7th EAI International Conference, Braga, Portugal, on 'Interactivity, Game Creation, Design, Learning, and Innovation'* (ed.) Springer, pp. 169–175. DOI: 10.1007/978-3-030-06134-0_18

Bartholomew, H. et al. (2011) "I am Natural, and I Do It for a Love!": Exploring Students' Accounts of Studying Mathematics. *International Journal of Mathematical Education in Science and Technology*, 42(7), pp. 915–924.

Bassey, M. (1999) *Case Study Research in Educational Settings*. Buckingham: Open University Press.

Bellotti, F., Kapralos, B. Lee, K. Moreno-Ger, P. and Berta, R. (2013) Assessment in and of Serious Games: An overview. *Advances in Human-Computer Interaction*, 2013, p. 1.

Belluigi, D.Z. (2009) Exploring the Discourses Around 'Creativity' and 'Critical Thinking' in a South African Creative Arts Curriculum. *Studies in Higher Education*, 34(6), pp. 699–717.

Bennedesen, J. and Caspersen, M.E. (2004) Programming in Context: A Model-First Approach to CS1. In *Proceedings of the 35th SIGCSE Technical Symposium on the Computer Science Education*, pp. 477–481. ACM Press.

Bevan, N. (2001) International Standards for HCl and Usability. *International Journal of Human Computer Studies*, 55(4), pp. 533–552.

Boaler, J. (2009) *The Elephant in the Classroom*. London: Souvenir Press.

British Education Research Association (BERA) (2011) BERA Ethical Guidelines for Educational Research. Available at: http://www.bera.ac.uk/ (Accessed 15/10/2016).

British Education Research Association (BERA) (2018) BERA Ethical Guidelines for Educational Research. Available at: https://www.bera.ac.uk/wp-content/uploads/2018/06/BERA-Ethical-Guidelines-for-Educational-Research_4thEdn_2018.pdf?noredirect=1 (Accessed 25/03/2019).

Brown, J.S., Collins, A. and Duguid, P. (1989) Situated Cognition and the Culture of Learning. *Educational Researcher*, 18(1), 32–42.

Bruner, J. (1973) *Going Beyond the Information Given*. New York: Norton.

Caillois, R. (1961) *Man, Play, and Games*. New York: Free Press.

Cardinot, A. and Fairfield, J.A. (2019) Game-Based Learning to Engage Students with Physics and Astronomy Using a Board Game. *International Journal of Game-Based Learning (IJGBL)*, 9(1), pp. 42–57.

Charmaz, K. (2006) *Constructing Grounded Theory*, 1st Edition. London: Sage Publications Ltd.

Charmaz, K. (2014) *Constructing Grounded Theory*, 2nd Edition. London: Sage Publications Ltd.

Chen, M.P., Wong, Y.T. and Wang, L.C. (2014) Effects of Type of Exploratory Strategy and Prior Knowledge on Middle School Students' Learning of Chemical Formulas from a 3D Roleplaying Game. *Educational Technology Research and Development*, 62, pp. 163–185. DOI: 10.1007/s11423-013-9324-3

Clark, D.B., Tanner-Smith, E.E. and Killingsworth, S.S. (2015) Digital Games, Design, and Learning a Systematic Review and Meta-Analysis. *Review of Educational Research*, 86(1), pp. 79–122.

Claxton, G. (2002) *Building Learning Power*, 1st Edition. Bristol: TLO Limited.

Conlon, C., Carney, G., Timonen, V. and Scharf, T. (2015) Emergent Reconstruction in Grounded Theory: Learning from Team-Based Interview Research. *Qualitative Research*, 15, pp. 39–56. DOI: 10.1177/1468794113495038

Cook, J. (1992) Negotiating the Curriculum: Programming for Learning. In *Negotiating the Curriculum*, pp. 15–31. Routledge.

Cooper, P.A. (1993) Paradigm Shifts in Designed Instruction: From Behaviorism to Cognitivism to Constructivism. *Educational Technology*, 33(5), pp. 12–19.

Corbin, J. and Strauss, A. (2015) *Basics of Qualitative Research: Techniques and Procedures for Developing Grounded Theory*, 4th Edition. Thousand Oaks, CA: Sage.

Cordova, D.I. and Lepper, M.R. (1996) Intrinsic Motivation and the Process of Learning: Beneficial Effects of Contextualization, Personalization, and Choice. *Journal of Educational Psychology*, 88, pp. 715–730.

Coulthard, G.J. (2009) A Review of the Educational Use and Learning Effectiveness of Simulations and Games [online]. PhD Thesis, Purdue University. Available at: http://www.coulthard.com/library/paper-simulation.html (Accessed 15/12/2015).

Crockett, L. (2018) Bogost's War on Gamification: Irony and Play on Grandma's Sofa. In: *12th European Conference on Games Based Learning (ECGBL 2018)*. Reading: Academic Conferences and Publishing International Ltd.

Draper, S. (2013) Social Constructivism. Available at: http://www.psy.gla.ac.uk/~steve/courses/archive/CERE12-13-safari-archive/topic3/webarchive-index.html#social%20con%201 (Accessed 23/12/2019).

Dweck, C.S. (2006) *Mindset: The New Psychology of Success*. United States: Ballantine Books Trade Paperback Edition.

Dwyer, J. and Slyman, S. (2016) *An Introduction to Quantitative Methods for Business Research*. London: Algana Publishing. 200 pages; ISBN−13: 9781907934094.

Early, R.E. (1992) The Alchemy of Mathematical Experience: A Psychoanalysis of Student Writings. *For the Learning of Statistics*, 12(1), pp. 15–20.

Faria, A.J. (2001) The Changing Nature of Business Simulation/Gaming Research. *Simulation and Gaming*, 32(1), pp. 97–110. DOI: 10.1177/104687810103200108

Faria, A.J., Hutchinson, D. Wellington, W.J. and Gold, S. (2009) Developments in Business Gaming: A Review of the Past 40 Years. *Simulation and Gaming*, 40(4), pp. 464–487. DOI: 10.1177/1046878108327585

Faria, A.J. and Wellington, W.J. (2004) 'A Survey of Simulation Game Users, Former-Users, and Never-Users. *Simulation and Gaming*, 35(2), pp. 178–207. DOI: 10.1177/1046878104263543

Federoff, M.A. (2002) Heuristics and Usability Guidelines for the Creation and Evaluation of Fun in Video Games. Master Thesis, Indiana University, Bloomington.

Fienberg, S.E. (1992) A Brief History of Statistics in Three and One-Half Chapters: A Review Essay. Available at: https://www.jstor.org/stable/2246307?casa_token=Vek3Rmd5Mk0AAAAA%3AJ0zRZC5AxskwvTNY7GlrgsRV6oLrE20DY3FdeaPad0JvNSNrZDav8uZpX-YqY1hXTDtaz6zM87S4vwkQdflDLSzmrLCNlEsaBqWnykepc0jlYXsQXA

Forest, E. (2014) ADDIE Model. Available at: http://educationaltechnology.net/the-addie-model-instructional-design/ (Accessed 20/04/2016).

Fosnot, T.C. (1989) *Enquiring Teachers, Enquiring Learners: A Constructivist Approach for Teaching*. New York: Teachers College Press.

Fosnot, T.C. and Perry, R.S. (2005) Constructivism: A Psychological Theory of Learning. Available at: http://beyondbitsandatoms.org/readings/fosnot_constructivism_1996.pdf (Accessed 20/07/2019).

Garris, R., Ahlers, R. and Driskell, E.J. (2002) Games, Motivation, and Learning: A Research and Practice Model. *Simulation and Gaming*, 33(4), pp. 441–467.

Gee, J.P. (2003a) *What Video Games Have to Teach Us About Learning and Literacy*. New York: Palgrave Macmillan.

Gee, J.P. (2003b) Good Video Games and Good Learning. Available at: https://academiccolab.org/resources/documents/Good_Learning.pdf (Accessed 03/01/2021).

Gee, J.P. (2004) *Situated Language and Learning: A Critique of Traditional Schooling*. London: Routledge.

Gee, J.P. (2005a) Why are Video Games Good for Learning? Retrieved January 1, 2018, from http://academiccolab.org/resources/documents/MacArthur.pdf (Accessed 20/04/2016).

Gee, J.P. (2005b) *Why Video Games Are Good for Your Soul: Pleasure and Learning*. Melbourne: Common Ground.

Gee, J.P. (2007) *Good Video Games and Good Learning: Collected Essays on Video Games, Learning, and Literacy*. New York: Peter Lang.

Gee, J.P. (2008) *Getting over the Slump: Innovation Strategies to Promote Children's Learning*. New York: The Joan Ganz Cooney Center at Sesame Workshop.

Gee, J.P., Hayes, E., Torres, R.J., Games, I.A., Squire, K. and Salen, K. (2008) Playing to Learn Game Design Skills in A Game Context. In: *Proceeding ICLS'08 Proceedings of the 8th International Conference on International Conference for the Learning Sciences*, Vol. 3, pp. 368–374.

Gee, J.P. and Hayes, E.R., 2011. *Language and Learning in the Digital Age*. Routledge.

Gibbons, J. and Gray, M. (2004) Critical Thinking as Integral to Social Work Practice. *Journal of Teaching in Social Work*, 24, pp. 19–37.

Glaser, B.G. and Strauss, A.L. (1967) *The Discovery of Grounded Theory*. Chicago: Aldine.

Gomes, A. and Mendes, A.J. (2007) Learning to Program-Difficulties and Solutions. In *International Conference on Engineering Education–ICEE*, Vol. 7.

Granic, I., Lobel, A. and Engels, R.C. (2014) The Benefits of Playing Video Games. *American Psychological Association*, 69(1), pp. 66–78. DOI: 10.1037/a0034857

Hackett, G. and Betz, N.E. (1989) An Exploration of the Mathematics Self-Efficacy/Mathematics Performance Correspondence. *Journal for Research in Mathematics Education*, 20(3), pp. 261–273.

Hamari, J., Shernoff, D.J., Rowe, E., Coller, B., Asbell-Clarke, J. and Edwards, T. (2016) Challenging Games Help Students Learn: An Empirical Study on Engagement, Flow and Immersion in Game-Based Learning. *Computers in Human Behavior*, 54, pp. 170–179.

Hamari, J. and Tuunanen, J. (2014) Player Types: A Meta-Synthesis. *Transactions of the Digital Games Research Association*, 1(2), pp. 29–53.

Hargreaves, A. (2003) *Teaching in the Knowledge Society: Education in the Age of Insecurity*. New York: Teachers' College Press.

Holbert. N., Penney, L. and Wilensky, U. (2010) *Bringing Constructionism to Action Game-Play, Constructionism*. Paris. Available at: https://www.google.co.uk/search?q=Bringing+Constructionism+to+Action+Game-Playandoq=Bringing+Constructionism+to+Action+Game-Playandaqs=chrome..69i57.7170j0j8andsourceid=chromeandie=UTF-8 (Accessed 20/03/2017).

Hoover, W.A. (1996) The Practice Implications of Constructivism. *SEDL Letter*, 9(3), pp. 1–2.

Horn, B., Hoover, A.K. Barnes, J. Folajimi, Y. Smith, G. and Harteveld, C. (2016) Opening the Black Box of Play: Strategy Analysis of An Educational Game. In *Proceedings of the 2016 Annual Symposium on Computer-Human Interaction in Play*, pp. 142–153.

Hsu, C.-Y., Tsai, C.-C. and Wang, H.-Y. (2012) Facilitating Third Graders' Acquisition of Scientific Concepts Through Digital Game-Based Learning: The Effects of Self-Explanation Principles. *Asia-Pacific Education Researcher*, 21(1), pp. 71–82.

Illeris, K. (2007) *How We Learn: Learning and Non-Learning in School and Beyond*. London: Routledge.

Jagger, S., Siala, H. and Sloan, D. (2016) It's All in The Game: A 3D Learning Model for Business Ethics. *Journal of Business Ethics*, 137(2), pp. 383–403.

Jameson, M.M. and Fusco, B.R. (2014) Math Anxiety, Math Self-Concept, and Math Self-Efficacy in Adult Learners Compared to Traditional Undergraduate Students. *Adult Education Quarterly*, 64(4), pp. 306–322.

Jeffrey, B. and Craft, A. (2004) Teaching Creatively and Teaching for Creativity: Distinctions and Relationships. *Educational Studies*, 30(1), pp. 77–87.

Jenson, J. and Droumeva, M. (2015) Making Games with Game Maker: A Computational Thinking Curriculum Case Study. In *ECGBL 2015–9th European Conference on Games Based Learning: ECGBL2015*. Academic Conferences and Publishing Limited, p. 260.

Jones, M.G. (1998) Creating Engagement in Computer-Based Learning Environments. Available at: http://itech1.coe.uga.edu/itforum/paper30/paper30.htmI (Accessed 12/04/2016).

Kafai, Y. (2006) *Playing and Making Games for Learning: Instructionist and Constructionist Perspectives for Game Studies. Games and Culture*, 1(1), pp. 36–40.

Kafai, Y.B. and Resnick, M. eds. (2012) *Constructionism in Practice: Designing, Thinking, and Learning in a Digital World*. Routledge. Available at: https://www.google.co.uk/books/edition/Constructionism_in_Practice/KGkIEQAAQBAJ?hl=en&gbpv=1&dq=Kafai,+Y.B.+and+Resnick,+M.+eds.+(2012)+Constructionism+in+Practice:+Designing,+Thinking,+and+Learning+in+a+Digital+World.+Routledge.&pg=PP1&printsec=frontcover

Kirriemuir, J. and Mcfarlane, A. (2004) Literature Review in Games and Learning. Available at: https://telearn.archives-ouvertes.fr/hal-00190453/document (Accessed 12/04/2020).

Kolb, A.Y. and Kolb, D.A. (2008) The Learning Way: Meta-Cognitive Aspects of Experiential Learning. *Simulation and Gaming*, 40(3), pp. 297–327.

Kolb, D.A. (1984) *Experiential Learning: Experience as the Source of Learning and Development*, Vol. 1. Englewood Cliffs, NJ: Prentice-Hall.

Korhonen, H. and Koivisto, E.M.I. (2007) Playability Heuristics for Mobile Multi-Player Games. *Proceedings of the 2nd International Conference on Digital Interactive Media in Entertainment and Arts*, pp. 28–35.

Kraiger, K., Ford, J.K. and Salas, E. (1993) Application of Cognitive, Skill-based, and Affective Theories of Learning Objectives to New Methods of Training Evaluations, *Journal of Applied Psychology*, 78(2), pp. 311–328.

Lave, J. and Wenger E. (1991) *Situated Learning: Legitimate Peripheral Participation*. Cambridge University Press. Available at: https://www.google.co.uk/books/edition/Situated_Learning/CAVIOrW3vYAC?hl=en&gbpv=1&dq=Lave,+J.+and+Wenger+E.+(1991)+Situated+Learning:+Legitimate+Peripheral+Participation.+Cambridge+University+Press&pg=PA11&printsec=frontcover

Lee, J.S., Keil, M. and Kasi, V. (2012) The Effects of an Initial Budget and Schedule Goal on Software Project Escalation. *Journal of Management Information System*, 29(1), pp. 53–77. DOI: 10.2753/MIS0742-1222290102

Lenhart, A., Kahne, J., Middaugh, E. Macgill, A.R., Evans, C. and Vitak, J. (2008) Teens, Video Games, and Civics: Teens' Gaming Experiences Are Diverse and Include Significant Social Interaction and Civic Engagement. *Pew Internet & American Life Project*. Available at: https://www.google.co.uk/search?q=Lenhart+et+al.%2C+2008andoq=Lenhart+et+al.%2C+2008andaqs=chrome..69i57j0.1244j0j8andsourceid=chromeandie=UTF-8 (Accessed 20/8/2017).

Li, F.W.B. and Watson, C. (2011) *Game-Based Concept Visualization for Learning Programming*. Scottsdale, Arizona, USA: ACM. 978-1-4503-0994-3/11/12.

Li, Z.Z., Cheng, Y.B. and Liu, C.C. (2013) A Constructionism Framework for Designing Game-Like Learning Systems: Its Effect on Different Learners. *British Journal of Educational Technology*, 44(2), pp. 208–224.

Lister, R., Adams, E.S. Fitzgerald, S. Fone, W. Hamer, J. Lindholm, M. McCartney, R. Moström, J.E. Sanders, K. Seppälä, O. and Simon, B. (2004) A Multi-National Study of Reading and Tracing Skills in Novice Programmers. In *ACM SIGCSE Bulletin*, 36(4), pp. 119–150. ACM.

Loftus, C., Thomas, L. and Zander, C. (2011) Can Graduating Students Design: Revisited. In *Proceedings of the 42nd ACM Technical Symposium on Computer Science Education*, pp. 105–110.

Lyons, I.M. and Beilock, S.L. (2012) When Math Hurts: Math Anxiety Predicts Pain Network Activation in Anticipation of Doing Math. *PLoS One*, 7(10), p. e48076.

Malone, T.W. (1980) What Makes Things Fun to Learn? A Study of Intrinsically Motivating Computer Games. Available at: http://cci.mit.edu/malone/Malone%20computer%20games%20bibliography.htm

Malone, T.W. and Lepper, M.R. (1987) 'Making Learning Fun: A Taxonomy of Intrinsic Motivations for Learning. Aptitude, Learning, and Instruction': Volume 3: Cognitive and Affective Process Analyses. R. E. Snow and M. J. Farr. Hilsdale, USA, Erlbaum.

Maloney, E.A., Schaeffer, M.W. and Beilock, S.L. (2013) Mathematics Anxiety and Stereotype Threat: Shared Mechanisms, Negative Consequences and Promising Interventions. *Research in Mathematics Education*, 15(2), pp. 115–128.

Marciano, J.N., Miranda, L.C. and Miranda, E.E.C. (2014) Evaluating Multiple Aspects of Educational Computer Games. Available at: https://www.hindawi.com/journals/ijcgt/2014/139205/ (Accessed 27/08/2017).

Marklund, B.B. and Taylor, A.S.A. (2016) Creating Games in the Classroom: From Native Gamers to Reflective Designers. In: Munkvold, R. (ed.). *Games Based learning for Researchers, Teachers and Students*, Vol. 1. Reading: Academic Conferences and Publishing International Ltd. ISBN: 978-1-911278-12-890000.

Mayer, E.R. (2014) *Computer Games for Learning: An Evidence Based Approach*. London: MIT.

Mayer, I., Warmelink, H. and Bekebrede, G. (2013) Learning in a Game-Based Virtual Environment: A Comparative Evaluation in Higher Education. *European Journal of Engineering Education*, 38(1), pp. 85–106.

Mayo, M. (2009) Video Games: A Route to Large-Scale STEM education? *Science*, 323, pp. 79–82.

Mccrindle, M. (2018) What Comes after Generation Z? Introducing Generation Alpha. Available at: https://mccrindle.com.au/insights/blogarchive/what-comes-after-generation-z-introducing-generation-alpha/ (Accessed 15/10/2018).

McFarlane, A., Sparrowhawk, A., and Heald, Y. (2002) *Report on the Educational Use of Games: An Exploration by TEEM on the Contribution Which Games Can Make to the Educational Process*. Cambridge, UK: TEEM. http://www.teem.org.uk/publications/teem_gamesined_full.pdf (Accessed 06/12/2015).

McGonigal, J. (2015) *Superbetter: A Revolutionary Approach to Getting Happier Stronger and Happier*. Braver and More Resilient, New York: Penguin Press.

Monroy-Hernandez, A. and Resnick, M. (2008) Empowering Kids to Create and Share Programmable Media. *Interactions*, 15(2), pp. 50–53. Retrieved August 30, 2015.

Moreno-Ger, P., Burgos, D. Martínez-Ortiz, I. Sierra, J.L. and Fernández-Manjón, B. (2008) Educational Game Design for Online Education. *Computers in Human Behavior*, 24(6), pp. 2530–2540. DOI: 10.1016/j.chb.2008.03.012

Moreno-Ger, P., Torrente, J. Hsieh, Y.G. and Lester, W.T. (2012) Usability Testing for Serious Games: Making Informed Design Decisions with User Data. *Advances in Human-Computer Interaction*, 2012. DOI: 10.1155/2012/369637

Naik, N., 2017. The Use of GBL to Teach Mathematics in Higher Education. *Innovations in Education and Teaching International*, *54*(3), pp. 238–246.

Pajares, F. and Kranzler, J. (1995) Self-Efficacy Beliefs and General Mental Ability in Mathematical Problem-Solving. *Contemporary Educational Psychology*, 20, pp. 426–443.

Pelletier, C., 2009. Games and Learning: What's the Connection. *International Journal of Learning and Media*, 1(1), pp. 83–101.

Pinelle, D., Wong, N., Stach, T. and Gutwin, C. (2009) Usability Heuristics for Networked Multiplayer Games. *Proceedings of the ACM 2009 International Conference on Supporting Group Work*, pp. 169–178.

Preconception. (2020) In: Dictionary.com. [Online] Available at: https://www.dictionary.com/browse/preconception (Accessed 01/08/2020).

Prensky, M. (2002) The Motivation of Gameplay: The Real Twenty-First Century Learning Revolution. *On the Horizon*, 10(1), pp. 5–11.

Prensky, M. (2005) Listen to Natives. *Association for Supervision and Curriculum Development Educational Leadership*, 63(4), pp. 9–13.

Prensky, M. (2006) *'Don't Bother Me Mom, I'm Learning!': How Computer and Video Games Are Preparing Your Kids for the Twenty-First Century*. St Paul, MN: Paragon House.

Pritchard, A. and Woollard, J. (2010) *Psychology for the Classroom: Constructivism and Social Learning*. Oxon: Routledge.

Punch, K.F. (2009) *Introduction to Research Methods in Education*. London: Sage Publications Ltd.

Resnick, M. (2017) *Lifelong Kindergarten: Cultivating Creativity Through Projects, Passion, Peers, and Play*. London, England: MIT Press.

Robins, A., Runtree, J. and Rountree, N. (2003) Learning and Teaching Programming: A Review and a Discussion. *Computer Science Education*, 13(2), pp. 137–172.

Roth, W.M. (2000) Authentic School Science: Intellectual Traditions In R. McCormick and C. Paechter (ed.), *Learning & Knowledge*. London, UK: Paul Chapman Publishing, pp. 6–20.

Russ, T., and Drury-Grogan, M.L. (2013) Assessing the Impacts of a Business Communication Simulation on Students' Self-Perceptions. *Eastern Communication Association*, 61(5), pp. 584–595. DOI: 10.1080/01463373.2013.822404

Salen, K. and Zimmerman, E. (2011) The Game Designer Reader, A Rules of Play Anthology. Roger Caillois: The Definition of Play. *The Classification of Games*. Cambridge, Massachusetts. London, England: The MIT Press.

Salisbury, J.H. and Cole. T. (2016) Grounded Theory in Games Research: Making the Case and Exploring the Options. Available at: https://www.academia.edu/27624170/Grounded_Theory_in_Games_Research_Making_the_Case_and_Exploring_the_Options (Accessed 15/10/2015).

Schell, J. (2015) *The Art of Game Design: A Book of Lenses*, 2nd Edition. USA: CRC Press.

Schell, J. (2020) *The Art of Game Design: A Book of Lenses*, 3rd Edition. USA: CRC Press.

Senge, P. (1990) *The Fifth Discipline: The Art and Practice of Learning Organization*. United Kingdom: Random House Business Books.

Shell, D.F., Murphy, C.C. and Bruning, R.H. (1989) Self-efficacy and Outcome Expectancy Mechanisms in Reading and Writing Achievement. *Journal of Educational Psychology*, 81(1), p. 91.

Shunk, D.H. (2000) *Learning Theories: An Educational Perspective*, 3rd Edition. Upper Saddle River, NJ: Prentice-Hall.

Siegel, R.G., Galassi, J.P. and Ware, W.B. (1985) A Comparison of Two Models for Predicting Mathematics Performance: Social Learning Versus Math Aptitude-Anxiety. *Journal of Counseling Psychology*, 32, pp. 531–538.

Sindre, G., Natvig, L. and Jahre, M. (2009) Experimental Validation of the Learning Effect for A Pedagogical Game on Computer Fundamentals. *IEEE Transactions on Education*, 52 (1), pp. 10–18.

Skemp, R.R. (1986) *The Psychology of Learning Mathematics Great*. Britain: Penguin Books Ltd.

Skemp, R.R. (1989) *Mathematics in the Primary School*. Bristol: Routledge.

Skilbeck, A. (2017) Dewey on Seriousness, Playfulness and the Role of the Teacher. *Education Sciences*, 7(1), p. 16.

Slyman, S. (2014) I Hate Maths! An Investigation into Students' Attitudes Towards Mathematics. MA Thesis, University of Roehampton, London, United Kingdom.

Slyman, S. (2018) *Games Based Learning in Universities, a Life Learning Experience*. New York: Business Expert Press Expert Insights Publishing. ISBN:978-1-94819-801-1.

Slyman, S., Gillies, M. and Jessel, J. (2019) Acumenous or Inquisitional? Towards a New Theoretical Lens within Games Learning. In: *13th European Conference on Games Based Learning*, University of South Denmark, Odense, Denmark. ISBN:978-1-912764-38-9 or E-book: ISSN:2049-100X.

Sobke, H., Chan, E. Buttlar, R. Grobe-Wortmann, J. and Londong, J. (2014) Cat King's Metamorphosis. The Reuse of an Educational Game in a Further Technical Domain. In: Gobel, S., and Wiemeyer, J. (ed.), *Games for Training Education, Health and Sports, 4th Conference on Serious Games, GameDays 2014*, Darmstadt, Germany, April 1–5, 2014, Springer.

Song, S. and Lee, J. (2007) Key Factors of Heuristic Evaluation for Game Design: Towards massively multi-player online Roleplaying game. *International Journal of Human-Computer Studies*, 65, pp. 709–723.

Strauss, A. and Corbin, J. (1990) *Basics of Qualitative Research: Grounded Theory Procedures and Techniques*, Newbury Park, CA: Sage.

Sweetser, P., Johnson, D. and Wyeth, P. (2012) Revisiting the Game Flow Model with Detailed Heuristics. *Journal of Creative Technologies*, 2012(3), pp. 1–16.

Sweetser, P. and Wyeth, P. (2005) GameFlow: A Model for Evaluating Player Enjoyment in Games. *Computers in Entertainment*, 3(3), p. 3. DOI: 10.1145/1077246.1077253

Tamura, H. (2007) Foundational Value of Mathematics Education for Management Curriculum. *International Statistical Review*, 75(3), pp. 397–405. DOI: 10.1111/j.1751-5823.2007.00032.x

Tao, Y.H., Cheng, C.J. and Sun, S.Y. (2012) Alignment of Teacher and Students Perceptions on the Continued Use of Business Simulation Games. *Educational Technology and Society*, 15(3), pp. 177–189.

Vlachopoulos, D. and Makri, A. (2017) The Effect of Games and Simulations on Higher Education: A Systematic Literature Review. *International Journal of Educational Technology in Higher Education*, 14(1), pp. 22. DOI: 10.1186/s41239-017-0062-1

Vos, L. (2014) Marketing Simulation Games: A Review of Issues in Teaching and Learning. *The Marketing Review*, 14(1), p. 6796. DOI: 10.1362/146934714X1394890 9473220

Vos, L. and Brennan, R. (2010) Marketing Simulation Games: Student and Lecturer Perspectives. *Marketing Intelligence and Planning*, 28(7), pp. 882–897.

Walonick, D.S. (2000) *Designing and Using Questionnaire in Survival Statistic*. MN: StatPacInc. Available at Internet WWW page at URL: http://www.statpac.com/surveys/surveys.pdf (Accessed 02/04/2016).

Whitton, N. (2011) Games-Based Experiences for Learning. Futurelab Innovation in Education. Available at: https://www.nfer.ac.uk/media/1775/futl11.pdf (Accessed 02/04/2016).

Wolfe, J. (1997) The Effectiveness of Business Games in Strategic Management Coursework. *Simulation and Gaming*, 28(4), pp. 360–376. DOI: 10.1177/104687 8197284003

Women in STEM – 'Facts and Statistics' (Reporting period 2012–2015) IET. Available at: https://communities.theiet.org/files/8042 (Accessed 20/06/2017).

Yin, R.K. (1994) *Case Study Research: Design and Methods*, 2nd Edition. London: Sage.

Young, M. (2008) *"Powerful Knowledge" and "Knowledge of the Powerful": Some Reflections*. London: Presentation at the Institute of Education.

Yusoff, A., Crowder, R. and Gilbert, L. (2010) Validation of Serious Games Attributes Using the Technology Acceptance Model [online]. In: IEEE (ed.), *2010 Second International Conference on Games and Virtual Worlds for Serious Applications*, Braga, Portugal, 25–26 March 2010, pp. 45–51. Available at: http://eprints.soton.ac.uk/270809/ (Accessed 10/05/2015).

Zimmerman, B.J. (2000) Self-Efficacy: An Essential Motive to Learn. *Contemporary Educational Psychology*, 25(1), pp. 82–91. DOI: 10.1006/ceps.1999.1016

10 Acumenous and Inquisitional Theory

This chapter delves into the intricate connections between Gee's 13 principles, *Acumenous and Inquisitional Theory*, and how these theoretical frameworks inform the research findings. The theoretical concepts underpinning *Acumenous and Inquisitional Theory* are presented in detail, providing a comprehensive understanding of their significance.

Moreover, this chapter ventures beyond academic discourse and examines the implications and recommendations for policy and practice, particularly within the realm of education. Leveraging 20 years of teaching experience, the author puts forth insights and potential impacts of the research findings on curriculum, pedagogy, and educational practices.

In a final section, Chapter 10 takes a practical turn and provides guidance on enhancing employability. It outlines the next steps for readers in relation to *Acumenous and Inquisitional* Theory, showcasing how this new theoretical contribution aligns with the AC_3 Teaching and Learning framework. The chapter offers valuable insights on how readers can leverage this framework and theoretical approach to enhance their employability in today's dynamic professional landscape.

In summary, this chapter, Chapter 10, serves as a comprehensive conclusion to this book, encapsulating the main contributions, discussing theoretical frameworks, proposing implications for policy and practice, and providing guidance for readers to enhance their employability. This chapter offers a holistic perspective on the research conducted, tying together theoretical and practical dimensions to inspire positive change in education and empower individuals to thrive in their professional journeys.

As previously articulated, the main key contributions of this book are:

- **Impacts of GBL**.
- **Acumenous and Inquisitional Theory**.

The purpose of this **Final Chapter** is to **discuss** my key **findings and contributions together**, addressing this time the question of how can SGs *acumenify*

DOI: 10.4324/9781032723426-13

students' self-efficacy in Mathematics and Statistics Education, transforming their fears and empowering their confidence? Next, I will discuss how Gee's 13 principles are linked to *Acumenous and Inquisitional Theory*, moving on to implications for policy and practice, followed by AC_3 teaching and learning framework.

An outline of the Main Findings and My Theoretical Contribution is:

- **Impacts of GBL**.
- Positive Impacts of GBL (Chapters 6–9).
- Further Problems with GBL: MEGs and Overall Mathematics and Statistics (Chapters 7–9).
- Development and Application of the ESGAE Model (Chapters 6 and 8).
- Inquisitional Learning Can Cause Disengagement with MEGs and Overall Mathematics and Statistics (Chapters 7 and 9).
- The Pedagogical Benefits of SGs: Mathematical and Statistical Learning, Creative Thinking, Decision-Making, and Problem-Solving (Chapters 6, 8, and 9).
- Acumenous GBL in SGs Can Increase Self-Efficacy in Mathematics and Statistics Learning (Chapters 6, 8, and 9).
- **Acumenous and Inquisitional Theory**.

Impacts of GBL

The positive impacts of GBL were some of the emerging insights from interviewees' responses about reasons behind engagement with games such as perceived usefulness and fun, confidence boosting, engagement, social interaction, agent, identity, competition, more learning, focus and concentration, fantasy, challenge, goals/storylines, control, and game mechanics. This can be linked to previous research such as Malone (1980), Malone and Lepper (1987), Jones (1998), Harel and Papert (1990), Faria (2001), Garris et al. (2002), Gee (2003a, 2003b, 2004, 2005a, 2005b, 2007, 2008, 2011), Lazzaro (2004), Squire (2005), Gee et al. (2008), Moreno-Ger et al. (2008), Whitton (2011), Li et al. (2013), Mayer et al. (2013), Granic et al. (2014), Malaka (2014), Vos (2014), McGonigal (2015), Jagger et al. (2016), Arnab and Clarke (2017), Cardinot and Fairfield (2019), and Schell (2020). They emphasize the several benefits of games on learning (see Chapters 2–5 for further details). However, none of the existing literature mentioned above indicated that there are further problems with mathematics educational games (MEGs). This enlightening unique book, therefore, has identified further problems with some MEGs.

Further Problems with GBL

As previously articulated, interviewees noted that not all games especially MEGs were considered to be fun or a good tool for entertainment and learning. Educational games were reported as disengaging and dull (see Chapters 7 and 9). MEGs have to change because currently they are putting learners (our interviewees in particular) off Mathematics and hence are likely to have detrimental impacts on their learning rather than enhancing it. Since participants illustrated that they have preconceptions of educational games that there are not going to be fun, these are the motives and perceptions (not useful and fun) that drove their self-efficacy and therefore their preconceptions, leading to negative experiences being added to their learning journey. This is because games can transform players for the better or worse, and when fulfilment is achieved, when users are left to investigate and have fun playing games, they learn new skills without even realizing. But when they are restricted/limited, according to participants, they do not learn. There should be more creative educational games, relevant to everyday life, so that students can use them (Chapter 7).

Development and Application of the ESGAE Model

Previous findings indicate that there are issues with existing game evaluation heuristics models as they tend to be either based on entertainment or educational games perspectives but not both as previously argued in Chapter 4. Furthermore, this book's findings of Chapter 6 show the significance of gameplay, game story and storylines, game design, game developments (mechanics, aesthetics, and technology) and usability and HCI principles in de/increasing self-efficacy and fear, leading players to feel confident to return to play the game or discouraged. This gaming experience can have significant impacts on players' experiences and therefore their self-efficacy. This therefore necessitates the development of the ESGAE. The proposed game evaluation heuristics model: the ESGAE (Chapter 6: Table 6.1) is designed to fill this gap. The aim of the ESGAE is thus to generate a comprehensive model for analyzing and evaluating ESGs. As I explained above, this model has been applied to two games: "Numbas" and "SimCity" in order to obtain a detailed understanding of game design and development systems and thus create a positive game experience. Having analyzed and evaluated these two games from the ESGAE perspectives, both games have similarities and differences. However, they cannot detect high/low self-efficacy in Mathematical and Statistical learning, picking up signals and customizing learning (see Chapter 6 for more details).

Inquisitional Learning Can Cause Disengagement with MEGs and Overall Mathematics and Statistics

According to my interviewees, MEGs are disengaging because there is lack of surprise, rewards, creativity, problem-solving, social interaction, and challenging environment. MEGs are boring. MEGs were found to be dominant when participants were young (from primary/elementary to secondary/high schools) because most students found their experience of Mathematics, which involved memory learning in the way it was taught hard as a subject. Mathematics just becomes different and difficult as learners progress to a higher level (upper secondary to A level). MEGs at a younger age were interactive but as Mathematics becomes advanced, MEGs curriculum become complex. This is because the type of Mathematics they contain is disengaging or *Inquisitional*. "*Inquisitional* GBL" (Chapters 7 and 9) is what I have referred to as "quiz-like" type of learning where learners face fewer multiple options to choose from. It is not interactive, nor appealing, and it just becomes hard with no amazements, no competition, no animation, no self-satisfaction, no mission, no variety, no direction, and no hints (Chapter 7). In Chapter 9, interviewees shared this *Inquisitional* learning experience as being complex, difficult to understand or remember and irrelevant to real life. It is manual and intense where learners must carry out all the steps. It is theoretical, just formulas, and based on memorization. This further indicates that this is how Mathematics and Statistics was taught to my interviewees for instance. My interviewees indicated that they do not feel at ease with Mathematics at an advanced level, or in our case (and in particular) MEGs at a higher level, implying that they have lack of self-efficacy in Mathematical and Statistical Learning, that is fear. According to the ***Acumenous and Inquisitional Theory***, external variables (perceived not useful and fun) can affect how users perceive the world and IT, which in turn affect their preconceptions and therefore their self-efficacy. The Mathematics and Statistics curriculum that my cohort of interviewees experienced is Inquisitional type of learning (Chapters 7 and 9). This thus can have detrimental impacts on our learners' experiences, leading to low self-efficacy in Mathematics and Statistics. Participants explained that their generation would like to have control over their learning and more autonomy. They prefer *Acumenous GBL (AGBL)*. This furthermore suggests, according to the ***Acumenous and Inquisitional Theory***, that learners' perceptions of IT as a fun and useful can have significant impacts on their preconceptions and thus their self-efficacy. Game design with its characteristics positively impacts learners' self-efficacy, behavior, and interaction with the game and therefore a significant part of the learning experience.

The Pedagogical Benefits of SGs: Mathematical and Statistical Learning, Creative Thinking, Decision-Making, and Problem-Solving Skills

As previously articulated, my interviewees reported that SGs have many pedagogical benefits, they can be adopted as a learning tool for improving various transferable skills in HE like Mathematical and Statistical learning, creative thinking, confidence boosting, leadership, interpersonal skills to visual attention, multi-tasking and neural processing to performance analysis. SGs have many possibilities and real-life scenarios with real-time feedback (Chapters 6–9). These findings are parallel to the theories of Faria (2001), Mayo (2009), Vos and Brennan (2010), Vos (2014), McGonigal (2015), Jagger et al. (2016), Hamari et al. (2016), Vlachopoulos and Makri (2017), and Cardinot and Fairfield (2019). They emphasize the several potential benefits of using SGs in teaching and learning (e.g., employability and leadership skills; see Chapters 2 and 3).

My participants indicated that they liked the SimCity gameplay, a good example, particularly the following features: the element of trial and error; enhancing problem-solving; making decisions, planning, creative thinking; and having a safe environment where they can make mistakes without the fear of failure. They suggested (my participants) that to improve Mathematical and Statistical skills, we can use this type of SGs and relate it to real-world scenarios and the economy (Chapter 9).

According to the **Acumenous and Inquisitional Theory**, participants regard the SG as useful and fun and these are the motives and perceptions that increased their self-efficacy and confidence, drove positively their preconceptions, and their interaction with the game as they seemed to be motivated, focused, and stimulated during the *AGBL* process (Chapters 7 and 9). This demonstrates the significance of *AGBL*, as participants felt they were on a mission, in a scenario, acting as mayors/leaders/heroes, and this gives them a sense of relevancy for what they were doing, that is, *AGBL*. SGs gave participants a sense of purpose, clear instructions, and a way to improve and make things better. Participants prefer to learn and live life from other players' perspectives because with every new game, they play as somebody else and learn from their perspective, there is something new to learn and there are many other skills to learn (Chapters 7 and 9). This is also because transferable skills such as Mathematical and Statistical learning, creative thinking, decision-making, and problem-solving are what stimulate participants to play SGs. This technique ***"AGBL"*** in SGs is what stimulates participants' self-efficacy to continuously engage with these SGs which stretch their imagination and therefore increase their self-efficacy and overall, their learning experience.

Acumenous GBL in SGs Can Increase Self-Efficacy in Mathematics and Statistics

As previously explained, simulation GBL approach to teaching was employed, that is, "SimCity" gameplay scenario with Excel, with the aim of testing participants' confidence in Statistical learning before and after the game (using self-efficacy questionnaires) (see Chapters 5 and 9).

Findings show (in particular, quantitative results) that participants' confidence increased significantly after playing the SG, as pairwise t-test results demonstrated (see Chapter 5 for further details).

Findings also indicate that participants liked the many pedagogical benefits of SGs such as a "sense of mission," "purpose" and "ownership," building confidence, solving issues, "more motivation to learn" and "absorb learning quicker rather than passively taking in information," as well as a chance to experience and investigate "real-life examples with real-time feedback." Findings suggest that *AGBL* (discussed next) can play a beneficial role in enhancing Mathematical and Statistical learning since players can see the outcomes of their decisions straightaway, and as they have an opportunity to try it again, they are more likely to create the best outcomes (Chapters 7 and 9). This is because interviewees felt they can "digest information more easily rather than absorbing knowledge passively." This engagement and ability to absorb the information quickly is what I have associated with *AGBL* since this ability to understand, reason and decide, think creatively and critically is what stimulates participants to rapidly absorb knowledge. Interviewees suggested that *AGBL* could simplify learning. They could "put a stop" to students' "struggles" or "fear" and thus facilitate the advancement of Mathematical and Statistical learning through the *AGBL* process. This is because simulation GBL approach to learning Statistics was perceived as useful and great fun (as aforementioned). These (usefulness and fun) drove interviewees' preconceptions leading to high self-efficacy that acts as a force changing positively their behavior. Participants interacted with the game scenario in playful ways. They felt excited, relaxed, and were more engaged. Participants were understanding, reasoning, and deciding. They were constantly thinking creatively and critically, collaborating, managing, and leading their learning, using the SG, that is, **acumenifying their learning** (*AGBL*), and subsequently developing new skills added to their learning journey. This *AGBL* process was apparent in some of the participants' responses demonstrated earlier.

Acumenous GBL Can Improve MEGs and Overall Mathematics and Statistics

MEGs can be engaging if we embody them with the following features: Encourage high emphasis on problem-solving, decision-making, creative thinking, and Statistical learning; competition; storylines; and a choice of outcomes. Learners also liked to investigate and see what will happen, why and what the impacts of their decision are. It is important that they can see the outcome of their actions straightaway. What these generations, Z and Alpha, would like is more autonomy and control over their learning. They prefer *Acumenous GBL* where the focus is on building, financing, profit-making, and stock markets, a chance to trial and experiment, illustrating "trends, patterns" and "future predications," giving users a "sense of purpose," creating the "best outcomes" and seeing the outcome straightaway. They need a "sense of ownership," a "sense of control," and "mission" to "create artefacts the way they want" and imagine it. These transferable skills in SGs are what I have referred to as "*AGBL*." This is because transferable skills and investigation come from SGs. *AGBL* was found to be stimulating and students reported that it was "easy to remember" some of things they learnt (Chapters 7 and 9). There are several explanations or many stories of why users are using it. According to the **Acumenous and Inquisitional Theory**, and from cognitive perspectives, there are enormous perceptions that IT is fun and useful; this has significant impacts on users' preconceptions, which subsequently affect learners' behavior and interaction with the SGs, leading to transferable skills, that is, *AGBL*. *AGBL* originated from Study 2 (Chapter 7). It then re-emerged and was further developed in Study 4 (see Chapter 9). Consequently, AGBL can be utilized to improve Mathematical and Statistical learning and teaching in HE Business/Computing Schools (see implication for policy and practice for further recommendations).

Acumenous and Inquisitional Theory

In Chapters 7 and 9, I presented, defined, and argued for **Acumenous and Inquisitional Theory** in relation to educational theories that become prominent in computer game-based learning. I explained in these chapters how these theoretical concepts were coming out from the research data: Chapters 7 and 9. I have also discussed the new contributions of each study and how these make up the ***Acumenous and Inquisitional Theory***. In this section, I will present a summary of this theory (for further details see Chapters 7 and 9). Next, I will discuss how this theory is building on Gee's principles. In particular, empowering learners (agent/codesign and identity), problem-based (sequenced well-ordered problems), and system thinking (deep understanding).

Acumenous and Inquisitional Theory is a perceived game-based theory which I have associated with *Acumenous* (AGBL) or *Inquisitional* (IGBL). "Acumenous" or "Inquisitional" are two types of GBL.

IGBL is part of a perceived game-based theory. Inquisitional refers to a "quiz-like" type of learning that can occur in conversations or discourses such as police investigations or courtroom cross-examinations: Questions and answers.

In participants' discourse *Inquisitional* was defined as having "fewer multiple options," more text involved (Chapters 7 and 9). It consists of a set of instructions. It has "multiple-choice type of answers." Inquisitional consists of "formulas with numbers, all variables Xs and Ys". *Inquisitional* has "letters and numbers" (Chapter 9). It relies on memorization and "formulas" where learners have to "do all steps" and "calculations." It is "theoretical"/feels theoretical (Chapter 9).

According to participants, *Inquisitional* is hard to understand its concepts. It is "too complex" (Chapter 9). It has "no imagination/creativity" (Chapter 7). It is "[n]ot as interactive." Not "appealing," "boring," and has "no animation." *Inquisitional* is characterized as "manual and intense," and it is "difficult" because learners are "passively taking knowledge" (Chapter 9). *Inquisitional* has more text involved, and if it is "too hard," it "discourage[s] people off" (Chapters 7 and 9). It can be "quite an exhausting process" for listeners/learners since it "requires a considerable amount of effort" and it is "not personal" (Chapter 9). *Inquisitional* can deter learners who do "not know where all information is coming" from, and "why they are required to understand it". This is because students "do not see its relevance outside classroom," and therefore it is "theoretical"/feels theoretical (Chapter 9).

The second type of GBL which was investigated within participants' discourse is what I have referred as *AGBL* (from acumen). *AGBL* is the ability to understand, reason, and decide "why something is occurring" as I previously discussed in Chapters 7 and 9. From the research data of Chapter 7, *Acumenous Game-Based Learning (AGBL)* emerges as a new theoretical lens, and I have defined it. In Chapter 9, *AGBL* re-emerges with new contributions, explanations, and further developments, discussed next (see Chapter 9 for further details). What I term *AGBL* is part of a perceived game-based theory. *AGBL* is a cognitive process, a thinking cap that goes through game players' minds while interacting with the games. *AGBL* is a science of games learning, and it is a way of thinking. "*AGBL*" is a high-level synergy of concentration, focus, and alertness.

In participants' discourse *AGBL* was defined as having the ability to see "[w]hat will happen?", why, and "create the best outcome" (Chapter 7). Participants indicated that they could see all the "X[s], Y[s] and X^2[s]" and "understand that they all fit together" (Chapter 9). *AGBL* has "many possibilities that you can pick." This "sort of participation" and "explain[ed] feature[s]" are important (Chapter 9). *AGBL* involves "applying knowledge sub-consciously," and there are "several" stories of "why learners are using it or why it occurs" (Chapter 9). This ability to "break" down "problems into logical steps," I suggest, is what I have referred to as *AGBL* since one has to understand, reason, and apply the rules. This is computational learning (programming) that is applied in GBL and which can, I propose, be used to improve Mathematical and Statistical learning within MEGs, as well as challenging any preconceptions and perhaps how students perceive MEGs, transforming their fears and empowering their confidence.

AGBL was defined as a type of learning that "stimulates" intelligence, "making connections" (Chapter 9). It is "easy to learn" and it simplifies learning since learners "can see how it can be applied to everyday life" (Chapter 7). Parallel to this, other interviewees (Chapter 9) valued the following features: the "relevance of real-life scenarios" with "real-time feedback" and "rewards;" the acumen or **"acumenifying"** process; the creative and critical thinking/way of solving issues differently; the step-by-step problem-solving aspects of it; the "reasoning" and many explanations and "possibilities of why they are using" it; the "collaboration" and "interactivity" (Chapter 9). Participants noted that they can see the "necessity" of this learning and how they obtain their "decisions." *AGBL* have "steps," "makes sense," "fun" and learning "hides behind it" (Chapter 9).

Participants indicated that "when looking back" *AGBL* "makes the learning quite easy" than "going straight on without knowing" "where it comes from." It is "stimulat[ing]" and "reward[ing]" where the learners are "not passively taking information." In other words, obtaining "information is not just a chore" (Chapter 9). *AGBL* gives learners "a chance to trial and make your decisions." It makes learners "feel like [they] have a chance." They "can always redo" their decisions and it can "help people who may have issues" with "confidence knowing that [they] can change [their] decisions." In participants' discourse, *AGBL* was described as "more engaging way of learning" and "easy to remember." "It gives [learners] more real-time feedback" with "real-time data" and a chance to "learn something" [for instance, market shares] "because there is no way [they] can do this in real life" (Chapter 9). *AGBL* is "relevan[t] to everyday life" and learners know "how it is applied in real-life experiments" with "[r]eal-life scenarios" and it is "personal." *AGBL* is "interactive" and "creative." This is because users are "applying" the learning "immediately" and "perhaps subconsciously." *AGBL* give learners "connection[s]" in their "brains that [they] make," since they "apply the knowledge subconsciously." Participants also liked the "collaboration" and "many possibilities that they can use." Participants felt they can learn complex ideas in fun easy ways rather than "absorbing knowledge passively" (Chapter 9). They were **"acumenifying"** their learning. They were constantly thinking creatively and critically, managing, and leading their learning.

I will now discuss how Gee's (2003a, 2003b, 2004, 2005a, 2005b, 2007, 2011) 13 principles (see Chapter 2) are linked to **Acumenous and Inquisitional Theory** (for further details about how *AGBL* is linked to other existing theories, see Chapter 9).

Gee's 13 Principles and Acumenous and Inquisitional Theory

In addressing how Gee's 13 principles are linked to my theory. I found that Gee's 13 principles (Chapter 2) can be linked in some aspect to my theoretical contribution (e.g., AGBL) but not to others (e.g., *Inquisitional*) as I will explain. For example, "agent/codesign" and "identity" (in empowering learners' principles in Gee's work) can be linked to self-efficacy and my game experiment (Chapters 5 and 9). What the game did, as a learning tool, is empower participants to undertake Mathematics and Statistics activities with confidence/self-efficacy. It empowers participants giving them a sense of purpose, a mission and roleplay, involving them and making them part of the experience/activity. The game gives them a sense of control, a sense of being active agents/co-designers, leaders, and managers to embark on a journey of problem-solving activities (sequenced problems) and deep understanding. This led participants' preconceptions and fears toward Mathematics and Statistics to change and to their confidence to super increase. The game, as a way of social interaction, has challenged participants' preconceptions. They realized and developed a social identity as leaders and managers. The game therefore demonstrated (as indicated in Chapters 7–9) that self-efficacy as a psychological construct can be amended and fixed. My participants thus became highly efficacious in problem-solving; they learned Mathematical and Statistical skills with ease. The game thus assisted participants, giving them a sense of responsibility, to unpack Mathematics and Statistics as a system, thinking through it and understanding it. They understood the problem at hand and thought of a solution as in system thinking principle. They *acumenify* their learning as in deep understanding (Gee's system thinking principle). These principles (problem-solving and deep understanding) represent two of the most fundamental elements that make up *Acumenous GBL* (as explained earlier; see Chapters 7, 9, and 10).

In addressing how Gee's 13 principles are not linked to my ***Acumenous and Inquisitional Theory.*** I found that Gee's work cannot be linked to *Inquisitional Theory*. *Inquisitional Learning* (as defined and argued earlier; Chapters 7–9) is totally the opposite of what Gee is advocating because Gee's 13 principles are not adhered to (e.g., empowering learners: agent/codesign, identity, problem-solving skills, and deep understanding) in these games. This can also explain the above-mentioned problems with MEGs. Hence, preconceptions and fears toward Mathematical and Statistical learning due to *Inquisitional Learning* are not challenged.

In the next final part of this chapter, I will examine Implications for Policy and Practice.

In this final section, I will discuss implications and recommendations for policy and practice, that is, the potential impacts of my research on HE education including curriculum and pedagogy. These are based on my findings (Chapters 6–9) and on my 20 years' teaching experience as a senior lecturer.

Implications for Policy and Practice

As previously explained (Chapter 9), and in terms of making recommendations for policy and practice, many participants felt that SGs should be integrated in Education, Business, and Computing Schools (as a part of the HE curriculum) for teaching complex subjects such as Mathematical and Statistical learning, using *AGBL* (as argued in Chapters 7 and 9). This could be generalized to other STEM subjects such as Physics and Engineering since most people find memory learning (scientific subjects) hard and complex leading to preconceptions and therefore negative emotions and fears toward these stimulating subjects. Similarly using *Acumenous GBL* learners can have confidence in learning these subjects, as noted in Chapters 7 and 9.

AC_3 Framework and Employability Skills

AGBL can be summarized in the acronym **AC_3** (acumen, creativity, critical thinking, and collaboration). Since these skills are the learning skills I found in game-based learning. *AC_3* is a *new T&L* (teaching and learning) *Repertoire*, a stock of skills aimed at guiding lecturers/educators into what skills they should focus on when teaching complex subjects such as Statistics Education and what game educators and designers should integrate into the design of educational games. *AC_3* is another contribution this book has uniquely produced. These transferable skills can be related to those for employability (World Economic Forum, 2016, 2021, 2030). They are skills for the 21st century and thus the skills and outcomes that graduates should possess by graduation, that is employability skills.

It is worth noting that in order to educate students we must "see through the academic illusion to their real abilities" (Robinson, 2001:93), which can be solved by employing *AGBL* in HE (Chapters 6, 7, and 9), i.e., a simulation game-based approach to teaching and learning. I would like to lead a nationwide initiative on comprehensive university reform. This would focus on preparing students to meet the needs of a changing society through *Acumenous GBL* in SGs. I recommend we will sustain motivation, engagement, and increase students' confidence, performances, attainment, and retention. Undergraduates and postgraduates will therefore achieve their potentials, become employable, and be work-force ready. To make further progress in this area, and considering my research findings, I recommend that Education, Business, and Computing Schools would benefit from:

- Integrating gameplay learning/*AGBL* into lectures as part of T&L, that is, seminars/lab work and as indicated in Chapters 7 and 9. However, we are also required to take into account that some students may not pay attention to lectures and therefore lecturers should have some authority/control over the game so that they can limit the amount of time that is spent on it, and, if they are using it during a lecture, they should be able to stop it if necessary (see Chapters 7 and 9).
- Providing lab/game space where undergraduates can visit and obtain hands-on experience in experimenting with *AGBL*. Thus, SGs should be accessible (Chapters 7 and 9) so that under/postgraduates can access these resources.
- Exploring games/SGs technologies and what is available practically and theoretically, informing and influencing curriculum development for the 21st century.
- Training IT staff, managers, and lecturers to incorporate and implement SGs into the curriculum, and thus we are required to think about funding, IT, and human resources.
- Acknowledging educational games as a new medium and motivating tool for increasing self-efficacy, transforming fears, and therefore increasing students' potentials, progression, and retention. However, we are also required to reflect and think about different types of available useful and accessible games (Chapters 7 and 9). I recommend therefore choosing open SGs, such as "SimCity," that could be integrated creatively in lectures. Yet educators should also think about game engines, installation, available software and hardware, IT resources, staff training, and the length of preparation and planning time as it could take longer than expected, and thus I recommend allowing extra time when planning to integrate any SGs into teaching and learning.

Bibliography

Acumen (2017) In: Cambridge University Press. [Online] Available at: http://dictionary.cambridge.org/dictionary/english/acumen (Accessed 30/05/2017).

Adams, A., Lunt, P. and Cairns, P. (2008) A Qualitative Approach to HCI Research. In: Cairns, P. and Cox, A. (ed.). *Research Methods for Human-Computer Interaction*. Cambridge, UK: Cambridge University Press, pp. 138–157. Available at: http://www.cambridge.org/catalogue/catalogue.asp?isbn=9780521870122andss=toc (Accessed 20/11/2016).

ADDIE Model (2016) Available at: http://www.learning-theories.com/addie-model.html (Accessed 20/04/2016).

Anderson, S. (2020) Quantitative Analysis. Available at: https://www.investopedia.com/terms/q/quantitativeanalysis.asp (Accessed 20/02/2021).

Anon (2018) Efficacy. In: Cambridge University Press. [Online] Available at: https://dictionary.cambridge.org/dictionary/english/efficacy (Accessed 30/09/2018).

Applefield, J.M., Huber, R. and Moallem, M. (2000) Constructivism in Theory and Practice: Toward a Better Understanding. *The High School Journal*, 84(2), pp. 35–53.

Arnab, S. and Clarke, S., 2017. Towards a Trans-disciplinary Methodology for a Game-based Intervention Development Process. *British Journal of Educational Technology*, 48(2), pp. 279–312.

Arthur, J., Waring, M. Coe, R. and Hedges, L.V. (2012) *Research Methods and Methodologies in Education*. California: Sage Publications Inc.

Atkins, L. and Wallace, W. (2012) *Qualitative Research in Education*. Los Angels, London, New Delhi, Singapore, Washington DC: Sage: Publications Ltd.

Bandura, A. (1986) *Social Foundations of Thought and Action: A Social Cognitive Theory*. Englewood Cliffs, NJ: Prentice-Hall.

Bandura, A. (1994) Self-efficacy. In Ramachaudra, V.S. (ed.). *Encyclopedia of Human Behaviour*, 4, pp. 71–81. New York: Academic Press. (Reprinted in H. Friedman (Ed.) (1998). Encyclopedia of Mental Health. San Diego: Academic Press).

Bandura, A. (1997) *Self-efficacy: The Exercise of Control*. New York: W.H. Freeman and Company.

Bandura, A. and Schunk, D. (1981) Cultivating Competence, Self-Efficacy, and Intrinsic Interest Through Proximal Self-Motivation. *Journal of Personality and Social Psychology*, 41(3), pp. 586–598.

Barbosa, T., Lopes, S.P. Leao. C. Soares, F. and Carvalho, V. (2019) Serious Game for Teaching Statistics in Higher Education: Storyboard Design. In: *7th EAI International Conference, Braga, Portugal, on 'Interactivity, Game Creation, Design, Learning, and Innovation'* (ed.) Springer, pp. 169–175. DOI: 10.1007/978-3-030-06134-0_18

Bartholomew, H. et al. (2011) "I am Natural, and I Do It for a Love!": Exploring Students' Accounts of Studying Mathematics. *International Journal of Mathematical Education in Science and Technology*, 42(7), pp. 915–924.

Bassey, M. (1999) *Case Study Research in Educational Settings*. Buckingham: Open University Press.

Bellotti, F., Kapralos, B. Lee, K. Moreno-Ger, P. and Berta, R. (2013) Assessment in and of Serious Games: An overview. *Advances in Human-Computer Interaction*, 2013, p. 1.

Belluigi, D.Z. (2009) Exploring the Discourses Around 'Creativity' and 'Critical Thinking' in a South African Creative Arts Curriculum. *Studies in Higher Education*, 34(6), pp. 699–717.

Bennedesen, J. and Caspersen, M.E. (2004) Programming in Context: A Model-First Approach to CS1. In *Proceedings of the 35th SIGCSE Technical Symposium on the Computer Science Education*, pp. 477–481. ACM Press.

Bevan, N. (2001) International Standards for HCl and Usability. *International Journal of Human Computer Studies*, 55(4), pp. 533–552.

Boaler, J. (2009) *The Elephant in the Classroom*. London: Souvenir Press.

British Education Research Association (BERA) (2011) BERA Ethical Guidelines for Educational Research. Available at: http://www.bera.ac.uk/ (Accessed 15/10/2016).

British Education Research Association (BERA) (2018) BERA Ethical Guidelines for Educational Research. Available at: https://www.bera.ac.uk/wp-content/uploads/2018/06/BERA-Ethical-Guidelines-for-Educational-Research_4thEdn_2018.pdf?noredirect=1 (Accessed 25/03/2019).

Brown, J.S., Collins, A. and Duguid, P. (1989) Situated Cognition and the Culture of Learning. *Educational Researcher*, 18(1), 32–42.

Bruner, J. (1973) *Going Beyond the Information Given*. New York: Norton.

Bryant, A. (2017) *Grounded Theory and Grounded Theorizing: Pragmatism in Research Practice*. New York, NY: Oxford University Press.

Bryman, A. (2008) *Social Research Methods*, 3rd Edition. Oxford: Oxford University Press (first published: 2001).

Bryman, A. (2012) *Social Research Methods*. 4th Edition. United States: Oxford University Press.

Bryman, A. (2016) *Social Research Methods*. 5th Edition. United States: Oxford University Press.

Caillois, R. (1961) *Man, Play, and Games*. New York: Free Press.

Cardinot, A. and Fairfield, J.A. (2019) Game-Based Learning to Engage Students with Physics and Astronomy Using a Board Game. *International Journal of Game-Based Learning (IJGBL)*, 9(1), pp. 42–57.

Charmaz, K. (2006) *Constructing Grounded Theory*, 1st Edition. London: Sage Publications Ltd.

Charmaz, K. (2014) *Constructing Grounded Theory*, 2nd Edition. London: Sage Publications Ltd.

Chen, M.P., Wong, Y.T. and Wang, L.C. (2014) Effects of Type of Exploratory Strategy and Prior Knowledge on Middle School Students' Learning of Chemical Formulas from a 3D Roleplaying Game. *Educational Technology Research and Development*, 62, pp. 163–185. DOI: 10.1007/s11423-013-9324-3

Clark, D.B., Tanner-Smith, E.E. and Killingsworth, S.S. (2015) Digital Games, Design, and Learning a Systematic Review and Meta-Analysis. *Review of Educational Research*, 86(1), pp. 79–122.

Claxton, G. (2002) *Building Learning Power*, 1st Edition. Bristol: TLO Limited.

Cohen, L., Manion, L. and Morrison, K. (2011) *Research Methods in Education*. London: Routledge.

Cohen, L., Manion, L. and Morrison, K. (2018) *Research Methods in Education*. London: Routledge.

Conlon, C., Carney, G., Timonen, V. and Scharf, T. (2015) Emergent Reconstruction in Grounded Theory: Learning from Team-Based Interview Research. *Qualitative Research*, 15, pp. 39–56. DOI: 10.1177/1468794113495038

Cook, J. (2005) Negotiating the Curriculum: Programming for Learning. In *Negotiating the Curriculum*, pp. 15–31. Routledge.

Cooper, P.A. (1993) Paradigm Shifts in Designed Instruction: From Behaviorism to Cognitivism to Constructivism. *Educational Technology*, 33(5), pp. 12–19.

Corbin, J. and Strauss, A. (2015) *Basics of Qualitative Research: Techniques and Procedures for Developing Grounded Theory*, 4th Edition. Thousand Oaks, CA: Sage.

Cordova, D.I. and Lepper, M.R. (1996) Intrinsic Motivation and the Process of Learning: Beneficial Effects of Contextualization, Personalization, and Choice. *Journal of Educational Psychology*, 88, pp. 715–730.

Coulthard, G.J. (2009) A Review of the Educational Use and Learning Effectiveness of Simulations and Games [online]. PhD Thesis, Purdue University. Available at: http://www.coulthard.com/library/paper-simulation.html (Accessed 15/12/2015).

Craft, A. (1999) Creative Development in the Early Years: Some Implications of Policy for Practice. *The Curriculum Journal*, 10(1), pp. 135–150.

Craft, A. (2002) *Creativity and Early Years Education*. London: Continuum.

Crawford, C. (1984) *The Art of Computer Game Design*. Berkeley: California, McGraw-Hill/Osborne Media.

Creswell, I.W. (2009) *Research Design: Qualitative, Quantitative, and Mixed Methods Approaches*, 3rd Edition. London: Sage.

Crockett, L. (2018) Bogost's War on Gamification: Irony and Play on Grandma's Sofa. In: *12th European Conference on Games Based Learning (ECGBL 2018)*. Reading: Academic Conferences and Publishing International Ltd.

Cronan, T.P. and Douglas, D.E. (2012) A Student ERP Simulation Game: A Longitudinal Study. *Journal of Computer Information Systems*, 53(1), pp. 3–13.

Csikszentmihalyi, M. (1990) Psychology Today: Finding Flow. Available at: https://www.researchgate.net/publication/200026151_Finding_Flow_The_Psychology_of_Engagement_With_Everyday_Life/link/55ad6c2f08aed9b7dcdae222/download (Accessed 03/05/2016).

Csikszentmihalyi, M. (1997) *Creativity, Flow and the Psychology of Discovery and Invention*. New York: Harper.

Dann, W.P., Cooper, S. and Pausch, R. (2008) *Learning to Program with Alice*. Prentice Hall Press.

Data Protection Act (1998). Available at: http://www.legislation.gov.uk/ukpga/1998/29/pdfs/ukpga_19980029_en.pdf (Accessed 10/12/2016).

Data Protection Act (2018). Available at: http://www.legislation.gov.uk/ukpga/2018/12/pdfs/ukpga_20180012_en.pdf (Accessed 10/09/2019).

Davis, F.D. (1989) Perceived Usefulness, Perceived Ease of Use, and User Acceptance of Information Technology. *MIS Quarterly*, 13(3), pp. 319–340.

De Bono, E. (1992) *Teach Your Child How to Think*. England: McQuaig Group Inc.

Dede, C. (2005) Planning for Neomillennial Learning Styles. *Educause Quarterly*, 1.

Derryberry, A. (2007) Serious Games: Online Games for Learning. White Paper. [Online]. Available at: http://www.adobe.com/resources/elearning/pdfs/serious_games_wp.pdf (Accessed 10/12/2015).

Desurive, H., Caplan, M. and Toth, J. (2004) Using Heuristics to Evaluate the Playability of Games. *CHI Conference*, 32 (1), pp. 97–110. DOI: 10.1177/104687810103200108

Desurvire, H. and Wiberg, C. (2009) Game Usability Heuristics (PLAY) for Evaluation and Designing Better Games: The Next Iteration. *CHI Conference*, pp. 557–566.

Desurvire, H. and Wiberg, C. (2010) User Experience Design for Inexperienced Gamers: GAP – Game Approachability Principles. In: Regina Bernhaupt (ed.). *Evaluating User Experience in Games*. Springer, pp. 131–148. DOI: 10.1007/978-1-84882-963-3_1

Dondlinger, J.M. (2007) Educational Video Game Design: A Review of the Literature. *Journal of Applied Educational Technology*, 4(1), Spring/Summer, pp. 21–31. Available at: http://reserachgate.net/publication/238444705 (Accessed 12/12/2015).

Draper, S. (2013) Social Constructivism. Available at: http://www.psy.gla.ac.uk/~steve/courses/archive/CERE12-13-safari-archive/topic3/webarchive-index.html#social%20con%201 (Accessed 23/12/2019).

Dunne, C. (2011) The Place of The Literature Review in Grounded Theory Research. *International Journal of Social Research Methodology*, 14, pp. 111–124. DOI: 10.1080/13645579.2010.494930

Dweck, C.S. (2006) *Mindset: The New Psychology of Success*. United States: Ballantine Books Trade Paperback Edition.

Dwyer, J. and Slyman, S. (2016) *An Introduction to Quantitative Methods for Business Research*. London: Algana Publishing. 200 pages; ISBN−13: 9781907934094.

Early, R.E. (1992) The Alchemy of Mathematical Experience: A Psychoanalysis of Student Writings, for the Learning of Statistics, 12(1), pp. 15–20.

Ejsing-Duun, S. and Karoff, H.S. (2016) Creativity and Playfulness: Producing Games as a Pedagogical Strategy Case studies. In: Munkvold, R. (ed.). *Games Based learning for Researchers, Teachers and Students*, Vol. 1. Reading: Academic Conferences and Publishing International Ltd. ISBN:978-1-911278-12-890000.

Fairweather, E. and Cramond, B. (2010) Infusing Creative and Critical Thinking into the Curriculum Together. In: Beghetto, A. and Kaufman, J.C. (ed.). *Nurturing Creativity in the Classroom*. United States: Cambridge University Press.

Falstein, N. and Barwood, H. (2001) The 400 Project. Available at: http://theinspiracy.com/400_project.htm. (Accessed 20/04/2016).

Faria, A.J. (2001) The Changing Nature of Business Simulation/Gaming Research. *Simulation and Gaming*, 32(1), pp. 97–110. DOI: 10.1177/104687810103200108

Faria, A.J., Hutchinson, D. Wellington, W.J. and Gold, S. (2009) Developments in Business Gaming: A Review of the Past 40 Years. *Simulation and Gaming*, 40(4), pp. 464–487. DOI: 10.1177/1046878108327585

Faria, A.J. and Wellington, W.J. (2004) 'A Survey of Simulation Game Users, Former-Users, and Never-Users. *Simulation and Gaming*, 35(2), pp. 178–207. DOI: 10.1177/1046878104263543

Federoff, M.A. (2002) Heuristics and Usability Guidelines for the Creation and Evaluation of Fun in Video Games. Master Thesis, Indiana University, Bloomington.

Fienberg, S.E. (1992) A Brief History of Statistics in Three and One-Half Chapters: A Review Essay.

Forest, E. (2014) ADDIE Model. Available at: http://educationaltechnology.net/the-addie-model-instructional-design/ (Accessed 20/04/2016).

Fosnot, T.C. (1989) *Enquiring Teachers, Enquiring Learners: A Constructivist Approach for Teaching*. New York: Teachers College Press.

Fosnot, T.C. and Perry, R.S. (2005) Constructivism: A Psychological Theory of Learning. Available at: http://beyondbitsandatoms.org/readings/fosnot_constructivism_1996.pdf (Accessed 20/07/2019).

Garris, R., Ahlers, R. and Driskell, E.J. (2002) Games, Motivation, and Learning: A Research and Practice Model. *Simulation and Gaming*, 33(4), pp. 441–467.

Gee, J.P. (2003a) *What Video Games Have to Teach Us About Learning and Literacy*. New York: Palgrave Macmillan.

Gee, J.P. (2003b) Good Video Games and Good Learning. Available at: https://academiccolab.org/resources/documents/Good_Learning.pdf (Accessed 03/01/2021).

Gee, J.P. (2004) *Situated Language and Learning: A Critique of Traditional Schooling*. London: Routledge.

Gee, J.P. (2005a) Why are Video Games Good for Learning? Retrieved January 1, 2018, from http://academiccolab.org/resources/documents/MacArthur.pdf (Accessed 20/04/2016).

Gee, J.P. (2005b) *Why Video Games Are Good for Your Soul: Pleasure and Learning*. Melbourne: Common Ground.

Gee, J.P. (2007) *Good Video Games and Good Learning: Collected Essays on Video Games, Learning, and Literacy*. New York: Peter Lang.

Gee, J.P. (2008) *Getting over the Slump: Innovation Strategies to Promote Children's Learning*. New York: The Joan Ganz Cooney Center at Sesame Workshop.

Gee, J.P., Hayes, E., Torres, R.J., Games, I.A., Squire, K. and Salen, K. (2008) Playing to Learn Game Design Skills in A Game Context. In: *Proceeding ICLS'08 Proceedings of the 8th International Conference on International Conference for the Learning Sciences*, Vol. 3, pp. 368–374.

Gee, J.P. and Hayes, E.R., 2011. *Language and Learning in the Digital Age*. Routledge.

Gibbons, J. and Gray, M. (2004) Critical Thinking as Integral to Social Work Practice. *Journal of Teaching in Social Work*, 24, pp. 19–37.

Glaser, B.G. and Strauss, A.L. (1967) *The Discovery of Grounded Theory*. Chicago: Aldine.

Gomes, A. and Mendes, A.J. (2007) Learning to Program-Difficulties and Solutions. In *International Conference on Engineering Education–ICEE*, Vol. 7.

Granic, I., Lobel, A. and Engels, R.C. (2014) The Benefits of Playing Video Games. *American Psychological Association*, 69(1), pp. 66–78. DOI: 10.1037/a0034857

Hackett, G. and Betz, N.E. (1989) An Exploration of the Mathematics Self-Efficacy/Mathematics Performance Correspondence. *Journal for Research in Mathematics Education*, pp. 261–273.

Hamari, J., Shernoff, D.J., Rowe, E., Coller, B., Asbell-Clarke, J. and Edwards, T. (2016) Challenging Games Help Students Learn: An Empirical Study on Engagement, Flow and Immersion in Game-Based Learning. *Computers in Human Behavior*, 54, pp. 170–179.

Hamari, J. and Tuunanen, J. (2014) Player Types: A Meta-Synthesis. *Transactions of the Digital Games Research Association*, 1(2), pp. 29–53.

Harel, I. and Papert, S. (1990) *Software Design as a Learning Environment*. Massachusetts Institute of Technology Cambridge, MA: MIT Media Laboratory.

Harel, I. and Papert, S. (1991) *Constructionism*. Norwood, NY: Ablex Publishing Corporation.

Hargreaves, A. (2003) *Teaching in the Knowledge Society: Education in the Age of Insecurity*. New York: Teachers' College Press.

Holbert. N., Penney, L. and Wilensky, U. (2010) *Bringing Constructionism to Action Game-Play, Constructionism*. Paris. Available at: https://www.google.co.uk/search?q=Bringing+Constructionism+to+Action+Game-Playandoq=Bringing+Constructionism+to+Action+Game-Playandaqs=chrome..69i57.7170j0j8andsourceid=chromeandie=UTF-8 (Accessed 20/03/2017).

Hoover, W.A. (1996) The Practice Implications of Constructivism. *SEDL Letter*, 9(3), pp. 1–2.

Horn, B., Hoover, A.K. Barnes, J. Folajimi, Y. Smith, G. and Harteveld, C. (2016) Opening the Black Box of Play: Strategy Analysis of An Educational Game. In *Proceedings of the 2016 Annual Symposium on Computer-Human Interaction in Play*, pp. 142–153.

Hsu, C.-Y., Tsai, C.-C. and Wang, H.-Y. (2012) Facilitating Third Graders' Acquisition of Scientific Concepts Through Digital Game-Based Learning: The Effects of Self-Explanation Principles. *Asia-Pacific Education Researcher*, 21(1), pp. 71–82.

Hunicke, R., LeBlanc M. and Zubek, R. (2004) MDA: A Formal Approach to Game Design and Game Research. Available at: https://www.google.co.uk/search?q=MDA%3A+A+Formal+Approach+to+Game+Design+and+Game+Researchandoq=MDA%3A+A+Formal+Approach+to+Game+Design+and+Game+Researchandaqs=chrome..69i57j69i58.3868j0j8andsourceid=chromeandie=UTF-8 (Accessed 8/10/2016).

Illeris, K. (2007) *How We Learn: Learning and Non-Learning in School and Beyond*. London: Routledge.

ISO (1998) Available at: http://www.standardsglossary.com/iso1.htm (Accessed 8/10/2016).

Jackson, N. and Shaw, M. (2007) Developing Perspectives on Creativity in Higher Education. In: N. Jackson, M. Oliver, M. Shaw, and J. Wisdom (ed.). *Developing Creativity in Higher Education: An imaginative curriculum*. London: Routledge.

Jagger, S., Siala, H. and Sloan, D. (2016) It's All in The Game: A 3D Learning Model for Business Ethics. *Journal of Business Ethics*, 137(2), pp. 383–403.

Jameson, M.M. and Fusco, B.R. (2014) Math Anxiety, Math Self-Concept, and Math Self-Efficacy in Adult Learners Compared to Traditional Undergraduate Students. *Adult Education Quarterly*, 64(4), pp. 306–322.

Jarvis, M. (2005) *'Motivation' in Psychology of Learning and Teaching*. Cheltenham: Nelson Thornes.

Jeffrey, B. and Craft, A. (2004) Teaching Creatively and Teaching for Creativity: Distinctions and Relationships. *Educational Studies*, 30(1), pp. 77–87.

Jenson, J. and Droumeva, M. (2015) Making Games with Game Maker: A Computational Thinking Curriculum Case Study. In *ECGBL 2015–9th European Conference on Games Based Learning: ECGBL2015*. Academic Conferences and Publishing Limited, p. 260.

Jones, M.G. (1998) Creating Engagement in Computer-Based Learning Environments. Available at: http://itech1.coe.uga.edu/itforum/paper30/paper30.htmI (Accessed 12/04/2016).

Kafai, Y. (2006) *Playing and Making Games for Learning: Instructionist and Constructionist Perspectives for Game Studies. Games and Culture*, 1(1), pp. 36–40.

Kafai, Y.B. and Resnick, M. eds. (2012) *Constructionism in Practice: Designing, Thinking, and Learning in a Digital World*. Routledge. Available at: https://www.google.co.uk/books/edition/Constructionism_in_Practice/KGkIEQAAQBAJ?hl=en&gbpv=1&dq=Kafai,+Y.B.+and+Resnick,+M.+eds.+(2012)+Constructionism+in+Practice:+Designing,+Thinking,+and+Learning+in+a+Digital+World.+Routledge.&pg=PP1&printsec=frontcover

Kato, P.M., Cole, S. Bradlyn, A.S. and Pollock, B.H. (2008) Video Games Improves Behavioural Outcomes in Adolescents and Young Adults with Cancer: A Randomized Trial. *Pediatrics*, 122(2), pp. e305–e217.

Kessner, M., Wood, J. Dillon, R.F. and West, R.L. (2001) On the Reliability of Usability Testing. In *Proceedings of the Extended Abstracts on Human Factors in Computing Systems (CHI '01)*, p. 97.

Kim, B. (2001) Social Constructivism. *Emerging Perspectives on Learning, Teaching, and Technology*, 1(1), p. 16.

Kirriemuir, J. and Mcfarlane, A. (2004) Literature Review in Games and Learning. Available at: https://telearn.archives-ouvertes.fr/hal-00190453/document (Accessed 12/04/2020).

Knight, P.T. (2001) A Briefing on Key Concepts – Formative and Summative, Criterion and Norm Referenced Assessment. *Learning and Teaching Support Network Generic Centre*, Vol. 30. http://www.heacademy.ac.uk

Kodugamelab (2015) Available at: https://www.kodugamelab.com/ (Accessed 8/11/2018).

Kolb, A.Y. and Kolb, D.A. (2008) The Learning Way: Meta-Cognitive Aspects of Experiential Learning. *Simulation and Gaming*, 40(3), pp. 297–327.

Kolb, D.A. (1984) *Experiential Learning: Experience as the Source of Learning and Development*, Vol. 1. Englewood Cliffs, NJ: Prentice-Hall.

Korhonen, H. and Koivisto, E.M.I. (2007) Playability Heuristics for Mobile Multi-Player Games. *Proceedings of the 2nd International Conference on Digital Interactive Media in Entertainment and Arts*, pp. 28–35.

Korris, J. (2004) Full Spectrum Warrior: How the Institute for Creative Technologies Built a Cognitive Training Tool for The Xbox. Available at: http://scholar.google.co.uk/scholar?hl=enandq=FULL+SPECTRUM+WARRIOR%3A+HOW+THE+INSTITUTE+FOR+CREATIVE+TECHNOLOGIESandbtnG=andas_sdt=1%2C5andas_sdtp= (Accessed 6/10/2015).

Koster, R. (2005) *A Theory of Fun for Game Design, Electronic Resource*, Published Scottsdale, AZ: Paraglyph Press. Available at: http://capitadiscovery.co.uk/roehampton/items/510379?resultsUri=http%3A%2F%2Fcapitadiscovery.co.uk%2Froehampton%2Fitems%3Fquery%3Dtheory%2Bof%2Bfun%26target%3Dcatalogue%26facet%255B0%255D%3Dfulltext%253Ayes#availability (Accessed 29/12/2015).

Kraiger, K., Ford, J.K. and Salas, E. (1993) Application of Cognitive, Skill-based, and Affective Theories of Learning Objectives to New Methods of Training Evaluations, *Journal of Applied Psychology*, 78(2), pp. 311–328.

Kuss, D.J. and Griffiths, M.D. (2012) Internet Gaming Addiction: A Systematic Review of Empirical Research. *International Journal of Mental Health and Addiction*, 10(2), pp. 278–296.

Lazzaro, N. (2004) Why We Play Games: Four Keys to More Emotion without Story. Available at: http://www.xeodesign.com/xeodesign_whyweplaygames.pdf (Accessed 28/12/2015).

Lee, J.S., Keil, M. and Kasi, V. (2012) The Effects of an Initial Budget and Schedule Goal on Software Project Escalation. *Journal of Management Information System*, 29(1), pp. 53–77. DOI: 10.2753/MIS0742-1222290102

Lenhart, A., Kahne, J., Middaugh, E. Macgill, A.R., Evans, C. and Vitak, J. (2008) Teens, Video Games, and Civics: Teens' Gaming Experiences Are Diverse and Include

Significant Social Interaction and Civic Engagement. *Pew Internet & American Life Project*. Available at: https://www.google.co.uk/search?q=Lenhart+et+al.%2C+2008andoq=Lenhart+et+al.%2C+2008andaqs=chrome..69i57j0.1244j0j8andsourceid=chromeandie=UTF-8 (Accessed 20/8/2017).
Levine, D.M., Krehbiel, T.C. and Berenson, M.L. (2003) *Business Statistics: A First Course*, 3rd Edition. New Jersey: Pearson Education Ltd.
Lewis, G. (2013) Emotion and Disaffection with School Mathematics. *Research in Mathematics Education*, 15(1), pp. 70–86.
Li, F.W.B. and Watson, C. (2011) *Game-Based Concept Visualization for Learning Programming*. Scottsdale, Arizona, USA: ACM. 978-1-4503-0994-3/11/12.
Li, Z.Z., Cheng, Y.B. and Liu, C.C. (2013) A Constructionism Framework for Designing Game-Like Learning Systems: Its Effect on Different Learners. *British Journal of Educational Technology*, 44(2), pp. 208–224.
Lister, R., Adams, E.S. Fitzgerald, S. Fone, W. Hamer, J. Lindholm, M. McCartney, R. Moström, J.E. Sanders, K. Seppälä, O. and Simon, B. (2004) A Multi-National Study of Reading and Tracing Skills in Novice Programmers. In *ACM SIGCSE Bulletin*, 36(4), pp. 119–150. ACM.
Litosseliti, L. (2005) *Using Focus Groups in Research*. London: Continuum.
Loftus, C., Thomas, L. and Zander, C. (2011) Can Graduating Students Design: Revisited. In *Proceedings of the 42nd ACM Technical Symposium on Computer Science Education*, pp. 105–110.
Lyons, I.M. and Beilock, S.L. (2012) When Math Hurts: Math Anxiety Predicts Pain Network Activation in Anticipation of Doing Math. *PLoS One*, 7(10), p. e48076.
Majgaard, G. (2016) Creating Games in the Classroom: From Native Gamers to Reflective Designers. In: Munkvold, R. (ed.). *Games Based Learning for Researchers, Teachers and Students*, Vol. 1. Reading: Academic Conferences and Publishing International Ltd. ISBN:978-1-911278-12-890000.
Malaka, R. (2014) How Computer Games Can Improve Your Health and Fitness. In: Gobel, S., and Wiemeyer, J., (ed.). *Games for Training Education, Health and Sports, 4th Conference on Serious Games, GameDays 2014*, Darmstadt, Germany, April 1–5, 2014, Springer.
Malone, T.W. (1980) What Makes Things Fun to Learn? A Study of Intrinsically Motivating Computer Games. Available at: http://cci.mit.edu/malone/Malone%20computer%20games%20bibliography.htm
Malone, T.W. and Lepper, M.R. (1987) 'Making Learning Fun: A Taxonomy of Intrinsic Motivations for Learning. Aptitude, Learning, and Instruction': Volume 3: Cognitive and Affective Process Analyses. R. E. Snow and M. J. Farr. Hillsdale, USA, Erlbaum.
Maloney, E.A., Schaeffer, M.W. and Beilock, S.L. (2013) Mathematics Anxiety and Stereotype Threat: Shared Mechanisms, Negative Consequences and Promising Interventions. *Research in Mathematics Education*, 15(2), pp. 115–128.
Marciano, J.N., Miranda, L.C. and Miranda, E.E.C. (2014) Evaluating Multiple Aspects of Educational Computer Games. Available at: https://www.hindawi.com/journals/ijcgt/2014/139205/ (Accessed 27/08/2017).
Marklund, B.B. and Taylor, A.S.A. (2016) Creating Games in the Classroom: From Native Gamers to Reflective Designers. In: Munkvold, R. (ed.). *Games Based learning for Researchers, Teachers and Students*, Vol. 1. Reading: Academic Conferences and Publishing International Ltd. ISBN: 978-1-911278-12-890000.
Maslow, A.H. (1970) *Motivation and Personality*, 2nd Edition. New York; London: Harper and Row.
Mayer, E.R. (2014) *Computer Games for Learning: An Evidence Based Approach*. London: MIT.
Mayer, I., Warmelink, H. and Bekebrede, G. (2013) Learning in a Game-Based Virtual Environment: A Comparative Evaluation in Higher Education. *European Journal of Engineering Education*, 38(1), pp. 85–106.

Mayo, M. (2009) Video Games: A Route to Large-Scale STEM education? *Science*, 323, pp. 79–82.

McConnell, D. (2000) *Implementing Computer Supported Cooperative Learning*, 2nd Edition. London: Kogan Page.

Mccrindle, M. (2018) What Comes after Generation Z? Introducing Generation Alpha. Available at: https://mccrindle.com.au/insights/blogarchive/what-comes-after-generation-z-introducing-generation-alpha/ (Accessed 15/10/2018).

McFarlane, A., Sparrowhawk, A., and Heald, Y. (2002) *Report on the Educational Use of Games: An Exploration by TEEM on the Contribution Which Games Can Make to the Educational Process*. Cambridge, UK: TEEM. http://www.teem.org.uk/publications/teem_gamesined_full.pdf (Accessed 06/12/2015).

McGonigal, J. (2015) *Superbetter: A Revolutionary Approach to Getting Happier Stronger and Happier*. Braver and More Resilient, New York: Penguin Press.

Milane, I. and Rowe, G. (2002) Difficulties in Learning and Teaching Programming-Views of Students and Tutors. *Education and Information Technologies*, 7(1), pp. 55–66.

Monroy-Hernandez, A. and Resnick, M. (2008) Empowering Kids to Create and Share Programmable Media. *Interactions*, 15(2), pp. 50–53. Retrieved August 30, 2015.

Moreno-Ger, P., Burgos, D. Martínez-Ortiz, I. Sierra, J.L. and Fernández-Manjón, B. (2008) Educational Game Design for Online Education. *Computers in Human Behavior*, 24(6), pp. 2530–2540. DOI: 10.1016/j.chb.2008.03.012

Moreno-Ger, P., Torrente, J. Hsieh, Y.G. and Lester, W.T. (2012) Usability Testing for Serious Games: Making Informed Design Decisions with User Data. *Advances in Human-Computer Interaction*, 2012. DOI: 10.1155/2012/369637

Morgan, D.L. (1988) *Focus Groups as Qualitative Research*. Beverly Hills, CA: Sage.

Morrison, K. (2002) *School Leadership and Complexity Theory*. London: Routledge Falmer.

Muijs, D. (2011) *Doing Quantitative Research in Education with SPSS*. London: Sage Publication Ltd.

Mullins, L. (2013) *Management and Organisational Behaviour*, 10th Edition. Harlow: Pearson Education.

Mullins, L. (2019) *Organisational Behaviour in The Workplace*, 12th Edition. Harlow: Pearson Education.

Multon, K.D., Brown, S.D. and Lent, R.W. (1991) Relation of Self-Efficacy Beliefs to Academic Outcomes: A Meta-Analytic Investigation. *Journal of Counselling Psychology*, 18, pp. 30–38.

Naik, N. (2017) The Use of GBL to Teach Mathematics in Higher Education. *Innovations in Education and Teaching International*, 54(3), pp. 238–246. DOI: 10.1080/14703297.2015.1108857

Newby, P. (2010) *Research Methods for Education*, Harlow: Pearson Education Ltd.

Nicholls, D., De Charon, L. and Hutkin, R. (2010) Development of Critical Thinking and Creativity: Practical Guidelines for the Postsecondary Classroom. *ATEA Journal*, 38(1), p. 12.

Nielsen, J. (1995) Usability Inspection Methods, *CHI'95 Mosaic of Creativity*, pp. 7–11.

Nielsen, J. (2000) *Designing Web Usability*. Indianapolis, IN, USA: New Riders Publishing.

Nygård, S., Kolås, L. and Sigurdardottir, H. (2015) Teachers' Experiences Using KODU as a Teaching Tool. In: *European Conference on Games Based Learning*, Academic Conferences International Limited. p. 416.

Otkay, J.S. (2012) *Grounded Theory*. Oxford, England: Oxford University Press.

Paavilainen, J. (2010) Critical Review on Video Game Evaluation Heuristics: Social Games Perspective. *Proceedings of the International Academic Conference on the Future of Game Design and Technology*, pp. 56–65. DOI: 10.1145/1920778.1920787

Pachler, N. (2007) *Mobile Learning: Towards a Research Agenda*. London: WLE Centre, IoE. In: Kress, G. and Pachler, N (ed.) (2007) Thinking about the 'm' in m-learning.

Pajares, F. and Kranzler, J. (1995) Self-Efficacy Beliefs and General Mental Ability in Mathematical Problem-Solving. *Contemporary Educational Psychology*, 20, pp. 426–443.

Palincsar, A.S. (1998) Social Constructivist Perspectives on Teaching and Learning. *Annual Review of Psychology*, 49, pp. 345–375.

Papert, S. (1993) *Mindstorms: Children, Computers and Powerful Ideas*. New York: Basic Books.

Paras, B. and Bizzocchi, J. (2005) Game, Motivation, and Effective Learning: An Integrated Model for Educational Game. In *Proceedings of DiGRA Conference on Changing Views–Worlds of Play*. Available at: https://www.researchgate.net/publication/221217604_Game_Motivation_and_Effective_Learning_An_Integrated_Model_for_Educational_Game_Design (Accessed 27/09/2016).

Park, H.J. and Kim, S.B. (2013) Guidelines of Serious Game Accessibility for the Disabled. In *Proceedings of the International Conference on Information Science and Applications (ICISA '13)*, pp. 1–3, IEEE.

Pinelle, D., Wong, N., Stach, T. and Gutwin, C. (2009) Usability Heuristics for Networked Multiplayer Games. *Proceedings of the ACM 2009 International Conference on Supporting Group Work*, pp. 169–178.

Pollard, A., Anderson, J., Maddock, M. Swaffield, S., Warin, J. and Warwick, P. (2008) *Reflective Teaching*, 3rd Edition. London: Continuum International Publishing Group.

Preconception. (2020) In: Dictionary.com. [Online] Available at: https://www.dictionary.com/browse/preconception (Accessed 01/08/2020).

Preece, Y., Rogers, Y. and Sharp, H. (2015) *Interaction Design: Beyond Human-Computer Interaction*, 4th Edition. Chichester: John Wiley and Sons Ltd.

Prensky, M. (2002) The Motivation of Gameplay: The Real Twenty-First Century Learning Revolution. *On the Horizon*, 10(1), pp. 5–11.

Prensky, M. (2005) Listen to Natives. *Association for Supervision and Curriculum Development Educational Leadership*, 63(N4), pp. 9–13.

Prensky, M. (2006) *'Don't Bother Me Mom, I'm Learning!': How Computer and Video Games Are Preparing Your Kids for the Twenty-First Century*. St Paul, MN: Paragon House. Available at: https://www.scirp.org/reference/referencespapers?referenceid=633456

Pritchard, A. and Woollard, J. (2010) *Psychology for the Classroom: Constructivism and Social Learning*. Oxon: Routledge.

Punch, K.F. (2009) *Introduction to Research Methods in Education*. London: Sage Publications Ltd.

Resnick, M. (2017) *Lifelong Kindergarten: Cultivating Creativity Through Projects, Passion, Peers, and Play*. London, England: MIT Press.

Robins, A., Runtree, J. and Rountree, N. (2003) Learning and Teaching Programming: A Review and a Discussion. *Computer Science Education*, 13(2), pp. 137–172.

Robson, C. (2010) *Real World Research*, 3rd Edition. Chichester: John Wiley and Sons Limited.

Robson, S. (2006) *Developing Thinking and Understanding in Young Children: An Introduction for Students*. London: Routledge Ltd.

Rogoff, B. (1998) Cognition as a Collaborative Process. In W. Damon, D. Kuhn and R.S. Seigler (ed.), *Handbook of Child Psychology*, 5th Edition. New York: Wiley.

Roth, W.M. (2000) Authentic School Science: Intellectual Traditions In R. McCormick and C. Paechter (ed.), *Learning & Knowledge*. London, UK: Paul Chapman Publishing, pp. 6–20.

Russ, T., and Drury-Grogan, M.L. (2013) Assessing the Impacts of a Business Communication Simulation on Students' Self-Perceptions. *Eastern Communication Association*, 61(5), pp. 584–595. DOI: 10.1080/01463373.2013.822404

Salen, K. and Zimmerman, E. (2011) The Game Designer Reader, A Rules of Play Anthology. Roger Caillois: The Definition of Play. *The Classification of Games*. Cambridge, Massachusetts. London, England: The MIT Press.

Schell, J. (2015) *The Art of Game Design: A Book of Lenses*, 2nd Edition. USA: CRC Press.

Schell, J. (2020) *The Art of Game Design: A Book of Lenses*, 3rd Edition. USA: CRC Press.

Senge, P. (1990) *The Fifth Discipline: The Art and Practice of Learning Organization*. United Kingdom: Random House Business Books.

Senge, P.M. (2006) *The Fifth Discipline: The Art and Practice of Learning Organisation*. London: Random House Business Books.

Shell, D.F., Murphy, C.C. and Bruning, R.H. (1989) Self-efficacy and Outcome Expectancy Mechanisms in Reading and Writing Achievement. *Journal of Educational Psychology*, 81(1), p. 91.

Shunk, D.H. (2000) *Learning Theories: An Educational Perspective*, 3rd Edition. Upper Saddle River, NJ: Prentice-Hall.

Siegel, R.G., Galassi, J.P. and Ware, W.B. (1985) A Comparison of Two Models for Predicting Mathematics Performance: Social Learning Versus Math Aptitude-Anxiety. *Journal of Counseling Psychology*, 32, pp. 531–538.

Silverman, B.G., Johns, M., Weaver, R. and Mosley, J. (2003) Authoring Edutainment Stories for Online Players (AESOP): Introducing Game play into Interactive Dramas. In: *International Conference on Virtual Storytelling*, Springer.

Silverman, D. (1993) *Interpreting Qualitative Data: Methods for Analysing Talk, Text, and Interaction*. London: Sage Publications Ltd.

Silverman, D. (2014) *Interpreting Qualitative Data*, 5th Edition. London: Sage Publications Ltd.

Sindre, G., Natvig, L. and Jahre, M. (2009) Experimental Validation of the Learning Effect for A Pedagogical Game on Computer Fundamentals. *IEEE Transactions on Education*, 52 (1), pp. 10–18.

Skemp, R.R. (1986) *The Psychology of Learning Mathematics Great*. Britain: Penguin Books Ltd.

Skemp, R.R. (1989) *Mathematics in the Primary School*. Bristol: Routledge.

Skilbeck, A. (2017) Dewey on Seriousness, Playfulness and the Role of the Teacher. *Education Sciences*, 7(1), p. 16.

Slyman, S. (2014) I Hate Maths! An Investigation into Students' Attitudes Towards Mathematics. MA Thesis, University of Roehampton, London, United Kingdom.

Slyman, S. (2018) *Games Based Learning in Universities, a Life Learning Experience*. New York: Business Expert Press Expert Insights Publishing. ISBN:978-1-94819-801-1.

Slyman, S., Gillies, M. and Jessel, J. (2019a) Acumenous or Inquisitional? Towards a New Theoretical Lens within Games Learning. In: *13th European Conference on Games Based Learning*, University of South Denmark, Odense, Denmark. ISBN:978-1-912764-38-9 or E-book: ISSN:2049-100X.

Slyman, S., Gillies, M. and Jessel, J. (2019b) Game-Based Learning to Engage students with Applied Statistics using a Simulation Roleplay Game. In: *9th Irish Conference on Games Based Learning*, Cork, Ireland. Available at: http://igbl-conference.com/2016/programme/generate_book_of_abstracts.php (Accessed 25/09/2019).

Smith Report (2004) Available at: http://www.mathsinquiry.org.uk/report/mathsInquiryFinalReport.pdf (Accessed 20/12/2014).

Smith Report (2017) Available online at: https://www.gov.uk/government/publications/smith-review-of-post-16-maths-report-and-government-response (Accessed 20/12/2018).

Sobke, H., Chan, E. Buttlar, R. Grobe-Wortmann, J. and Londong, J. (2014) Cat King's Metamorphosis. The Reuse of an Educational Game in a Further Technical Domain.

In: Gobel, S., and Wiemeyer, J. (ed.), *Games for Training Education, Health and Sports, 4th Conference on Serious Games, GameDays 2014*, Darmstadt, Germany, April 1–5, 2014, Springer.
Song, S. and Lee, J. (2007) Key Factors of Heuristic Evaluation for Game Design: Towards massively multi-player online Roleplaying game. *International Journal of Human-Computer Studies*, 65, pp. 709–723.
Squire, K., 2005. *Game-based Learning: Present and Future State of the Field.* New York, NY: Masie Center e-learning Consortium.
Strauss, A. and Corbin, J. (1990) *Basics of Qualitative Research: Grounded Theory Procedures and Techniques*, Newbury Park, CA: Sage.
Sweetser, P., Johnson, D. and Wyeth, P. (2012) Revisiting the Game Flow Model with Detailed Heuristics. *Journal of Creative Technologies*, 2012(3), pp. 1–16.
Sweetser, P. and Wyeth, P. (2005) GameFlow: A Model for Evaluating Player Enjoyment in Games. *Computers in Entertainment*, 3(3), p. 3. DOI: 10.1145/1077246.1077253
Tamura, H. (2007) Foundational Value of Mathematics Education for Management Curriculum. *International Statistical Review*, 75(3), pp. 397–405. DOI: 10.1111/j.1751-5823.2007.00032.x
Tao, Y.H., Cheng, C.J. and Sun, S.Y. (2012) Alignment of Teacher and Students Perceptions on the Continued Use of Business Simulation Games. *Educational Technology and Society*, 15(3), pp. 177–189.
Tondello, G.F., Kappen, D.L. Mekler, E.D. Ganaba, M. and Nacke, L.E. (2016) Heuristic Evaluation for Gameful Design. *CHI PLAY'16*. ACM. 978-1-4503-4458-6/16/10. DOI: 10.1145/2968120.2987729
van der Spek, E.D., van Oostendorp, H. and Meyer, J.-J.C.H. (2013) Introducing Surprising events can stimulate deep learning in a serious game. *British Journal of Educational Technology*, 44(1), 156–169.
Vlachopoulos, D. and Makri, A. (2017) The Effect of Games and Simulations on Higher Education: A Systematic Literature Review. *International Journal of Educational Technology in Higher Education*, 14(1), pp. 22. DOI: 10.1186/s41239-017-0062-1
Vos, L. (2014) Marketing Simulation Games: A Review of Issues in Teaching and Learning. *The Marketing Review*, 14(1), p. 6796. DOI: 10.1362/146934714X1394890 9473220
Vos, L. and Brennan, R. (2010) Marketing Simulation Games: Student and Lecturer Perspectives. *Marketing Intelligence and Planning*, 28(7), pp. 882–897.
Vygotsky, L.S. (1980) *Mind in Society. The Development of Higher Psychological Processes*. Edited by Cole, M., John-Steiner, V., Scribner, S. and Souberman, E. London: Harvard University Press.
Walonick, D.S. (2000) *Designing and Using Questionnaire in Survival Statistic*. MN: StatPacInc. Available at Internet WWW page at URL: http://www.statpac.com/surveys/surveys.pdf (Accessed 02/04/2016).
Whitton, N. (2011) Games-Based Experiences for Learning. Futurelab Innovation in Education. Available at: https://www.nfer.ac.uk/media/1775/futl11.pdf (Accessed 02/04/2016).
Wolfe, J. (1997) The Effectiveness of Business Games in Strategic Management Coursework. *Simulation and Gaming*, 28(4), pp. 360–376. DOI: 10.1177/1046878197284003
Women in STEM – 'Facts and Statistics' (Reporting period 2012–2015) IET. Available at: https://communities.theiet.org/files/8042 (Accessed 20/06/2017).
Yin, R.K. (1994) *Case Study Research: Design and Methods*, 2nd Edition. London: Sage.
Young, M. (2008) *"Powerful Knowledge" and "Knowledge of the Powerful": Some Reflections*. London: Presentation at the Institute of Education.
Yusoff, A., Crowder, R. and Gilbert, L. (2010) Validation of Serious Games Attributes Using the Technology Acceptance Model [online]. In: IEEE (ed.), *2010 Second*

International Conference on Games and Virtual Worlds for Serious Applications, Braga, Portugal, 25–26 March 2010, pp. 45–51. Available at: http://eprints.soton.ac.uk/270809/ (Accessed 10/05/2015).

Zimmerman, B.J. (2000) Self-Efficacy: An Essential Motive to Learn. *Contemporary Educational Psychology*, 25(1), pp. 82–91. DOI: 10.1006/ceps.1999.1016

Zyda, M. (2005) *From Visual Simulation to Virtual Reality to Games*. USC Information Sciences Institute, IEEE Computer Society Edition September, pp. 25–31. Available at: http://scholar.google.co.uk/scholar?hl=enandq=zyda+2005andbtnG=andas_sdt=1%2C5andas_sdtp= (Accessed 07/10/2015).

Index

Pages in *italics* refer to figures and pages in **bold** refer to tables.

Kolibri 360 GmbH
Ferdinand-Jühlke-Straße 7
99095 Erfurt, Deutschland
produktsicherheit@kolibri360.de